Combined Treatments and Therapies
to Cure Spinal Cord Injury

Combined Treatments and Therapies to Cure Spinal Cord Injury

Editor

Nicolas Guerout

Basel • Beijing • Wuhan • Barcelona • Belgrade • Novi Sad • Cluj • Manchester

Editor
Nicolas Guerout
Université Paris Cité
Paris
France

Editorial Office
MDPI AG
Grosspeteranlage 5
4052 Basel, Switzerland

This is a reprint of articles from the Special Issue published online in the open access journal *Biomedicines* (ISSN 2227-9059) (available at: https://www.mdpi.com/journal/biomedicines/special_issues/cure_SCI).

For citation purposes, cite each article independently as indicated on the article page online and as indicated below:

Lastname, A.A.; Lastname, B.B. Article Title. *Journal Name* **Year**, *Volume Number*, Page Range.

ISBN 978-3-7258-2161-7 (Hbk)
ISBN 978-3-7258-2162-4 (PDF)
doi.org/10.3390/books978-3-7258-2162-4

Contents

About the Editor

Nicolas Guérout

Nicolas Guérout earned his PhD in 2010 from the University of Rouen, specializing in the EA3830. His doctoral research focused on developing and characterizing olfactory ensheathing cell cultures for therapeutic applications and their transplantation in various peripheral nerve injury models. Following his PhD, he undertook 4 years and 6 months of postdoctoral training, including a 2-year stint in the Department of Neuroscience at the Karolinska Institutet in Stockholm, Sweden. During his time in Sweden, he gained advanced knowledge and skills, particularly in developmental biology and stem cell biology. His postdoctoral work concentrated on the development and recruitment of ependymal stem cells from the spinal cord.

In 2014, he successfully competed for the position of Associate Professor (Ass. Prof.) in Physiology. During his tenure as an Ass. Prof., he began exploring new research themes, particularly the effects of magnetic stimulation on the recruitment of endogenous stem cells in the spinal cord following injury. In May 2022, he passed a competitive exam to become a University Professor and joined Dr. Daniel Zytnicki's team at Université Paris Cité, a leading group in motoneurons and amyotrophic lateral sclerosis (ALS). Since his appointment, he has continued to focus on research related to traumatic spinal cord injuries, also expanding his work into ALS research. In January 2025, he will take on the role of director of this team, renamed the "Biology and Pathophysiology of the Spinal Cord."

Editorial

Combined Treatments and Therapies to Cure Spinal Cord Injury

Nicolas Guérout

Saints Pères Paris Institute for the Neurosciences, Université Paris Cité, CNRS UMR8003, 75006 Paris, France; nicolas.guerout@u-paris.fr

Traumatic injuries of the spinal cord (SCIs) are still pathologies with a disastrous outcome. In humans, they most often lead to permanent motor, sensitive, and sphincter deficits, and to this day, there is no available treatment. While a few decades ago, SCI mainly affected young adults, today, it also affects older people in developed countries. The most common causes are road accidents or falls, but they can also be inflicted by knives or firearms [1].

Although these injuries have been described as incurable since ancient Egypt, numerous research teams have been interested in developing treatments to repair the spinal cord and thus bring about at least a partial recovery of lost functions [2].

In parallel with this research into therapies to improve the management of SCIs, a great deal of work has gone into better describing and understanding the mechanisms that take place after them. Thus, even if there is still some debate within the scientific community, the events that follow the injury and contribute to the development of spinal scars are becoming better understood. Similarly, the cell populations that make up the scar and the role they play have been investigated and are now clearly defined [3].

The fundamental knowledge provided by these studies has led to the development of a number of therapeutic avenues, which have been tested after SCIs in recent decades. These include peripheral nerve grafts to bypass the damaged area, the use of enzymes to degrade certain molecules in the extracellular matrix that inhibit axonal regrowth, and the use of antibodies to block the action of inhibitory molecules [4–6].

However, although these different strategies have shown promising results in animals, particularly rodents, they remain either difficult to implement in humans or have failed to demonstrate their efficacy in clinical trials [7].

More recently, the contribution of fundamental knowledge provided by the use of transgenic mouse lines, the latest omics techniques, and optogenetics has enabled us to better characterize cell diversity both in the injured and uninjured spinal cord and in its connections with the cortex and brain stem [8–15]. As a result, research into treatments is focusing on modulating spinal scar and inflammation, inducing the regrowth/survival of the motor and sensitive axons, and promoting new functional connections between the brain and spinal cord [16–18]. As a result, a great deal of research is based on cell transplantation, neuromodulation, or physiotherapy [9,19].

It is in this context that the project for this Special Issue, entitled "Combined Treatments and Therapies to Cure Spinal Cord Injury", was designed. The Special Issue includes six research papers and five literature reviews dealing exclusively with cell transplantation, neuromodulation, and physiotherapy techniques.

Indeed, among the six research papers, one concerns the study of inflammation after SCIs, two deal with cell transplantation, and three are related to neuromodulation techniques.

Similarly, two of the reviews are devoted to cell transplantation, two to new physiotherapy techniques, and one to chronic pain pathologies.

Citation: Guérout, N. Combined Treatments and Therapies to Cure Spinal Cord Injury. *Biomedicines* **2024**, *12*, 1095. https://doi.org/10.3390/biomedicines12051095

Received: 9 May 2024
Accepted: 14 May 2024
Published: 15 May 2024

Sabirov et al., in their clinical study, investigated the level of different cytokines present in the blood and cerebrospinal fluid (CSF) at different times post-injury, allowing a characterization of the dynamics of post-SCI inflammation in patients [20].

Kumru et al. studied the effect of transcutaneous spinal cord stimulation (tSCS) applied at the thoracic and cervical level to increase respiratory capacity in SCI patients. They demonstrated that tSCS coupled with inspiratory muscle training (IMT) increases patients' respiratory capacity, whereas IMT alone does not increase the various parameters measured [21].

Neves Vialle et al. investigated the effects of human mesenchymal stem cell transplantation in a rat model of SCI and demonstrated that, in their model, these cells increased neuronal survival at the site of injury [22].

Georgelou et al. investigated the effects of a combined treatment involving the administration of small-molecule mimetics of endogenous neurotrophins and neural stem cells, demonstrating the synergistic effect of this combination in a mouse model of SCI by means of functional and histological studies [23].

Leszczyńska et al. compared the effects of different neuromodulation methods such as repetitive transcranial magnetic stimulation (rTMS) and peripheral electrotherapy with a physiotherapy method: kinesiotherapy. They particularly demonstrated that the two neuromodulation techniques lead to better patient outcomes than kinesiotherapy alone and that the best results are obtained in the groups combining electrotherapy with kinesiotherapy and rTMS with kinesiotherapy [24].

Tharu et al. compared the effects of a neuromodulation technique, transcutaneous electrical spinal cord stimulation, with a physiotherapy technique, conventional task-specific rehabilitation, in patients with spinal cord injuries, demonstrating that both treatments induce functional recovery [25].

Two reviews in this Special Issue address the contribution of cell transplantation after SCI: Wen et al. provide an overview of the use of dental-derived stem cells, while Reshamwala et al. present various clinical studies that have used olfactory ensheathing cell transplantation in humans [26,27].

Two reviews focus on the effects of rehabilitation in patients with SCIs. Indeed, He et al. present a review of the effects of physical exercise on the reorganization of communication between the brain and the spinal cord, while Stanciu et al. describe the reported effects of rehabilitation through hydrotherapy [28,29].

Finally, the last review by Foreman et al. focuses on lower back pain and clinically presents the various pathologies that can lead to these chronic pains. They also introduce new therapeutic approaches that could help manage these types of pathologies, including neuromodulation and cell transplantation [30].

This Special Issue provides an overview of the latest research on spinal cord injuries, both in animals and in SCI patients. The various published articles highlight promising techniques such as cell transplantation and neuromodulation. They also emphasize the crucial importance of physiotherapy and the various associated rehabilitation methods. These different studies and reviews demonstrate, if it were still necessary, that future treatments must combine various complementary approaches in order to effectively treat this devastating pathology.

Conflicts of Interest: The author declares no conflict of interest.

References

1. Golestani, A.; Shobeiri, P.; Sadeghi-Naini, M.; Jazayeri, S.B.; Maroufi, S.F.; Ghodsi, Z.; Dabbagh Ohadi, M.A.; Mohammadi, E.; Rahimi-Movaghar, V.; Ghodsi, S.M. Epidemiology of Traumatic Spinal Cord Injury in Developing Countries from 2009 to 2020: A Systematic Review and Meta-Analysis. *Neuroepidemiology* **2022**, *56*, 219–239. [CrossRef]
2. van Middendorp, J.J.; Sanchez, G.M.; Burridge, A.L. The Edwin Smith papyrus: A clinical reappraisal of the oldest known document on spinal injuries. *Eur. Spine J.* **2010**, *19*, 1815–1823. [CrossRef]
3. Guérout, N. Plasticity of the Injured Spinal Cord. *Cells* **2021**, *10*, 1886. [CrossRef]

4. Gauthier, P.; Réga, P.; Lammari-Barreault, N.; Polentes, J. Functional reconnections established by central respiratory neurons regenerating axons into a nerve graft bridging the respiratory centers to the cervical spinal cord. *J. Neurosci. Res.* **2002**, *70*, 65–81. [CrossRef] [PubMed]

5. Liebscher, T.; Schnell, L.; Schnell, D.; Scholl, J.; Schneider, R.; Gullo, M.; Fouad, K.; Mir, A.; Rausch, M.; Kindler, D.; et al. Nogo-A antibody improves regeneration and locomotion of spinal cord-injured rats. *Ann. Neurol.* **2005**, *58*, 706–719. [CrossRef]

6. Zhao, R.-R.; Andrews, M.R.; Wang, D.; Warren, P.; Gullo, M.; Schnell, L.; Schwab, M.E.; Fawcett, J.W. Combination treatment with anti-Nogo-A and chondroitinase ABC is more effective than single treatments at enhancing functional recovery after spinal cord injury. *Eur. J. Neurosci.* **2013**, *38*, 2946–2961. [CrossRef] [PubMed]

7. Kucher, K.; Johns, D.; Maier, D.; Abel, R.; Badke, A.; Baron, H.; Thietje, R.; Casha, S.; Meindl, R.; Gomez-Mancilla, B.; et al. First-in-Man Intrathecal Application of Neurite Growth-Promoting Anti-Nogo-A Antibodies in Acute Spinal Cord Injury. *Neurorehabil. Neural Repair* **2018**, *32*, 578–589. [CrossRef] [PubMed]

8. Asboth, L.; Friedli, L.; Beauparlant, J.; Martinez-Gonzalez, C.; Anil, S.; Rey, E.; Baud, L.; Pidpruzhnykova, G.; Anderson, M.A.; Shkorbatova, P.; et al. Cortico-reticulo-spinal circuit reorganization enables functional recovery after severe spinal cord contusion. *Nat. Neurosci.* **2018**, *21*, 576–588. [CrossRef]

9. Kathe, C.; Skinnider, M.A.; Hutson, T.H.; Regazzi, N.; Gautier, M.; Demesmaeker, R.; Komi, S.; Ceto, S.; James, N.D.; Cho, N.; et al. The neurons that restore walking after paralysis. *Nature* **2022**, *611*, 540–547. [CrossRef]

10. Barnabé-Heider, F.; Göritz, C.; Sabelström, H.; Takebayashi, H.; Pfrieger, F.W.; Meletis, K.; Frisén, J. Origin of new glial cells in intact and injured adult spinal cord. *Cell Stem Cell* **2010**, *7*, 470–482. [CrossRef]

11. Floriddia, E.M.; Lourenço, T.; Zhang, S.; van Bruggen, D.; Hilscher, M.M.; Kukanja, P.; Gonçalves Dos Santos, J.P.; Altınkök, M.; Yokota, C.; Llorens-Bobadilla, E.; et al. Distinct oligodendrocyte populations have spatial preference and different responses to spinal cord injury. *Nat. Commun.* **2020**, *11*, 5860. [CrossRef] [PubMed]

12. Göritz, C.; Dias, D.O.; Tomilin, N.; Barbacid, M.; Shupliakov, O.; Frisén, J. A pericyte origin of spinal cord scar tissue. *Science* **2011**, *333*, 238–242. [CrossRef] [PubMed]

13. Dias, D.O.; Kim, H.; Holl, D.; Werne Solnestam, B.; Lundeberg, J.; Carlén, M.; Göritz, C.; Frisén, J. Reducing Pericyte-Derived Scarring Promotes Recovery after Spinal Cord Injury. *Cell* **2018**, *173*, 153–165.e22. [CrossRef] [PubMed]

14. Sabelström, H.; Stenudd, M.; Réu, P.; Dias, D.O.; Elfineh, M.; Zdunek, S.; Damberg, P.; Göritz, C.; Frisén, J. Resident neural stem cells restrict tissue damage and neuronal loss after spinal cord injury in mice. *Science* **2013**, *342*, 637–640. [CrossRef] [PubMed]

15. Habib, N.; Li, Y.; Heidenreich, M.; Swiech, L.; Avraham-Davidi, I.; Trombetta, J.J.; Hession, C.; Zhang, F.; Regev, A. Div-Seq: Single-nucleus RNA-Seq reveals dynamics of rare adult newborn neurons. *Science* **2016**, *353*, 925–928. [CrossRef] [PubMed]

16. Squair, J.W.; Milano, M.; de Coucy, A.; Gautier, M.; Skinnider, M.A.; James, N.D.; Cho, N.; Lasne, A.; Kathe, C.; Hutson, T.H.; et al. Recovery of walking after paralysis by regenerating characterized neurons to their natural target region. *Science* **2023**, *381*, 1338–1345. [CrossRef] [PubMed]

17. Chalfouh, C.; Guillou, C.; Hardouin, J.; Delarue, Q.; Li, X.; Duclos, C.; Schapman, D.; Marie, J.-P.; Cosette, P.; Guérout, N. The Regenerative Effect of Trans-spinal Magnetic Stimulation After Spinal Cord Injury: Mechanisms and Pathways Underlying the Effect. *Neurotherapeutics* **2020**, *17*, 2069–2088. [CrossRef] [PubMed]

18. Llorens-Bobadilla, E.; Chell, J.M.; Le Merre, P.; Wu, Y.; Zamboni, M.; Bergenstråhle, J.; Stenudd, M.; Sopova, E.; Lundeberg, J.; Shupliakov, O.; et al. A latent lineage potential in resident neural stem cells enables spinal cord repair. *Science* **2020**, *370*, eabb8795. [CrossRef] [PubMed]

19. Delarue, Q.; Brodier, M.; Neveu, P.; Moncomble, L.; Hugede, A.; Blondin, A.; Robac, A.; Raimond, C.; Lecras, P.; Riou, G.; et al. Lesion-induced impairment of therapeutic capacities of olfactory ensheathing cells in an autologous transplantation model for treatment of spinal cord injury. *bioRxiv* **2024**. [CrossRef]

20. Sabirov, D.; Ogurcov, S.; Shulman, I.; Kabdesh, I.; Garanina, E.; Sufianov, A.; Rizvanov, A.; Mukhamedshina, Y. Comparative Analysis of Cytokine Profiles in Cerebrospinal Fluid and Blood Serum in Patients with Acute and Subacute Spinal Cord Injury. *Biomedicines* **2023**, *11*, 2641. [CrossRef]

21. Kumru, H.; García-Alén, L.; Ros-Alsina, A.; Albu, S.; Valles, M.; Vidal, J. Transcutaneous Spinal Cord Stimulation Improves Respiratory Muscle Strength and Function in Subjects with Cervical Spinal Cord Injury: Original Research. *Biomedicines* **2023**, *11*, 2121. [CrossRef]

22. Vialle, E.N.; Fracaro, L.; Barchiki, F.; Dominguez, A.C.; de Oliveira Arruda, A.; Olandoski, M.; Brofman, P.R.S.; Kuniyoshi Rebelatto, C.L. Human Adipose-Derived Stem Cells Reduce Cellular Damage after Experimental Spinal Cord Injury in Rats. *Biomedicines* **2023**, *11*, 1394. [CrossRef] [PubMed]

23. Georgelou, K.; Saridaki, E.-A.; Karali, K.; Papagiannaki, A.; Charalampopoulos, I.; Gravanis, A.; Tzeranis, D.S. Microneurotrophin BNN27 Reduces Astrogliosis and Increases Density of Neurons and Implanted Neural Stem Cell-Derived Cells after Spinal Cord Injury. *Biomedicines* **2023**, *11*, 1170. [CrossRef]

24. Leszczyńska, K.; Huber, J. The Role of Transcranial Magnetic Stimulation, Peripheral Electrotherapy, and Neurophysiology Tests for Managing Incomplete Spinal Cord Injury. *Biomedicines* **2023**, *11*, 1035. [CrossRef]

25. Tharu, N.S.; Alam, M.; Ling, Y.T.; Wong, A.Y.; Zheng, Y.-P. Combined Transcutaneous Electrical Spinal Cord Stimulation and Task-Specific Rehabilitation Improves Trunk and Sitting Functions in People with Chronic Tetraplegia. *Biomedicines* **2022**, *11*, 34. [CrossRef]

26. Wen, X.; Jiang, W.; Li, X.; Liu, Q.; Kang, Y.; Song, B. Advancements in Spinal Cord Injury Repair: Insights from Dental-Derived Stem Cells. *Biomedicines* **2024**, *12*, 683. [CrossRef]
27. Reshamwala, R.; Murtaza, M.; Chen, M.; Shah, M.; Ekberg, J.; Palipana, D.; Vial, M.-L.; McMonagle, B.; St John, J. Designing a Clinical Trial with Olfactory Ensheathing Cell Transplantation-Based Therapy for Spinal Cord Injury: A Position Paper. *Biomedicines* **2022**, *10*, 3153. [CrossRef] [PubMed]
28. He, L.-W.; Guo, X.-J.; Zhao, C.; Rao, J.-S. Rehabilitation Training after Spinal Cord Injury Affects Brain Structure and Function: From Mechanisms to Methods. *Biomedicines* **2023**, *12*, 41. [CrossRef] [PubMed]
29. Stanciu, L.E.; Iliescu, M.G.; Vlădăreanu, L.; Ciota, A.E.; Ionescu, E.-V.; Mihailov, C.I. Evidence of Improvement of Lower Limb Functioning Using Hydrotherapy on Spinal Cord Injury Patients. *Biomedicines* **2023**, *11*, 302. [CrossRef]
30. Foreman, M.; Maddy, K.; Patel, A.; Reddy, A.; Costello, M.; Lucke-Wold, B. Differentiating Lumbar Spinal Etiology from Peripheral Plexopathies. *Biomedicines* **2023**, *11*, 756. [CrossRef]

 *biomedicines*

Article

Comparative Analysis of Cytokine Profiles in Cerebrospinal Fluid and Blood Serum in Patients with Acute and Subacute Spinal Cord Injury

Davran Sabirov [1], Sergei Ogurcov [2], Ilya Shulman [2], Ilyas Kabdesh [1,*], Ekaterina Garanina [3], Albert Sufianov [4,5], Albert Rizvanov [1] and Yana Mukhamedshina [1,6]

[1] OpenLab "Gene and Cell Technologies", Institute of Fundamental Medicine and Biology, Kazan (Volga Region) Federal University, 420008 Kazan, Russia
[2] Neurosurgical Department No. 2, Republic Clinical Hospital, 420138 Kazan, Russia
[3] Department of Genetics, Institute of Fundamental Medicine and Biology, Kazan (Volga Region) Federal University, 420008 Kazan, Russia
[4] Department of Neurosurgery, Sechenov First Moscow State Medical University of the Ministry of Health of the Russian Federation (Sechenov University), 119991 Moscow, Russia
[5] The Research and Educational Institute of Neurosurgery, Peoples' Friendship University of Russia (RUDN), 117198 Moscow, Russia
[6] Department of Histology, Cytology and Embryology, Kazan State Medical University, 420012 Kazan, Russia
* Correspondence: ikabdesh@kpfu.ru or ikabdesh@gmail.com

Abstract: Background: Cytokines are actively involved in the regulation of the inflammatory and immune responses and have crucial importance in the outcome of spinal cord injuries (SCIs). Examining more objective and representative indicators of the patient's condition is still required to reveal the fundamental patterns of the abovementioned posttraumatic processes, including the identification of changes in the expression of cytokines. Methods: We performed a dynamic (3, 7, and 14 days post-injury (dpi)) extended multiplex analysis of cytokine profiles in both CSF and blood serum of SCI patients with baseline American Spinal Injury Association Impairment Scale grades of A. Results: The data obtained showed a large elevation of IL6 (>58 fold) in CSF and IFN-γ (>14 fold) in blood serum at 3 dpi with a downward trend as the post-traumatic period increases. The level of cytokine CCL26 was significantly elevated in both CSF and blood serum at 3 days post-SCI, while other cytokines did not show the same trend in the different biosamples. Conclusions: The dynamic changes in cytokine levels observed in our study can explore the relationships with the SCI region and injury severity, paving the way for a better understanding of the pathophysiology of SCI and potentially more targeted and personalized therapeutic interventions.

Keywords: traumatic spinal cord injury; cytokine profile; cerebrospinal fluid; blood serum; multiplex analysis

Citation: Sabirov, D.; Ogurcov, S.; Shulman, I.; Kabdesh, I.; Garanina, E.; Sufianov, A.; Rizvanov, A.; Mukhamedshina, Y. Comparative Analysis of Cytokine Profiles in Cerebrospinal Fluid and Blood Serum in Patients with Acute and Subacute Spinal Cord Injury. *Biomedicines* **2023**, *11*, 2641. https://doi.org/10.3390/biomedicines11102641

Academic Editor: Nicolas Guerout

Received: 28 August 2023
Revised: 12 September 2023
Accepted: 18 September 2023
Published: 26 September 2023

1. Introduction

Spinal cord injury (SCI) is a global health problem affecting tens of thousands of people every year, with profound consequences for those affected and significant societal costs. Neurological disorders resulting from SCI often lead to severe disability and dramatically change the lives of patients and their families [1]. While SCIs used to predominantly affect younger individuals, its incidence is now equally high among the elderly [2].

The pathogenesis of SCI involves a complex interplay of primary and secondary mechanisms. The primary injury is immediate, resulting from mechanical trauma, while secondary events, such as inflammation, ischemia, and excitotoxicity, further exacerbate the injury. This complexity has necessitated research into novel and often overlooked therapeutic strategies. For instance, the work of Turczyn et al. (2022) delves into the yet inconclusive but promising role of omega-3 fatty acids in SCI treatment. Similarly, the

DISCUS trial led by Saadoun et al. (2023) is pioneering in its evaluation of duroplasty for cervical SCI, a technique not yet mainstream but showing potential for significant impact [3,4].

Concurrently, the management of SCI imposes a significant burden on global healthcare systems. The lifetime care for an individual with SCI can escalate to several million dollars, depending on the severity of the injury. This is further complicated by the long-term pharmacological management required for pain, spasticity, and secondary complications. The pharmacoeconomic impact is substantial, as evidenced by ongoing trials and studies, including those investigating the protective effects of hydrogen gas against spinal cord ischemia-reperfusion injury [5].

Despite significant advances in medical, surgical, and rehabilitation care for people with SCI, measures and interventions to ensure neurological recovery remain limited. This limited recovery is primarily due to the central nervous system's (CNS) inherently low regenerative capacity, the complexity of the injury, and the poorly understood pathophysiological events after SCIs [6].

The complexity of SCI diagnosis and treatment is further highlighted by the potential for misdiagnosis with other spinal conditions. For instance, chronic non-bacterial osteomyelitis (CNO) is often confused with infectious spondylodiscitis or malignant lesions in the spine, complicating the diagnostic process and potentially leading to ineffective treatments [7]. This underscores the need for a comprehensive approach to spinal disorders, including SCI, to minimize treatment-related morbidity.

Evaluating new treatments for SCI presents a significant challenge. Demonstrating neurological recovery during the acute phase is particularly challenging due to the potential for spontaneous recovery [8]. This problem is exacerbated by the fact that in the clinical evaluation of new treatments, we rely on a functional neurological examination only [9]. Although tools, such as the International Standards for Neurological Classification of Spinal Cord Injury (ISNCSCI), provide a standardized assessment of neurological function, it has limitations when used in clinical trials [10]. This highlights the importance of examining other, more objective and representative indicators of the patient's condition, such as biomarkers, for example.

In light of the need to search for more reliable and stable biomarkers in SCI, it seems more representative to study the cerebrospinal fluid (CSF), which is in close contact with the CNS and can provide a more accurate picture of the activity of various molecules in the area of injury. However, it should be noted that blood sampling is more acceptable and safer, and blood serum analysis can also provide information on some key biomarkers, although it requires additional research to clarify its value as a possible diagnostic and prognostic tool in the context of SCI.

Analysis of changes in cytokine levels at different periods after injury and their correlation with the American Spinal Injury Association (ASIA) Impairment Scale score can provide valuable insights not only into the pathophysiology of SCI but also potentially guide therapeutic interventions [11,12]. By including more samples of SCI patients and using multiplex analysis, our aim was to determine how cytokine change correlates with time after severe injury, which may help develop new strategies for early diagnosis and personalized treatment of SCI.

2. Materials and Methods

2.1. Study Population

All patients (n = 40) were recruited at the Neurosurgical Department No. 2 of the Republican Clinical Hospital (Kazan, Russia). Before proceeding, each participant gave their written consent to have cerebrospinal fluid (CSF) and blood serum samples collected for the study. The Local Ethical Committee of Kazan Federal University granted study approval (Protocol No. 3, dated 23 March 2017).

For this prospective study, the patient selection criteria for acute and subacute SCI were as follows: individuals older than 18; SCI locations between vertebrae C3 and L3; capability for valid neurological assessment; and initial classification as ASIA grade A. An experienced research neurologist evaluated the baseline severity of neurological damage using ASIA grade assessments. Follow-up neurological checks took place at 1- and 2-weeks post-injury to determine ASIA conversion and possible motor and sensory score improvement. Among the SCI patients, injuries were located in the cervical (C) region for 18 patients, thoracic (Th) for 14 patients, and lumbar (L) for 8 patients. Of the 40 patients, 35 (87%) were male, and the mean age was 42.2 ± 15.1 years.

For the control group, serum samples were drawn from 16 physically healthy individuals who willingly consented to venous blood collection. Uninjured subjects were recruited from the Kazan Federal University student and academic population and hospital staff. The CSF samples were collected from conditionally healthy patients who sought medical attention due to lumbar disc herniations or stenosis, from whom we obtained written informed consent. The criteria for including uninjured control participants were being older than 18; no past spinal cord or brain injuries; absence of key indicators of acute or chronic inflammation and autoimmune diseases; and a standard full blood count.

2.2. Samples Collection and Storage

The CSF and venous blood samples were obtained from SCI patients at 3, 7, and 14 dpi. The results of blood serum multiplex analysis at 14 dpi were presented by us earlier [13]. The CSF samples from the uninjured control group were collected the day before lumbar spine surgery. Venous blood was collected via standard venipuncture in 6 mL vacuum test tubes (Apexlab, Moscow, Russia). Following a 30-min coagulation period, the blood underwent centrifugation at a speed of 3000 RPM. Subsequently, it was partitioned into 300 µL fractions and preserved at a temperature of $-80\,^{\circ}$C until the time of analysis.

The CSF samples were collected following a strict aseptic technique; a lumbar puncture with an atraumatic needle (Medispine, 20G, Indore, Madhya Pradesh, India) was performed at L3–4, and a 3 mL sample of CSF was collected. The CSF samples were divided into 500 µL aliquots, centrifuged at 1000 RPM for 10 min, and the supernatant was then immediately frozen and stored at $-80\,^{\circ}$C. All obtained samples from injured patients and the uninjured control group underwent identical procedures and were stored for comparable durations.

It is important to note that some patients with SCI may develop cerebrospinal fluid dynamics disorders, which can make it difficult to collect cerebrospinal fluid at various stages after injury. Because of this, we were unable to obtain data for all 40 patients at investigated time points after injury. This means that the number of samples used in the analysis may vary depending on the time of collection.

2.3. Multiplex Analysis

Within the scope of our research, we assessed fluctuations in the cytokine composition of cerebrospinal fluid (CSF) and blood serum from patients with spinal cord injuries (SCI) at different dpi. We employed xMAP Luminex technology for multiplex analysis. The Bio-Plex Pro™ #171AK99MR2 (Bio-Rad, Hercules, CA, USA) facilitated the simultaneous examination of a panel of 40 human cytokines (CCL21, CXCL13, CCL27, CXCL5, CCL11, CCL24, CCL26, CX3CL1, CXCL6, GMCSF, CXCL1, CXCL2, CCL1, IFN-γ, IL1b, IL2, IL4, IL6, CXCL8/IL8, IL10, IL16, CXCL10, CXCL11, MCP-1, MCP-2/CCL8, MCP-3, MCP-4, CCL22, MIF, MIG/CXCL9, MIP-1a/CCL3, MIP-1b/CCL4, MIP-3a/CCL20, MIP-3b/CCL19, MPIF-1/CCL23, CXCL16, SDF-1/CXCL12, CCL17, CCL25 and TNFα), using just 50 µL of the sample. All serum specimens were collectively analyzed in a single assay setup. All the CSF and blood serum samples were analyzed simultaneously within a single assay.

2.4. Statistical Analysis

Data analysis was conducted using version 3.6.3 of R (R Foundation for Statistical Computing, Vienna, Austria). Quantitative variables are illustrated through their mean and standard deviation, as well as their median and interquartile ranges. Before choosing the appropriate statistical tests, we utilized the Pearson test to evaluate data normality.

The Kruskal–Wallis test was used to check for an overall difference in median levels of cytokines between groups, while the Dunn's correction was used for multiple comparisons. We used the Benjamini–Hochberg method for adjustments related to multiple comparisons.

3. Results

3.1. Dynamics of Cerebrospinal Fluid Cytokine Profile after Spinal Cord Injury

We analyzed 40 cytokines in the CSF of 35 patients at various days post-SCI, revealing significant changes in cytokine concentrations at 3, 7, and 14 dpi (Figures 1 and 2). In particular, we found a significant increase in the levels for CCL22 (2.7-fold, P.adj < 0.009), CCL26 (6.6-fold, P.adj < 0.017), IL8 (6.7-fold, P.adj < 0.013), CCL23 (8.2-fold, P.adj < 0.04), and IL6 (58.8-fold, P.adj < 0.001) at 3 dpi compared to uninjured control samples (Figure 3, Supplementary Table S1). At 7 and 14 dpi we continued to observe elevated levels of the abovementioned cytokines, although their concentrations were somewhat lower than at 3 dpi.

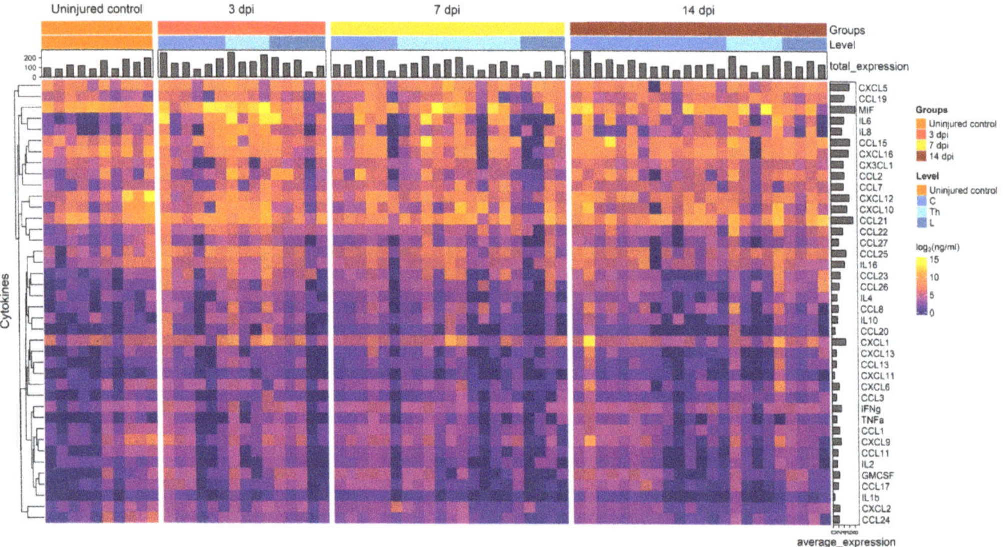

Figure 1. Graphical representation showing log2 cytokine concentrations (color keys), generated with the multiplex analysis of the CSF collected at 3 dpi (*n* = 15), 7 dpi (*n* = 21), 14 dpi (*n* = 23), or from uninjured control (*n* = 10). A dendrogram resulting from hierarchical clustering of cytokines is shown on the left.

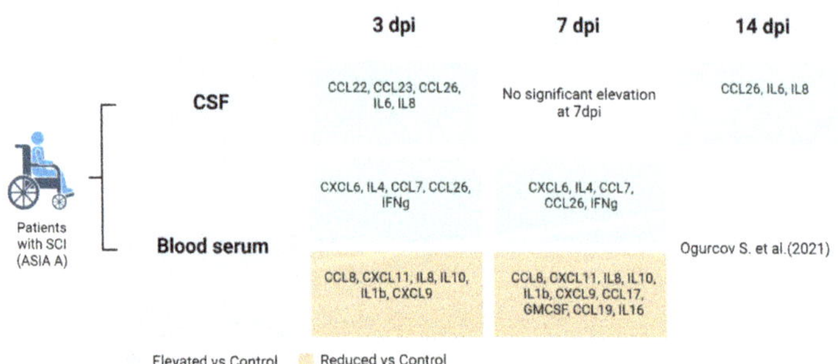

Figure 2. List of cytokines with statistically significant elevation or reduction in CSF and blood serum at 3 dpi, 7 dpi, and 14 dpi [13], compared to uninjured control.

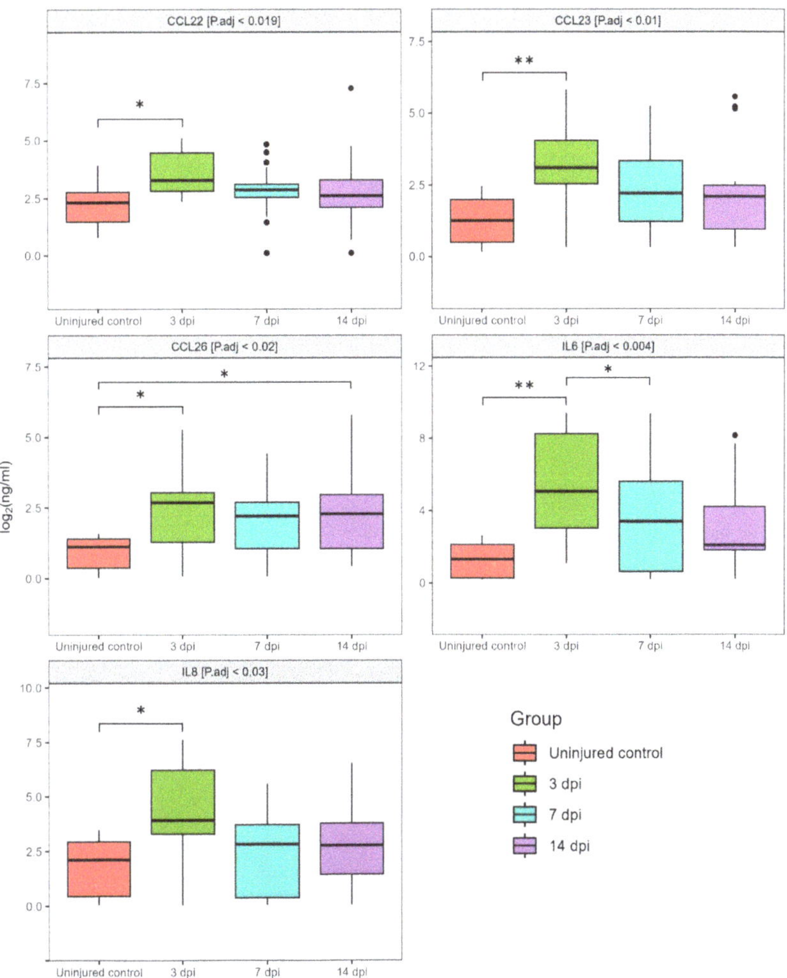

Figure 3. Log2-transformed CSF cytokine concentrations (ng/mL) between 3 dpi (green columns, *n* = 15), 7 dpi (blue columns, *n* = 21), 14 dpi (purple columns, *n* = 23) patients and uninjured subjects (red columns, *n* = 10). * P.adj < 0.05, ** P.adj < 0.01.

At 14 dpi, expression levels of CCL22 (1.3-fold), IL8 (2-fold), IL6 (2.6-fold), CCL23 (2.8-fold), and CCL26 (4.2-fold) were also increased compared to the uninjured control samples. However, it is important to note that the abovementioned changes were only statistically significant for CCL26 [SCI 8.69 (1.83–18.32) vs. the uninjured control 2.06 (0.44–3.06), P.adj < 0.012]. In addition, we found a significant difference in IL6 levels between the 3 and 7 dpi groups with a downward trend as the post-traumatic period increases.

We segregated cytokine levels based on the region of injury for determination of whether the region of SCI affects CSF cytokine levels. There were no discernible differences in cytokine levels among patients with varying SCI regions and the uninjured controls at 3 dpi. However, we observed that CCL17 levels in patients with C injury were significantly elevated compared to SCI at the L region and uninjured control subjects at 7 dpi, showing increases of 45-fold (P.adj < 0.005) and 10-fold (P.adj < 0.045), respectively (Figure 4A). SCI at the Th region had CCL3 levels in two-fold (P.adj < 0.05) and four-fold (P.adj < 0.003) higher compared to SCI at the C region and uninjured control subjects at 14 dpi, accordingly (Figure 4B).

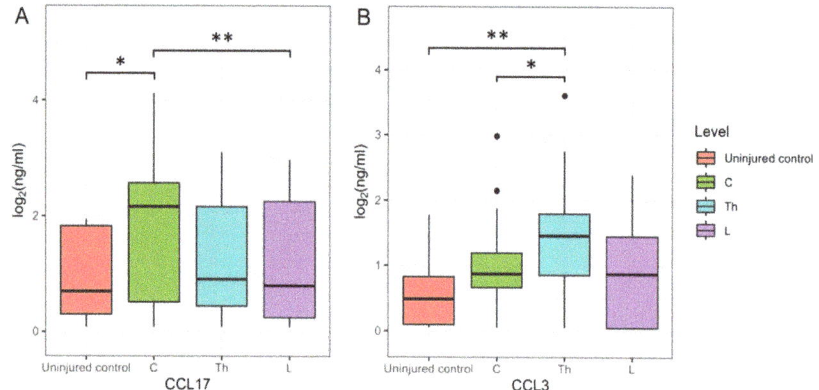

Figure 4. Log2-transformed CSF CCL17 and CCL3 concentrations (ng/mL) at 7 (**A**) and 14 (**B**) dpi, considering the cohort of cervical (C, *n* = 14), thoracic (Th, *n* = 11), and lumbar (L, *n* = 4) patients and uninjured subjects (red columns, *n* = 10). * P.adj < 0.05, ** P.adj < 0.01.

3.2. Dynamics of Blood Serum Cytokine Profile after Spinal Cord Injury

Based on the analysis of blood serum, we found significant changes in cytokine profiles after SCI. Out of the 40 cytokines we analyzed at 3 and 7 dpi, the levels of CCL26, CXCL6, GMCSF, IFN-γ, IL1b, IL4, IL8, IL10, IL16, CXCL11, CCL8, CCL7, CXCL9, CCL3, CCL19, CCL17 showed significant deviations when compared to uninjured control samples (Figure 5, Supplementary Table S1).

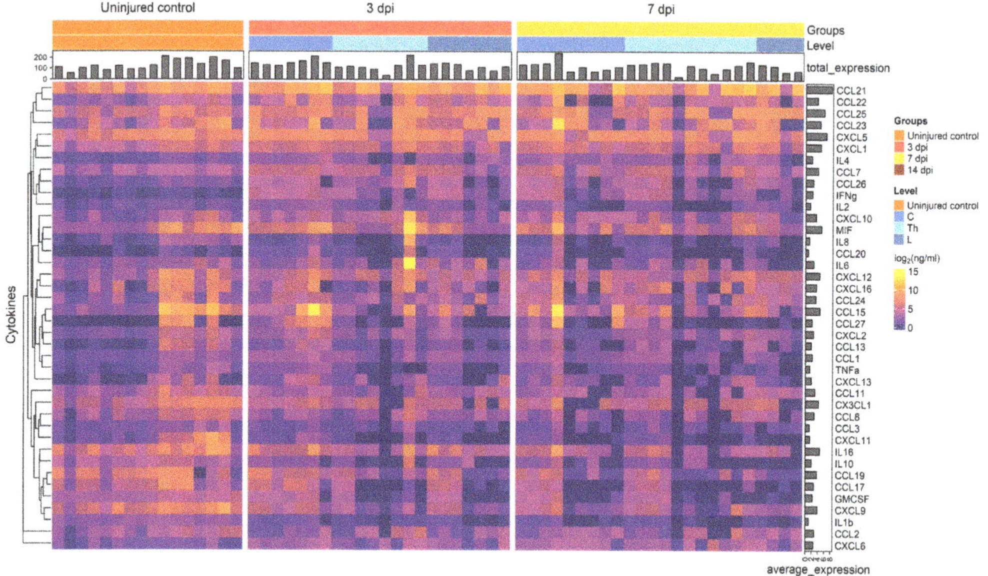

Figure 5. Graphical representation showing log2 cytokine concentrations (color keys), generated with the multiplex analysis of the blood serum collected at 3 dpi (*n* = 22), 7 dpi (*n* = 24), or from uninjured control (*n* = 16). A dendrogram resulting from hierarchical clustering of cytokines is shown on the left.

We found a significant increase in the levels for CXCL6 (2.8-fold, P.adj < 0.008), IL4 (4.5-fold, P.adj < 0.04), CCL7 (5.7-fold, P.adj < 0.01), CCL26 (6-fold, P.adj < 0.002), and IFN-γ

(26.9-fold, P.adj < 0.0001) at 3 dpi compared to uninjured control samples (Figure 6). At 7 dpi, the levels of the abovementioned cytokines also remained elevated: CXCL6 (4,2-fold, P.adj < 0.0002), IL4 (4,8-fold, P.adj < 0.0004), CCL7 (1.7-fold, P.adj < 0.014), CCL26 (5,7-fold, P.adj <0.01), and IFN-γ (14,9-fold, P.adj < 0.0005).

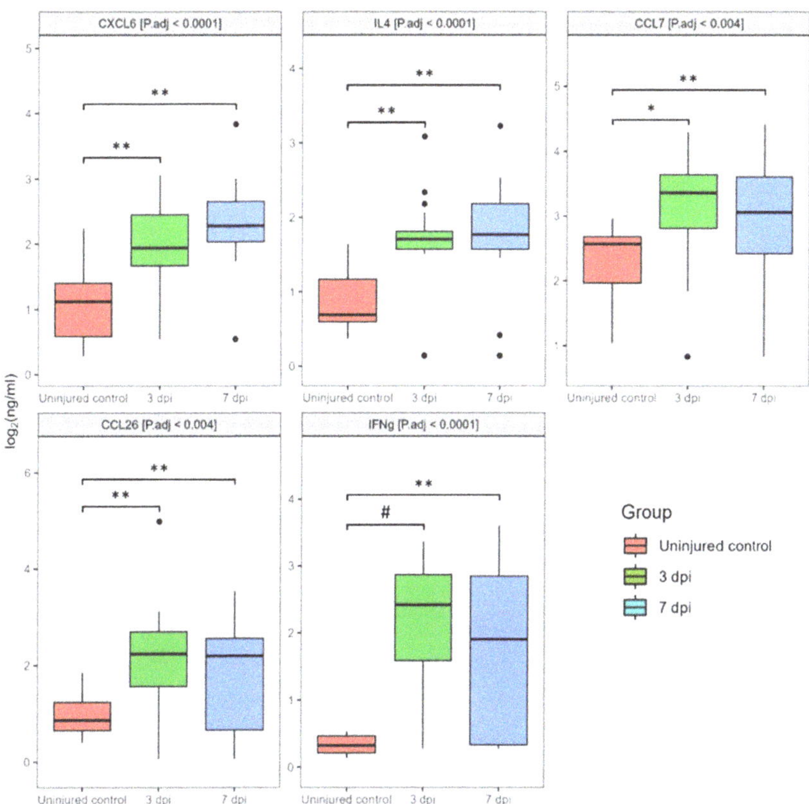

Figure 6. Log2-transformed blood serum cytokine concentrations (ng/mL) between 3 dpi (green columns, n = 22), 7 dpi (blue columns, n = 24) patients, and uninjured subjects (red columns, n = 16). * P.adj < 0.05, ** P.adj < 0.01, # P.adj < 0.0001.

We also detected a significant decrease in several other cytokines at 3 and 7 dpi, including CCL3, CCL8, IL8, CCL19, IL10, IL16, IL1b, CXCL9, CXCL11, GMCSF, CCL17 (Figure 7). In particular, the next cytokines were lower compared to the uninjured control at 3 dpi: CCL8 (2.5-fold, P.adj < 0.05), CXCL11 (2.9-fold, P.adj < 0.005), IL8 (2.9-fold, P.adj < 0.02), IL10 (4-fold, P.adj < 0.0005), IL1b (7-fold, P.adj < 0.0001), and CXCL9 (7.4-fold, P.adj < 0.0001). At 7 dpi we continued to observe decreased levels of the abovementioned cytokines: CCL8 (2.9-fold, P.adj < 0.007), CXCL11 (8.7-fold, P.adj < 0.0008), IL8 (2.6-fold, P.adj < 0.02), IL10 (21.4-fold, P.adj < 0.0001), IL1b (10.7-fold, P.adj < 0.0001), and CXCL9 (10.7-fold, P.adj < 0.0001). In addition, CCL17 (324.5-fold, P.adj < 0.0001), GMCSF (21.2-fold, P.adj < 0.009), CCL19 (19.9-fold, P.adj < 0.003), and IL16 (4.6-fold, P.adj < 0.004) all showed significant decreases in the SCI patient samples at 7 dpi compared to the uninjured control. It is important to note that we did not observe significant differences in the blood serum cytokine profile when comparing 3 and 7 dpi.

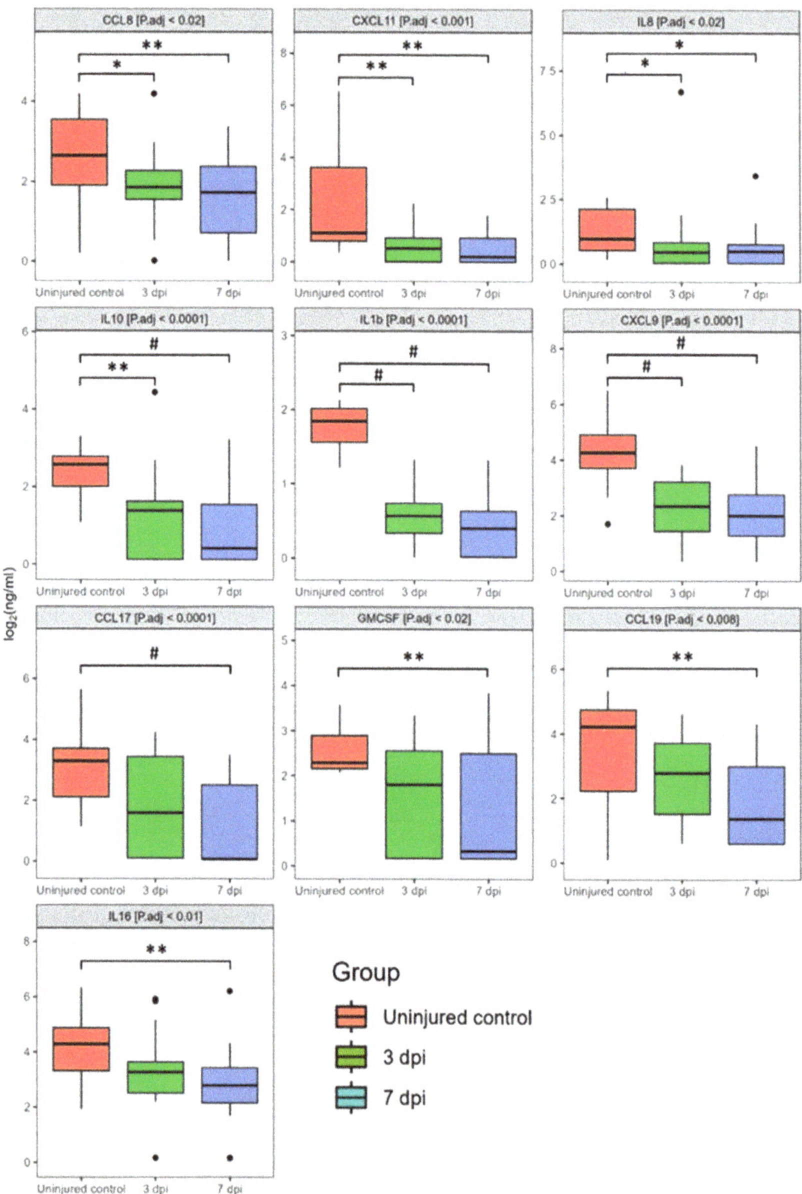

Figure 7. Log2-transformed blood serum cytokine concentrations (ng/mL) between 3 dpi (green columns, *n* = 22), 7 dpi (blue columns, *n* = 24) patients and uninjured subjects (red columns, *n* = 16). * P.adj < 0.05, ** P.adj < 0.01, # P.adj < 0.0001.

We also segregated cytokine levels based on the region of injury for determination of whether the region of SCI affects blood serum cytokine levels. It is interesting to note that changes in cytokine levels associated with the region of injury were detected only at 3 dpi. The CCL27 level in patients with C injury was elevated 560-fold (P.adj < 0.027) and 360-fold (P.adj < 0.034) compared to SCI at the L region and uninjured control subjects, accordingly (Figure 8). The concentration of IL2 was observed to significantly increase by 2.7-fold

(P.adj < 0.012) and 2.5-fold (P.adj < 0.001) in patients with C injury compared to those with SCI at the Th region and uninjured control subjects, respectively. While the concentrations of CCL17 [0.08 (0.08–4.34)], GMCSF [0.42 (0.19–4.58)], and TNFa [0.34 (0.02–0.84)] in patients with Th injury decreased compared to SCI at the C region in 566-fold, 28-fold, and 10-fold (CCL17 [45.34 (13.03–55.02)], GMCSF [11.97 (10.25–16.76)], TNFa [3.41 (2.84–5.95)]), and uninjured control subjects in 76-, 20-, and 3-folds (CCL17 [25.96 (7.71–39.98)], GMCSF [8.81 (7.62–17.18)], TNFa [1.12 (0.8–2.78)]).

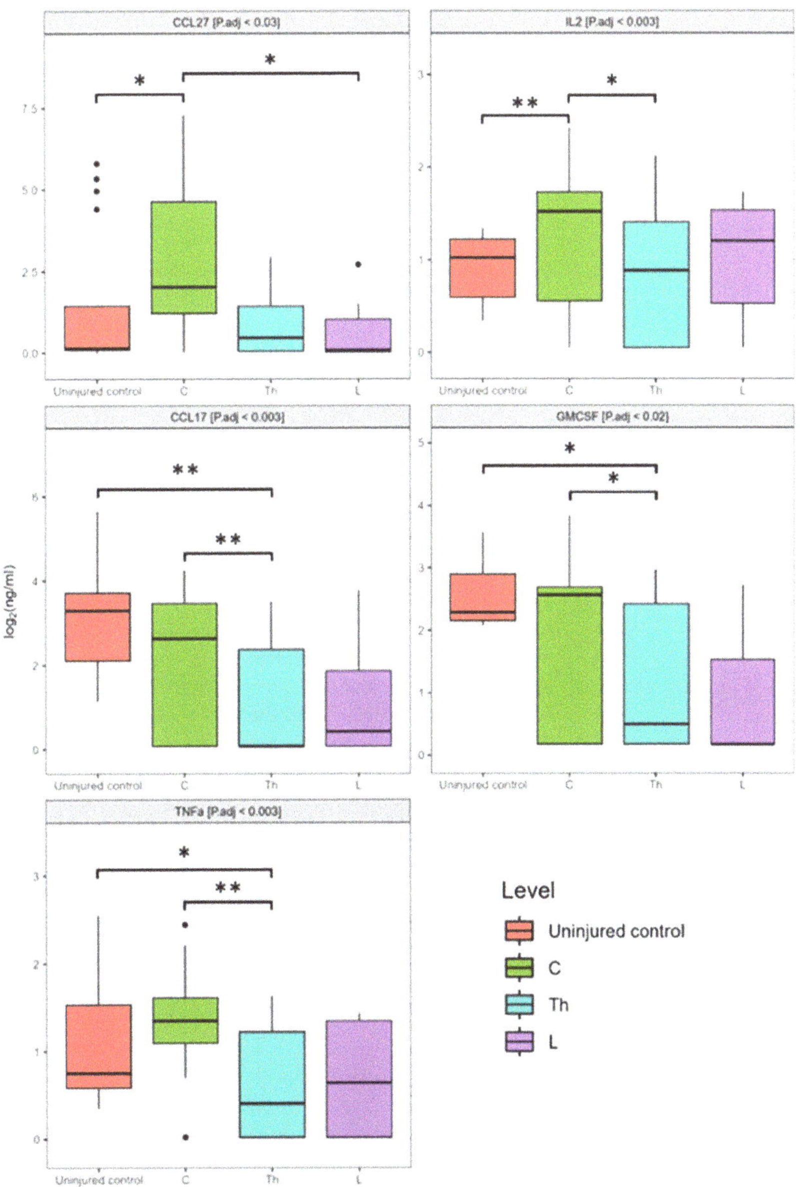

Figure 8. Log2-transformed blood serum cytokine concentrations (ng/mL) between patients at 3 dpi injury, considering the cohort of cervical (C, *n* = 7), thoracic (Th, *n* = 8), and lumbar (L, *n* = 7) patients, and uninjured control subjects (*n* = 16). * P.adj < 0.05, ** P.adj < 0.01.

4. Discussion

In our current study, we performed a multiplex analysis of 40 cytokines in the CSF and serum of patients with SCI at different time periods after injury (3, 7, and 14 dpi), comparing the obtained data with those of uninjured control subjects. We observed both upregulation and downregulation of various cytokines after SCI.

One of the most notable results was a sustained increase in CCL26 levels in SCI in both CSF by 3 and 14 dpi and in serum by 3 and 7 dpi. These results align with our previous study, suggesting a potentially universal role for CCL26 in the inflammatory response following traumatic SCI, regardless of the severity of the injury [13]. Elevated levels of CCL26, both in CSF and blood serum, have been observed in numerous diseases, including Alzheimer's disease [14], asthma [15], rheumatoid arthritis [16], cancer [17], and experimental autoimmune encephalomyelitis [18]. This suggests that CCL26 alone might not provide a definitive diagnostic tool due to its broad involvement in inflammatory responses; it could potentially be used in conjunction with other specific biomarkers to enhance the accuracy of early diagnosis and to tailor personalized treatment strategies for SCI patients.

In our study, we observed a significant increase in IL8 level in the CSF of patients following SCI, while a decrease in IL8 level was noted in the blood serum of the same patient group at 3 dpi. This contrasting pattern suggests a differential regulation of IL8 in the CNS and the systemic circulation following SCI, highlighting the complex and dynamic nature of the inflammatory and immune response to such injuries. Similar results on the level of expression of IL8 in CSF and blood serum were observed by Kossmann et al. (1997) examining patients with TBI. It has been suggested that IL8 release in the CSF after neurotrauma may be associated with BBB dysfunction and nerve growth factor production [19].

Elevated levels of IL6 were also found in the CSF of SCI patients at 3 dpi. The increase in the abovementioned cytokine is consistent with the results of a study by Kwon et al. (2010) where multiplex analysis found that IL6 levels in the CSF of SCI patients are extremely high, especially in patients with severe damage [20]. These results highlight the important role of IL6 and IL8 in the acute inflammatory response after SCI. IL6, for example, is significantly increased within hours of injury in both rodents and humans, and blocking its receptor (IL6R) has been shown to reduce glial scar, neutrophil and macrophage invasion, and improve functional recovery after SCI [21,22]. Using electrochemiluminescence, Fertleman et al. (2022) found that levels of many cytokines in CSF, including IL6 and IL8, are significantly increased after orthopedic surgery. The next day after the operation, their concentration even exceeds the figures recorded immediately after surgery [23]. Based on the above data, it can be assumed that after spinal decompression surgery for SCI, the levels of these cytokines may also increase. Apparently, IL6 and IL8 alone might not provide a definitive diagnostic tool but can be used in combined biomarker approaches.

In our study, we also observed a significant increase in the CCL22 level in the CSF of patients with SCI at 3 dpi. Functionally, CCL22 is a potent chemoattractant for antigen-testing but non-resting T-lymphocytes, as well as for NK cells and monocytes. CCL22 levels in CSF have been reported to be elevated in women with multiple sclerosis [24]. These data, combined with our observations, highlight the potential importance of CCL22 as an inflammatory marker not only in neurodegenerative diseases, such as multiple sclerosis, but also in acute SCI.

We observed a sharp decrease in the pro-inflammatory cytokine IL1b at 3 and 7 dpi in blood serum, mirroring trends in our previous SCI study in rats [25]. In the study performed by Biglari et al., IL1b concentration fluctuated greatly between 4 h and 7 dpi. However, between 1 and 4 weeks after injury, the level of IL1b in blood serum significantly decreased only in those patients in whom the improvement was less significant [11].

A significant decrease in blood serum chemokines CXCL9 and CXCL11 was found by our investigation after SCI at 3 and 7 dpi. Three structurally related chemokines (CXCL9, CXCL10, CXCL11) form the subgroup, the interferon (IFN)-inducible non-ELR

CXC chemokine [26]. The chemokines CXCL9 and CXCL11 function as central modulators of the immune response, predominantly in inflammatory cascades. Stein et al. (2013) in a pilot study on the simultaneous assessment of the level of 21 cytokines in the blood plasma of patients with chronic (>1 year from initial injury) SCI, on the contrary, found a significant increase in CXCL9 levels in patients with chronic SCI compared with uninjured control subjects [27].

We also found that the level of IFN-γ in the blood serum of patients was significantly increased by both 3 and 7 dpi, which may indicate inflammation and secondary damage that occurs after SCI [28]. IFN-γ has been previously shown to be involved in antimicrobial immunity and is elevated in various bacterial diseases [29]. This cytokine is also known for its role in antitumor immunity [30]. Monitoring of this cytokine can provide insight into the severity of injury and the body's response. However, since the level of IFN-γ is also increased in other pathological conditions, its use will only be justified along with other specific biomarkers.

Our data confirm that the inflammatory and immune response to SCI is a complex and multifaceted process involving various molecules. Unfortunately, the assessment of cytokine or chemokine levels separately cannot be a valuable tool for diagnosing and monitoring the status of patients with SCI. That is why the combined biomarker approach could offer a more comprehensive understanding of the individual patient's inflammatory profile, thereby improving the precision of therapeutic interventions. Despite the increased activity of researchers in this area, additional investigations are needed to better understand the role of injury-responsive cytokines in the context of SCI and their potential application in clinical practice.

5. Study Limitations

Although our study provides valuable insights, it is not without its shortcomings. This study was limited to just 40 patients with SCI, requiring larger sample sizes in future studies. The limited number of participants likely reduced the study's ability to detect significant cytokine patterns. We encountered challenges in simultaneously collecting blood and cerebrospinal fluid from all SCI patients, attributed to cerebrospinal fluid dynamics disorders arising in certain posttraumatic phases. It should also be noted that there is a need to confirm the results using other methods of molecular analysis.

6. Conclusions

In light of the need to find more reliable and stable SCI biomarkers, we conducted this study, which greatly expands on our previous analysis. In an earlier published paper, we identified several cytokines in blood serum that could potentially serve as biomarkers for the severity of SCI [13]. However, given the active involvement of cytokines in the regulation of the inflammatory and immune response and their possible crucial importance for the outcome of SCI [28], we hypothesized that analysis of blood serum alone may not fully reflect the actual activity of cytokines at the site of injury in the CNS. In this context, in the current study, we extended our analysis to cytokine profiles in both CSF and blood serum in more patients and at various stages after SCI. This allowed us to explore in more detail the dynamics of change and the relationship between cytokine levels, injury severity, and recovery time, paving the way for a better understanding of the pathophysiology of SCI and potentially more targeted and personalized therapeutic interventions.

Supplementary Materials: The following supporting information can be downloaded at: https://www.mdpi.com/article/10.3390/biomedicines11102641/s1, Table S1. Cytokine concentrations (ng/mL) in CSF and blood serum at 3, 7 and 14 days post-spinal cord injury in patients and uninjured control subjects.

Author Contributions: Conceptualization, S.O. and Y.M.; methodology, E.G.; software, D.S.; validation, Y.M., D.S. and A.R.; formal analysis, I.K., D.S., A.S. and S.O.; investigation, S.O., I.S., D.S., A.S. and Y.M.; resources, A.R.; data curation, A.R.; writing—original draft preparation, Y.M., I.K., S.O. and

D.S.; writing—review and editing, A.R.; visualization, S.O., I.S. and D.S.; supervision, A.R.; project administration, Y.M.; funding acquisition, A.R. All authors have read and agreed to the published version of the manuscript.

Funding: The study was funded by the subsidy allocated to Kazan Federal University for the state assignment No. FZSM-2023-0011 in the sphere of scientific activities.

Institutional Review Board Statement: The study was conducted according to the guidelines of the Declaration of Helsinki and approved by the Kazan Federal University Local Ethical Committee (Protocol No. 3, 23 March 2017).

Informed Consent Statement: Informed consent was obtained from all subjects involved in the study.

Data Availability Statement: The data presented in this study are available on request from the corresponding author. The data are not publicly available due to the evolving nature of the project.

Acknowledgments: This paper has been supported by the Kazan Federal University Strategic Academic Leadership Program (PRIORITY-2030).

Conflicts of Interest: The authors declare no conflict of interest.

References

1. Ahuja, C.S.; Wilson, J.R.; Nori, S.; Kotter, M.R.N.; Druschel, C.; Curt, A.; Fehlings, M.G. Traumatic Spinal Cord Injury. *Nat. Rev. Dis. Primers* **2017**, *3*, 17018. [CrossRef] [PubMed]
2. Singh, A.; Tetreault, L.; Kalsi-Ryan, S.; Nouri, A.; Fehlings, M.G. Global Prevalence and Incidence of Traumatic Spinal Cord Injury. *Clin. Epidemiol.* **2014**, *6*, 309–331. [CrossRef] [PubMed]
3. Turczyn, P.; Wojdasiewicz, P.; Poniatowski, Ł.A.; Purrahman, D.; Maślińska, M.; Żurek, G.; Romanowska-Próchnicka, K.; Żuk, B.; Kwiatkowska, B.; Piechowski-Jóźwiak, B.; et al. Omega-3 Fatty Acids in the Treatment of Spinal Cord Injury: Untapped Potential for Therapeutic Intervention? *Mol. Biol. Rep.* **2022**, *49*, 10797–10809. [CrossRef] [PubMed]
4. Saadoun, S.; Grassner, L.; Belci, M.; Cook, J.; Knight, R.; Davies, L.; Asif, H.; Visagan, R.; Gallagher, M.J.; Thomé, C.; et al. Duroplasty for Injured Cervical Spinal Cord with Uncontrolled Swelling: Protocol of the DISCUS Randomized Controlled Trial. *Trials* **2023**, *24*, 497. [CrossRef]
5. Kimura, A.; Suehiro, K.; Mukai, A.; Fujimoto, Y.; Funao, T.; Yamada, T.; Mori, T. Protective Effects of Hydrogen Gas against Spinal Cord Ischemia–Reperfusion Injury. *J. Thorac. Cardiovasc. Surg.* **2022**, *164*, e269–e283. [CrossRef]
6. Ahuja, C.S.; Nori, S.; Tetreault, L.; Wilson, J.; Kwon, B.; Harrop, J.; Choi, D.; Fehlings, M.G. Traumatic Spinal Cord Injury—Repair and Regeneration. *Neurosurgery* **2017**, *80*, S9–S22. [CrossRef]
7. Kubaszewski, Ł.; Wojdasiewicz, P.; Rożek, M.; Słowińska, I.E.; Romanowska-Próchnicka, K.; Słowiński, R.; Poniatowski, Ł.A.; Gasik, R. Syndromes with Chronic Non-Bacterial Osteomyelitis in the Spine. *Rheumatology* **2016**, *53*, 328–336. [CrossRef]
8. Fehlings, M.G.; Tetreault, L.A.; Wilson, J.R.; Kwon, B.K.; Burns, A.S.; Martin, A.R.; Hawryluk, G.; Harrop, J.S. A Clinical Practice Guideline for the Management of Acute Spinal Cord Injury: Introduction, Rationale, and Scope. *Glob. Spine J.* **2017**, *7*, 84S–94S. [CrossRef]
9. Kirshblum, S.C.; Burns, S.P.; Biering-Sorensen, F.; Donovan, W.; Graves, D.E.; Jha, A.; Johansen, M.; Jones, L.; Krassioukov, A.; Mulcahey, M.J.; et al. International Standards for Neurological Classification of Spinal Cord Injury (Revised 2011). *J. Spinal Cord Med.* **2011**, *34*, 535–546. [CrossRef]
10. Kirshblum, S.; Snider, B.; Eren, F.; Guest, J. Characterizing Natural Recovery after Traumatic Spinal Cord Injury. *J. Neurotrauma* **2021**, *38*, 1267–1284. [CrossRef]
11. Biglari, B.; Swing, T.; Child, C.; Büchler, A.; Westhauser, F.; Bruckner, T.; Ferbert, T.; Jürgen Gerner, H.; Moghaddam, A. A Pilot Study on Temporal Changes in IL-1β and TNF-α Serum Levels after Spinal Cord Injury: The Serum Level of TNF-α in Acute SCI Patients as a Possible Marker for Neurological Remission. *Spinal Cord* **2015**, *53*, 510–514. [CrossRef] [PubMed]
12. Chen, Y.; Wang, D.; Cao, S.; Hou, G.; Ma, H.; Shi, B. Association between Serum IL-37 and Spinal Cord Injury: A Prospective Observational Study. *Biomed. Res. Int.* **2020**, *2020*, 6664313. [CrossRef] [PubMed]
13. Ogurcov, S.; Shulman, I.; Garanina, E.; Sabirov, D.; Baichurina, I.; Kuznetcov, M.; Masgutova, G.; Kostennikov, A.; Rizvanov, A.; James, V.; et al. Blood Serum Cytokines in Patients with Subacute Spinal Cord Injury: A Pilot Study to Search for Biomarkers of Injury Severity. *Brain Sci.* **2021**, *11*, 322. [CrossRef] [PubMed]
14. Westin, K.; Buchhave, P.; Nielsen, H.; Minthon, L.; Janciauskiene, S.; Hansson, O. CCL2 Is Associated with a Faster Rate of Cognitive Decline during Early Stages of Alzheimer's Disease. *PLoS ONE* **2012**, *7*, e30525. [CrossRef]
15. Larose, M.-C.; Chakir, J.; Archambault, A.-S.; Joubert, P.; Provost, V.; Laviolette, M.; Flamand, N. Correlation between CCL26 Production by Human Bronchial Epithelial Cells and Airway Eosinophils: Involvement in Patients with Severe Eosinophilic Asthma. *J. Allergy Clin. Immunol.* **2015**, *136*, 904–913. [CrossRef]
16. Zhebrun, D.A.; Totolyan, A.A.; Maslyanskii, A.L.; Titov, A.G.; Patrukhin, A.P.; Kostareva, A.A.; Gol'tseva, I.S. Synthesis of Some CC Chemokines and Their Receptors in the Synovium in Rheumatoid Arthritis. *Bull. Exp. Biol. Med.* **2014**, *158*, 192–196. [CrossRef]

17. Chen, X.; An, Y.; Zhang, Y.; Xu, D.; Chen, T.; Yang, Y.; Chen, W.; Wu, D.; Zhang, X. CCL26 Is Upregulated by Nab-Paclitaxel in Pancreatic Cancer–Associated Fibroblasts and Promotes PDAC Invasiveness through Activation of the PI3K/AKT/MTOR Pathway. *Acta Biochim. Biophys. Sin.* **2021**, *53*, 612–619. [CrossRef]
18. Shou, J.; Peng, J.; Zhao, Z.; Huang, X.; Li, H.; Li, L.; Gao, X.; Xing, Y.; Liu, H. CCL26 and CCR3 Are Associated with the Acute Inflammatory Response in the CNS in Experimental Autoimmune Encephalomyelitis. *J. Neuroimmunol.* **2019**, *333*, 576967. [CrossRef]
19. Kossmann, T.; Stahel, P.F.; Lenzlinger, P.M.; Redl, H.; Dubs, R.W.; Trentz, O.; Schlag, G.; Morganti-Kossmann, M.C. Interleukin-8 Released into the Cerebrospinal Fluid after Brain Injury Is Associated with Blood–Brain Barrier Dysfunction and Nerve Growth Factor Production. *J. Cereb. Blood Flow Metab.* **1997**, *17*, 280–289. [CrossRef]
20. Kwon, B.K.; Stammers, A.M.T.; Belanger, L.M.; Bernardo, A.; Chan, D.; Bishop, C.M.; Slobogean, G.P.; Zhang, H.; Umedaly, H.; Giffin, M.; et al. Cerebrospinal Fluid Inflammatory Cytokines and Biomarkers of Injury Severity in Acute Human Spinal Cord Injury. *J. Neurotrauma* **2010**, *27*, 669–682. [CrossRef]
21. Okada, S.; Nakamura, M.; Mikami, Y.; Shimazaki, T.; Mihara, M.; Ohsugi, Y.; Iwamoto, Y.; Yoshizaki, K.; Kishimoto, T.; Toyama, Y.; et al. Blockade of Interleukin-6 Receptor Suppresses Reactive Astrogliosis and Ameliorates Functional Recovery in Experimental Spinal Cord Injury. *J. Neurosci. Res.* **2004**, *76*, 265–276. [CrossRef] [PubMed]
22. Nakamura, M.; Okada, S.; Toyama, Y.; Okano, H. Role of IL-6 in Spinal Cord Injury in a Mouse Model. *Clin. Rev. Allergy Immunol.* **2005**, *28*, 197–204. [CrossRef] [PubMed]
23. Fertleman, M.; Pereira, C.; Dani, M.; Harris, B.H.L.; Di Giovannantonio, M.; Taylor-Robinson, S.D. Cytokine Changes in Cerebrospinal Fluid and Plasma after Emergency Orthopaedic Surgery. *Sci. Rep.* **2022**, *12*, 2221. [CrossRef]
24. Galimberti, D.; Fenoglio, C.; Comi, C.; Scalabrini, D.; De Riz, M.; Leone, M.; Venturelli, E.; Cortini, F.; Piola, M.; Monaco, F. MDC/CCL22 Intrathecal Levels in Patients with Multiple Sclerosis. *Mult. Scler. J.* **2008**, *14*, 547–549. [CrossRef] [PubMed]
25. Mukhamedshina, Y.O.; Akhmetzyanova, E.R.; Martynova, E.V.; Khaiboullina, S.F.; Galieva, L.R.; Rizvanov, A.A. Systemic and Local Cytokine Profile Following Spinal Cord Injury in Rats: A Multiplex Analysis. *Front. Neurol.* **2017**, *8*, 581. [CrossRef]
26. Müller, M.; Carter, S.; Hofer, M.J.; Campbell, I.L. The Chemokine Receptor CXCR3 and Its Ligands CXCL9, CXCL10 and CXCL11 in Neuroimmunity—A Tale of Conflict and Conundrum. *Neuropathol. Appl. Neurobiol.* **2010**, *36*, 368–387. [CrossRef]
27. Stein, A.; Panjwani, A.; Sison, C.; Rosen, L.; Chugh, R.; Metz, C.; Bank, M.; Bloom, O. Pilot Study: Elevated Circulating Levels of the Proinflammatory Cytokine Macrophage Migration Inhibitory Factor in Patients with Chronic Spinal Cord Injury. *Arch. Phys. Med. Rehabil.* **2013**, *94*, 1498–1507. [CrossRef]
28. Hellenbrand, D.J.; Quinn, C.M.; Piper, Z.J.; Morehouse, C.N.; Fixel, J.A.; Hanna, A.S. Inflammation after Spinal Cord Injury: A Review of the Critical Timeline of Signaling Cues and Cellular Infiltration. *J. Neuroinflammation* **2021**, *18*, 284. [CrossRef]
29. Shtrichman, R.; Samuel, C.E. The Role of Gamma Interferon in Antimicrobial Immunity. *Curr. Opin. Microbiol.* **2001**, *4*, 251–259. [CrossRef]
30. Kursunel, M.A.; Esendagli, G. The Untold Story of IFN-γ in Cancer Biology. *Cytokine Growth Factor. Rev.* **2016**, *31*, 73–81. [CrossRef]

Article

Transcutaneous Spinal Cord Stimulation Improves Respiratory Muscle Strength and Function in Subjects with Cervical Spinal Cord Injury: Original Research

Hatice Kumru [1,2,3,*], Loreto García-Alén [1], Aina Ros-Alsina [1], Sergiu Albu [1,2,3], Margarita Valles [1,2,3] and Joan Vidal [1,2,3]

[1] Institut Guttmann, Institut Universitari de Neurorehabilitació Adscrit a la (UAB), 08916 Barcelona, Spain; loretogarcia@guttmann.com (L.G.-A.); superaina@gmail.com (A.R.-A.); salbu@guttmann.com (S.A.); mvalles@guttmann.com (M.V.); jvidal@guttmann.com (J.V.)
[2] Universitat Autònoma de Barcelona, Cerdanyola del Vallès, Bellaterra, 08193 Barcelona, Spain
[3] Fundació Institut d'Investigació en Ciències de la Salut Germans Trias i Pujol, 08916 Barcelona, Spain
* Correspondence: hkumru@guttmann.com

Abstract: (1) Background: Respiratory muscle weakness is common following cervical spinal cord injury (cSCI). Transcutaneous spinal cord stimulation (tSCS) promotes the motor recovery of the upper and lower limbs. tSCS improved breathing and coughing abilities in one subject with tetraplegia. Objective: We therefore hypothesized that tSCS applied at the cervical and thoracic levels could improve respiratory function in cSCI subjects; (2) Methods: This study was a randomized controlled trial. Eleven cSCI subjects received inspiratory muscle training (IMT) alone. Eleven cSCI subjects received tSCS combined with IMT (six of these subjects underwent IMT alone first and then they were given the opportunity to receive tSCS + IMT). The subjects evaluated their sensation of breathlessness/dyspnea and hypophonia compared to pre-SCI using a numerical rating scale. The thoracic muscle strength was assessed by maximum inspiratory (MIP), expiratory pressure (MEP), and spirometric measures. All assessments were conducted at baseline and after the last session. tSCS was applied at C3-4 and Th9-10 at a frequency of 30 Hz for 30 min on 5 consecutive days; (3) Results: Following tSCS + IMT, the subjects reported a significant improvement in breathlessness/dyspnea and hypophonia ($p < 0.05$). There was also a significant improvement in MIP, MEP, and forced vital capacity ($p < 0.05$). Following IMT alone, there were no significant changes in any measurement; (4) Conclusions: Current evidence supports the potential of tSCS as an adjunctive therapy to accelerate and enhance the rehabilitation process for respiratory impairments following SCI. However, further research is needed to validate these results and establish the long-term benefits of tSCS in this population.

Keywords: transcutaneous electrical spinal cord stimulation; cervical spinal cord injury; respiratory function; rehabilitation

Citation: Kumru, H.; García-Alén, L.; Ros-Alsina, A.; Albu, S.; Valles, M.; Vidal, J. Transcutaneous Spinal Cord Stimulation Improves Respiratory Muscle Strength and Function in Subjects with Cervical Spinal Cord Injury: Original Research. *Biomedicines* **2023**, *11*, 2121. https://doi.org/10.3390/biomedicines11082121

Academic Editors: Jacek M. Kwiecien and Nicolas Guerout

Received: 15 June 2023
Revised: 15 July 2023
Accepted: 22 July 2023
Published: 27 July 2023

1. Introduction

Impaired respiratory function is a common consequence of cervical spinal cord injury (cSCI) and may also occur following thoracic injuries [1–3]. A total of 84% of cervical (C1-4), 60% of C5-8, and 65% of thoracic (Th1-12) injuries had respiratory complications [1]. It affects ventilation and lung volumes, leading to breathlessness or dyspnea and an increased work of breathing [2,3]. A disruption of supraspinal input to intercostal and abdominal motor neurons can lead to reduced respiratory capacities, peak expiratory flow, and coughing strength. An impairment of the muscles of inspiration reduces vital capacity, prevents deep breaths, and may lead to dyspnea with exertion and/or collapse of the lungs (atelectasis). Dysfunctional expiratory muscles impair cough and secretion clearance, increase airways resistance and increase the susceptibility to and persistence of lower respiratory tract infections. Total lung capacity is usually reduced following SCI due to impaired inspiratory musculature and residual volume is relatively increased due to impaired expiratory musculature and subsequent reduced expiratory reserve volume [2–4]. Impaired respiratory function is a significant contributor to the overall morbidity and mortality of SCI individuals [5,6]. Thus, any intervention that could be shown to improve respiratory function and thereby prevent or alleviate these conditions would be of great benefit to the SCI population. On a day-to-day basis, if respiratory muscle training is effective for a person with SCI, it could mean breathing and speaking more easily, coughing and expelling mucus more effectively, and avoiding hospital admissions [2,4]. Each of these outcomes (both individually and in combination) could significantly improve such a person's quality of life.

Respiratory muscle training (RMT) is a non-invasive intervention that can improve respiratory muscle function and reduce the risk of respiratory complications in individuals with SCI, and is widely used in our hospital. RMT involves exercises that target the inspiratory and expiratory muscles to improve strength (maximal inspiratory and expiratory pressure: MIP and MEP), endurance, and coordination [2,3]. In addition to RMT, other non-invasive interventions in individuals with SCI include breathing exercises, airway clearance techniques, and positioning. These interventions aim at optimizing lung volume, reducing airway resistance, and promoting efficient breathing patterns [2,3]. The other current invasive neuromodulation techniques, including phrenic nerve pacemaker (PNP) [7] and a technique known as diaphragm motor point pacing (DMPP), have been developed and successfully tested [8] specifically in individuals with lesions above the C4 level [9]. DMPP is a technique that involves the implantation of electrodes near the motor points of the diaphragm to stimulate the diaphragm muscles, thereby improving respiratory function in people with spinal cord injury, used at the Guttmann Institute. In addition to these invasive techniques, epidural spinal cord stimulation has also shown promising results in improving respiratory function following SCI [10].

Transcutaneous electrical spinal cord stimulation (tSCS) is a non-invasive spinal cord stimulation technique promoting the functional and motor strength recovery of the upper [11–15] or lower extremities [16–18] and improving trunk stability [19,20] in SCI individuals. The most likely mechanisms of tSCS first involve acutely elevating the excitability and plasticity of residual spinal and supraspinal neuronal networks, through the recruitment of afferent fibers from the posterior roots of the spinal cord [11,19–23], and, secondly, and secondly, with training shapes to the more plastic network connectivity toward a more normal coordinated functional state guided via use-dependent mechanism [11–15].

Most tSCS studies so far have focused on lower [16–18] or upper extremity function [11–15], addressing both the proof of concept and elaborating on the mechanisms involved. However, much less is known about the feasibility of neuromodulating the brain-to-spinal-cord connectome controlling respiratory function. Mechanisms underlying the action of tSCS combined with physical training, although partially overlapping, may involve different and perhaps synergistic processes leading to a more efficient reorganization of neural circuits [11,22]. A recent report suggested a positive impact of tSCS over the cervical spinal cord on the breathing and coughing ability in a patient with chronic tetraplegia [24]. However, this was a single case study and further research is needed to establish the safety and efficacy of tSCS for improving respiratory function in larger patient populations [24].

Here, our study aimed to investigate the effects of tSCS combined with respiratory training on respiratory parameters in individuals with cervical SCI. The study sought to provide empirical evidence for this hypothesis by evaluating the impact of tSCS on inspiratory and expiratory muscle strength, pulmonary capacity, and subjective perception of respiratory function. The intention was to determine whether the combination of tSCS and respiratory training could lead to improvements in respiratory function more than respiratory training alone.

2. Materials and Methods

2.1. Subjects

Inclusion criteria were: (i) age between 18 and 70 years; (ii) SCI at the cervical level; (iii) time following SCI > 3 months because stabilization of respiratory function occurs during the first 3 months [25]; (iv) sensation of breathlessness/dyspnea scoring $\geq$ 3 according to a numerical rating scale (NRS; 0 = no changes, 10 = severe problem) compared to pre-SCI; (v) no change in medical treatment at least one week before and during the study; (vi) agreement to participate after signing a written informed consent form.

Exclusion criteria were: (i) respiratory problems before SCI; (ii) mechanically ventilated; (iii) unstable SCI or other health condition (cancer, unstable pulmonary or heart disease, etc.); (iv) contraindication for tSCS (e.g., pacemaker); (v) electrical stimulation intolerance; (vi) pregnancy.

SCI subjects were recruited from in-patients at Guttmann Institute (Badalona, Spain), where the study was conducted. The protocol was approved by the Ethics Committee of the Guttmann Institute and was carried out in accordance with the standards of the Declaration of Helsinki (protocol code 2019_317 and date of approval: 4 December 2019).

2.2. Clinical Assessment of Spinal Cord Injury

At baseline, the severity and level of SCI was assessed according to the American Spinal Cord Injury Association (ASIA) Impairment Scale (AIS) and the International Standards for Neurological Classification of Spinal Cord, which consists of 5 grades (A, B, C, D, E) [26]. We also collected height and weight of each subject.

Clinical assessment of SCI, weight, height, and neurophysiology was obtained at baseline only.

2.3. Assessment of Respiratory Function

The study used both subjective and objective measures to evaluate the effectiveness of the intervention. The subjective degree of breathlessness/dyspnea and hypophonia was scored by the subjects using an NRS (0 = no changes, 10 = severe problem) compared to pre-SCI. Objective measures included maximal inspiratory pressure (MIP), maximal expiratory pressure (MEP), and spirometric measures such as forced vital capacity (FVC), forced expiratory volume in 1 s (FEV1), peak expiratory flow (PEF), and forced expiratory flow (FEF25%, FEF50%, FEF75%). These measures are commonly used to assess respiratory motor function and to provide information on respiratory muscle weakness and function. MIP and MEP were measured using the MicroRPM Respiratory Pressure Meter (Micro Direct, Inc., Lewiston, ID, USA), while spirometric measures were performed using the Datospir Micro (Sibelmed; Barcelona, Spain). FVC is the volume of air that can forcibly be blown out after full inspiration; FEV1 is the volume of air that can forcibly be blown out in first 1 s after full inspiration; FEF is the flow (or speed) of air coming out of the lung during the middle portion of a forced expiration, and the usual discrete intervals are 25%, 50%, and 75%; PEF is the maximal flow (or speed) achieved during the maximally forced expiration initiated at full inspiration.

The measurements were conducted by the same physician (HK) for all subjects. The participants were seated, and any belts were loosened. They were also instructed to wear nose clips. The maneuver was demonstrated, and the participants were instructed to inhale or exhale completely (MIP/MEP) or "blast" the air out for 1 s (spirometric measures) following maximal inhalation. Participants were given the opportunity to practice before the final recordings were taken.

All subjects had their MIP, MEP, and spirometric measures recorded three times at baseline (pre-intervention) and three times after the last session (post-intervention).

2.4. Neurophysiological Assessment

In our previous research, we conducted a study that demonstrated the effectiveness of using 90% of the rest motor threshold (RT) of the hand muscle for transcutaneous spinal cord stimulation (tSCS) in modulating spinal cord and cortex excitability of upper extremity in healthy subjects. Building upon these findings, our aim in the current study was to apply a similar approach by utilizing 90% of the RT of the diaphragm for tSCS at the C3-4 spinal segment and 90% of the RT of the rectus abdominis muscle for tSCS at the Th9-10. In 4 young SCI subjects (age range: 18–23 years), in order to record the spinal motor response of the diaphragm, we placed the superficial electrodes in the parasternal line, and the monophasic rectangular 1 ms single pulses were delivered at the C3-4 level. For recording the spinal motor response of the rectus abdominis, the electrodes were placed at the umbilical level, and the monophasic rectangular 1 ms single pulses were delivered at the Th9-10 level. The monophasic rectangular 1 ms single pulses were delivered at C3-C4 and at Th9-Th10 using 2 cm diameter hydrogel adhesive electrodes as cathodes (Axelgaard, ValuTrode® Cloth) and two 5×12 cm^2 rectangular electrodes (Axelgaard, ValuTrode® Cloth) placed symmetrically over the iliac crests as anodes [22,23]. Intensity was applied up to 120 mA and RT was defined as the lowest intensity that elicited a spinal motor response ≥ 50 µV peak-to-peak amplitude. For recordings, we used routine electrodiagnostic equipment (Medelec Synergy, Oxford Instruments; Surrey, UK). Single electric stimulation was applied using the transcutaneous electrical stimulator BioStim-5 (Cosyma Inc., Moscow, Russia). However, we could not obtain any muscle response in diaphragm, nor in the rectus abdominal muscle. Finally, we decided to use biphasic rectangular 1 ms single pulses at C3-4 and C6-7 to record the RT of the abductor pollicis brevis (APB) muscle. These recorded 90% of RT values were then used as the intensity for tSCS at C3-4 and Th9-10, respectively. By using biphasic rectangular 1 ms pulses, we aimed to reduce the discomfort associated with the stimulation and ensure a more tolerable experience for the participants during the threshold determination and stimulation process, because the monophasic pulses were more painful.

2.5. Experimental Design

Initially, the experimental design was a randomized controlled trial that included two groups: (i) tSCS + IMT group, which was tSCS combined with IMT, and (ii) group of IMT alone (control group). Randomization was conducted using a computer-generated list to assign participants to their respective groups. Subjects who underwent IMT alone were given the opportunity to receive tSCS + IMT after completing the initial IMT (control) trial. This allowance was made in response to their request for tSCS, and it allowed them to experience the tSCS + IMT.

Each group underwent a total of five sessions, each lasting approximately 30 min, over the course of one week. Participant preparation for each session took approximately 15–25 min.

2.6. Interventions

Inspiratory muscle training (IMT): We used inspiratory muscle training (IMT) because this was feasible, safe, and a low-cost intervention that may be effective for cSCI subjects, either when performed alone or in conjunction with expiratory muscle training, then known as respiratory muscle training [27].

All patients received IMT using the Fruugo/Australia Lung Trainer 3 Chamber Breathing apparatus for approximately 30 min, five times per week. The inspiratory load was set at 30% of the mean of three trials of MIP. The training began at the smallest inspiratory pressure setting, and once subjects were able to perform the training sessions easily at a particular resistance setting, they progressed to 10 consecutive inspirations at 30% of MIP with a comfortable respiratory velocity, followed by 60 s of resting time, with a total of 15 repetitions (equivalent to 150 inspirations at 30% of MIP).

2.7. Transcutaneous Spinal Cord Stimulation

tSCS was applied at the respiratory muscle levels: at C3-4 to promote motor function of the diaphragm and intercostal muscles and at Th9-10 to promote motor function of the abdominal muscles. This approach could hold potential benefits, even if the diaphragm was not directly affected by SCI, because, in healthy individuals, tSCS has the potential to enhance the electrical activity and responsiveness of the spinal cord and brain areas involved in motor control, leading to improvements in motor function and potentially enhancing overall motor control, including in healthy subjects [22,23].

tSCS was carried out with a "BioStim-5" (Cosyma Inc., Moscow, Russia). Stimulation was delivered at C3-C4 and at Th9-Th10 using 2 cm diameter hydrogel adhesive electrodes as cathodes (Axelgaard, ValuTrode® Cloth) and two 5×12 cm^2 rectangular electrodes (Axelgaard, ValuTrode® Cloth) placed symmetrically over the iliac crests as anodes [22,23]. tSCS was delivered using biphasic rectangular 1 ms pulses at a frequency of 30 Hz, at 90% of rest motor threshold of APB with each pulse filled with a carrier frequency of 10 kHz.

tSCS was combined with IMT (as described above) and was turned on during training and turned off during resting periods.

Clinical, respiratory, and neurophysiological assessments were performed by a physician (HK), the treatment interventions were realized by a physiotherapist (LG). All assessments and interventions were conducted in the morning before 1:00 pm, at the Neurorehabilitation Hospital of the Guttmann Institute. The study took place between January 2020 and January 2023.

2.8. Data and Statistical Analysis

Data were collected for each subject after their respective assessment, and data analysis was performed after completing the assessment of the last subject. Means of MIP, MEP, and spirometric measures were calculated from three repetitions each. Group data were presented as mean ± standard deviation (SD) from individual means for each group separately. Data distribution was examined using the Kolmogorov–Smirnov test. To evaluate the response to the intervention in the two groups, we performed a repeated measure ANOVA on outcome variables, considering the variable "Time" (pre- and post-intervention) as the within-subject factor and the variable "Intervention" (IMT and tSCS + IMT) as the between-subject factor. For parametric data, paired t-tests were used to compare the data between pre- and post-intervention. When the data distribution was not normal (FVC), the Wilcoxon t-test was used instead. Statistical analyses were conducted using a commercial software package (IBM SPSS, version 21.0, SPSS Inc., Chicago, IL, USA). The significance level was set at $p < 0.05$.

3. Results

3.1. Subjects Clinical and Demographic Characteristics

Thirty-one patients with cervical SCI were recruited to participate in this study. Sixteen patients completed the inclusion criteria and signed a consent form. They were randomly allocated to two experimental groups: IMT alone (n = 11) and tSCS + IMT (n = 5 plus 6 patients who first completed IMT alone and then, after at least one week of break, received tSCS + IMT as shown in flow diagram in Figure 1). This approach was implemented to prevent any potential sustained effects of IMT from influencing the outcomes when the tSCS + IMT was introduced (Figure 1).

The mean age was 29.3 ± 10.1 years for tSCS + IMT and 37.0 ± 5.9 years for the IMT group. All subjects were male except one female in the tSCS + IMC group. The mean time since SCI was 8.1 ± 1.8 months for tSCS + IMT and 7.5 ± 1.7 for the IMT group. Lesion levels of SCI were C3/C4/C5/C6/C7, Lesion level of SCI was similar in both groups (at C3/C4/C5/C6/C7 in IMT group: n = 1/5/3/0/2; tSCS+IMT group: n = 1/5/2/1/2 respectively (Table 1). The severity of SCI according to AIS A/B/C/D was distributed in the IMT group 2/4/4/1, and in the tSCS + IMT group 2/4/4/1 (Table 1). The two groups were comparable in terms of age, time since SCI, lesion level, and SCI severity as assessed by the AIS scale. The statistical analysis showed no significant differences between the groups for these variables ($p > 0.05$ for each comparison, Table 1).

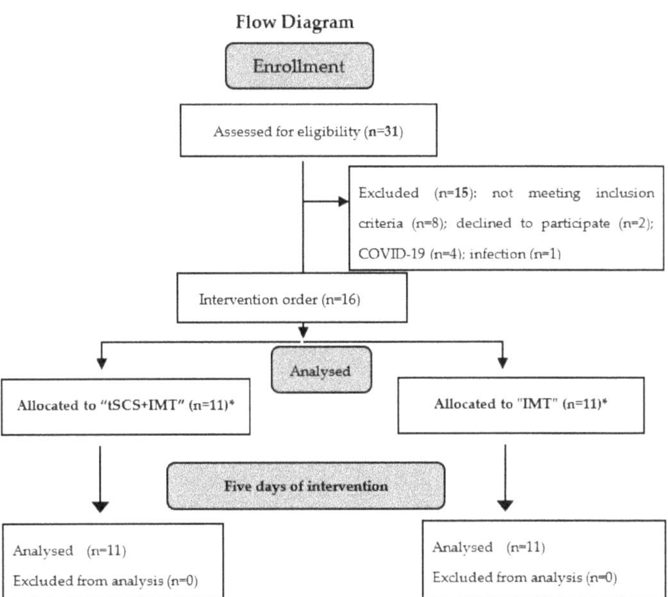

Figure 1. Participant flow throughout the trial duration. IMT: Inspiratory muscle training; tSCS: transcutaneous spinal cord stimulation; *: 6 patients first received IMT alone and, at least one week later, tSCS + IMT.

Table 1. Clinical and demographic characteristics of the subjects with spinal cord injury, heights, weight, and intensity of stimulation at C3-4 (C3) and Th9-10 (Th9) vertebral segments. The last two columns consist of the intensity of stimulation of each of the segment levels that received tSCS.

	Age	Sex	SCI Etiology	SCI Level	AIS	Time Since SCI (Month)	Height (cm)	Weight (kg)	tSCS Intensity at C3 (mA)	tSCS Intensity at Th9 (mA)
IMT	38 *	M	Trauma	C7	B	7	165	54	-	-
IMT	36	M	Trauma	C4	B	9	172	57	-	-
IMT	36	M	Trauma	C4	C	10	167	61	-	-
IMT	31	M	Trauma	C5	A	8	170	76	-	-
IMT	18 *	M	Trauma	C4	B	8	174	48	-	-
IMT	45	M	Trauma	C4	C	6	175	62	-	-
IMT	21 *	M	Trauma	C5	B	9	176	60	-	-
IMT	18 *	M	Trauma	C3	A	7	179	58	-	-
IMT	23 *	F	Trauma	C7	C	6	170	54	-	-
IMT	47	M	Trauma	C4	D	9	172	83.2	-	-
IMT	26 *	M	Trauma	C5	C	4	178	82.5	-	-
tSCS + IMT	25	M	Trauma	C4	A	9	186	76	67	60
tSCS + IMT	46	M	Trauma	C4	C	8	174	76	80	80
tSCS + IMT	28	M	Trauma	C6	A	7	176	71	77	90
tSCS + IMT	18 *	M	Trauma	C4	B	9	174	48	44	52
tSCS + IMT	35	M	Trauma	C4	C	7	175	62	34	40
tSCS + IMT	21 *	M	Trauma	C5	B	9	176	60	52	66
tSCS + IMT	18 *	M	Trauma	C3	B	10	179	58.5	66	70
tSCS + IMT	23 *	F	Trauma	C7	C	9	170	54	62	62
tSCS + IMT	26 *	M	Trauma	C5	C	5	178	82.5	57	61
tSCS + IMT	34	M	Trauma	C4	D	4	175	68.5	49	59
tSCS + IMT	38 *	M	Trauma	C7	B	8	165	54	78	82
p	0.08					1.00	0.20	0.79		

* SCI subjects who first received IMT and, at least one week later, tSCS + IMT. M: male; C: cervical; Th: thoracic.

All data (breathlessness/dyspnea, hypophonia, MIP, MEP, spirometric measures) were similar between both groups ($p > 0.05$) at baseline (pre-treatment) (Table 2).

Table 2. The data of subjective and objective assessment of respiratory changes.

		PRE	POST	p **
tSCS + IMT	breathlessness	5.09 ± 1.14	3.36 ± 1.36	F = 8.272, $p < $ **0.009**, η^2 = 0.293
IMT		4.91 ± 1.67	4.91 ± 1.64	TimexIntervention F = 19.449,
p *		0.336		$p < $ **0.001** η^2 = 0.493
tSCS + IMT	hypophonia	4.00 ± 2.32	2.27 ± 1.35	F = 18.06, $p < $ **0.001**, η^2 = 0.475
IMT		4.18 ± 1.17	3.91 ± 0.83	TimexIntervention F = 9.552,
p *		0.819		$p = $ **0.006**; η^2 = 0.323
tSCS + IMT	MIP	61.18 ± 28.01	69.64 ± 28.93	F = 4.452, $p < $ **0.048**, η^2 = 0.182
IMT		65.61 ± 26.01	65.03 ± 26.69	TimexIntervention F = 5.813,
p *		0.705		$p = $ **0.026**; η^2 = 0.225
tSCS + IMT	MEP	54.48 ± 27.53	66.80 ± 32.41	F = 15.240, $p < $ **0.01**, η^2 = 0.432
IMT		45.58 ± 25.15	48.09 ± 27.71	TimexIntervention F = 6.708,
p *		0.439		$p = $ **0.017**; η^2 = 0.251

spirometric measures

		PRE	POST	
tSCS + IMT	FVC (L)	2.28 ± 0.93	2.70 ± 1.37	$p^{\&}$ = **0.013**
IMT		2.24 ± 0.79	2.18 ± 0.98	$p^{\&}$ = 0.534
$p^{@}$		0.898		
tSCS + IMT	FEV$_1$ (L)	1.58 ± 0.57	1.75 ± 0.70	F = 0.067, p = 0.799, η^2 = 0.003
IMT		1.78 ± 0.46	1.57 ± 0.49	TimexIntervention F = 6.708,
p *		0.406		$p = $ **0.018**; η^2 = 0.251
tSCS + IMT	FEV$_1$/FVC (%)	72.30 ± 22.37	70.21 ± 20.92	F = 2.196, p = 0.154, η^2 = 0. 099
IMT		78.85 ± 12.51	75.92 ± 13.95	TimexIntervention F = 0.062,
p *		0.402		p = 0.806; η^2 = 0.003
tSCS + IMT	PEF (L/s)	2.90 ± 1.39	3.04 ± 1.32	F = 0.004, p = 0.950, η^2 = 0.000
IMT		2.94 ± 0.73	2.82 ± 0.85	TimexIntervention F = 0.840,
p *		0.772		p = 0.371; η^2 = 0.042
tSCS + IMT	FEF50% (L/s)	1.57 ± 0.68	1.58 ± 0.70	F = 0.232, p = 0.636, η^2 = 0.012
IMT		1.78 ± 0.49	1.67 ± 0.67	TimexIntervention F = 0.371,
p *		0.482		p = 0.550; η^2 = 0.019
tSCS + IMT	FEF25%/75% (L/s)	1.62 ± 0.83	2.46 ± 3.28	F = 0.851, p = 0.3670, η^2 = 0. 041
IMT		1.66 ± 0.46	1.57 ± 0.61	TimexIntervention F = 2.388,
p *		0.639		p = 0.268; η^2 = 0.061
tSCS + IMT	FEV$_1$/FEV0.5	1.44 ± 0.18	1.44 ± 0.18	F = 0.30, p = 0.864; η^2 = 0.001
IMT		1.45 ± 0.15	1.45 ± 0.17	TimexIntervention F = 0.74,
p *		0.880		p = 0.788, η^2 = 0. 004

* p. the differences between tSCS + IMT vs. IMT group at baseline condition according to t-test except FVC; $^{@}$: Mann–Whitney-U test; **: p. Repeated measure ANOVA between pre vs. post of tSCS + IMT vs. IMT group; except FVC, $^{\&}$ p value according to Wilcoxon-t-test. η^2 = effect size (η^2 = 0.01: small effect; η^2 = 0.06: medium effect; η^2 = 0.14: large effect); Maximum inspiratory and expiratory muscle pressure: MIP and MEP, respectively; FVC: forced vital capacity (FVC); FEV$_1$: forced expiratory volume in 1 s; PEF: peak expiratory flow; FEF: forced expiratory flow.

3.2. Respiratory Assessment

3.2.1. Subjective Evaluation

Subjects reported significant improvement in breathlessness/dyspnea post intervention (F = 8.272, $p < $ **0.009**; η^2 = 0.293) with a significant interaction of Time × Intervention (F = 19.449, $p < $ **0.001)** (Table 2; Figure 2). Breathlessness/dyspnea improved significantly compared to baseline in the tSCS + IMT group (p = 0.002, paired t-test) but not in the IMT group (p = 0.441, Figure 2).

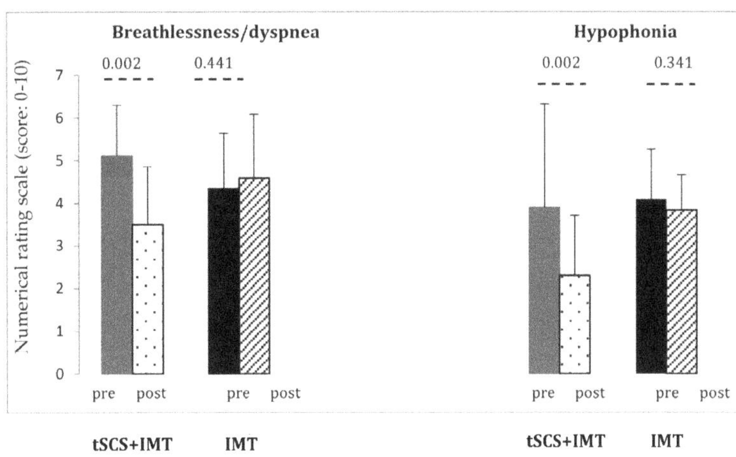

Figure 2. Changes in breathlessness/dyspnea and in hypophonia measured by numerical rating scale (NRS). IMT: Inspiratory muscle training; tSCS: transcutaneous spinal cord stimulation. NRS: 0 = no changes and 10 severe problems following SCI in comparison to pre-SCI. *p*-value according to paired *t*-test.

Hypophonia improved significantly (F = 18.06, $p < 0.001$, $\eta^2 = 0.475$) with a significant interaction of Time $\times$ Intervention (F = 9.552, $p = 0.006$, Table 2). Hypophonia improved in the tSCS + IMT significantly (paired *t*-test, $p = 0.002$) but not in the IMT group ($p = 0.341$, Figure 2).

3.2.2. Objective Evaluation

Inspiratory thorax muscle strength measured by MIP improved significantly compared to baseline (F= 4.452, $p = 0.048$, $\eta^2 = 0.182$), with a significant interaction of Time $\times$ Intervention (F = 5.813, $p = 0.026$, Table 2; Figure 3). After the last session, MIP improved significantly with respect to baseline in the tSCS + IMT group (paired *t*-test, $p = 0.004$) but not in the IMT group ($p = 0.814$, Figure 3).

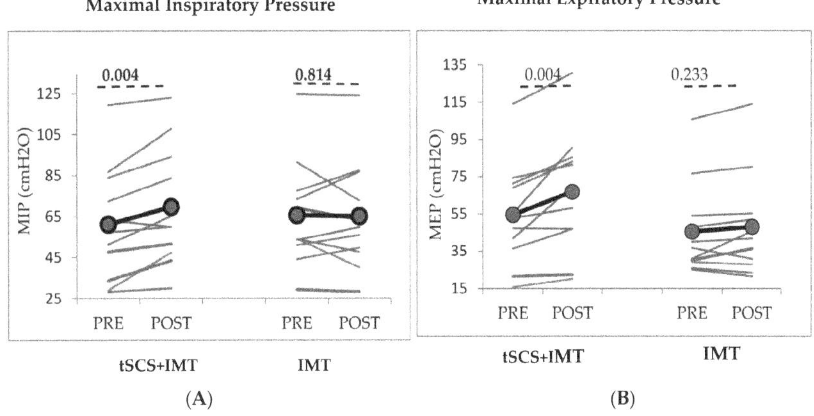

Figure 3. Changes: (**A**) in MIP (maximal inspiratory pressure) and, (**B**) in MEP (maximal expiratory pressure) with tSCS + IMT and with IMT. IMT: inspiratory muscle training; tSCS: transcutaneous spinal cord stimulation. The grey lines represent individual cSCI subjects and the black lines represent the group mean. Significant improvement post-tSCS + IMT in MIP and MEP. *p*-value according to paired *t*-test.

Expiratory thorax muscle strength (MEP) also improved significantly (F = 15.240, $p < 0.01$; $\eta^2 = 0.432$), with a significant interaction of Time × Intervention (F = 6.708, $p = 0.017$, Table 2; Figure 3). The improvement was significant in comparison to baseline in the tSCS + IMT group ($p = 0.004$, paired t-test) but not in the IMT group ($p = 0.233$, Figure 3).

Spirometric measures: FVC increased significantly compared to baseline following the last session of tSCS + IMT (Wilcoxon test, $p = 0.013$) but did not change significantly in the IMT group ($p = 0.534$) (Table 2, Figure 4). There were no significant changes in the other spirometric measures (FEV$_1$, FEV$_1$/FVC, PEF, FEF50%, FEV25%/75%, FEV$_1$/FEV0.5) in either group ($p > 0.05$, Table 2).

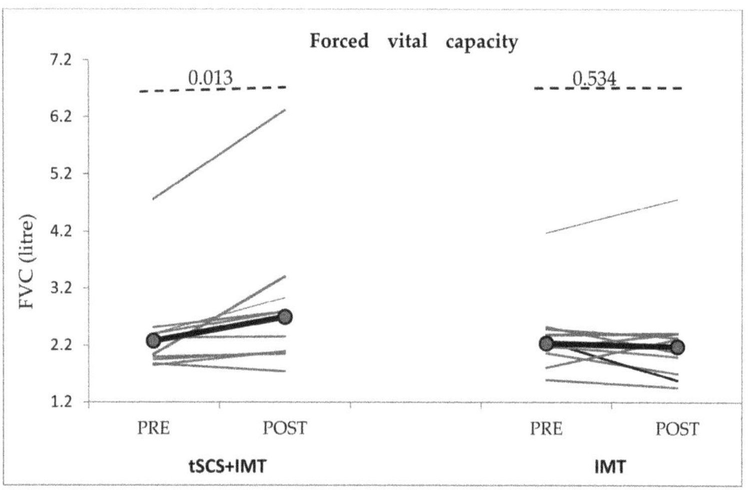

Figure 4. Changes in forced vital capacity (FVC) in both conditions. IMT: inspiratory muscle training; tSCS: transcutaneous spinal cord stimulation. The grey lines represent individual cSCI subjects and the black lines represent the group means. Significant improvement post-tSCS + IMT. p-value according to Wilcoxon-t-test.

Adverse effects: Seven subjects in the tSCS + IMT group complained of mild to moderate pain (range: 1–5) around the tSCS electrodes, particularly in the cervical segment, but no one left the study. None had significant changes in blood pressure during tSCS.

4. Discussion

This study demonstrates that tSCS combined with IMT improved respiratory function both through self-report measures such as breathlessness and hypophonia and through objective assessments such as inspiratory and expiratory muscle strength and pulmonary vital capacity. It is noteworthy that these improvements were observed after only one week of the combined intervention, indicating the potential for relatively rapid effects.

In the literature, only one article [24] reported potential therapeutic benefits of tSCS for improving respiratory function in a 39-year-old man with a complete SCI at the C5 level. The subjects received tSCS over the cervical spinal cord for 20 min twice a day in addition to respiratory muscle training, for a total of 8 weeks. The authors used biphasic pulses consisting of a 10 kHz carrier pulse at 30 Hz, applying the cathode electrodes at C3-4, C5-6, or Th1-2, and two anode electrodes over both shoulders. The authors selected the stimulation site with the lowest intensity that generated the greatest functional respiratory response, using inspiratory capacity and forced expiratory volume in 1 s (FEV1) as their indicator. Their results showed significant improvements in MIP, MEP, and FVC, as well as in subjective measures of dyspnea and cough effectiveness [24]. In our study, we used similar characteristic of tSCS (biphasic 10 kHz carrier pulses and at 30 Hz) in 11 cSCI subjects. On the other side, (i) we used tSCS at two spinal segments (C3-4 and Th9-10),

(ii) using the 90% of rest motor threshold in APB muscle. (iii) The cathode electrodes were placed at C3-4 and Th9-10, where each one was connected to an anode electrode placed symmetrically over the iliac crests, and (iv) the duration of the tSCS was 5 days.

Our subjects with cervical SCI did not present severe respiratory problems because we excluded those with mechanically ventilated cervical SCI and included subjects with stabilized respiratory function. However, we applied tSCS at C3-4 to exert a beneficial effect on respiratory parameters in the diaphragm, which was not affected severely in this study because, in our previous study, tSCS could promote the muscle function of the non-affected muscles in healthy subjects [22,23]. The diaphragm provides the major driving force for inspiration, which is innervated by the phrenic nerve, derived primarily from the C3 through C5 lower motor neurons (LMNs), most of which reside within the C4 spinal segment. High-cervical lesions may damage phrenic motor neurons and/or disrupt descending bulbospinal pathways [28]. However, lesions rostral to the C3 or C4 level spare phrenic LMNs but disrupt descending bulbospinal upper motor neuron signal transmission from pattern-generating respiratory centers in the medulla oblongata. Both these injury profiles typically result in a dependence on artificial, external respiratory replacement, traditionally a tracheostomy tube, and mechanical positive pressure ventilation. However, this was not the case in our SCI subjects, because such severe cases were not included in our study.

In this study, combined tSCS with IMT improved respiratory function both through self-report measures and through objective assessments. Additionally, the comparison with IMT alone suggests that tSCS may provide additional benefits beyond IMT alone. It was previously published that tSCS combined with muscle activity demonstrates a greater ability to modulate the muscle force, spinal cord, and cortex excitability compared to using intervention alone in healthy subjects [22]. This suggests that the combined approach may have synergistic effects in promoting neural modulation and plasticity [11,22]. Mechanisms underlying the action of tSCS combined with physical training, although partially overlapping, may involve different and perhaps synergistic processes leading to a more efficient reorganization of neural circuits [11,22,23].

tSCS was applied at two spinal segments: at the C3-4 level to exert a beneficial effect on respiratory parameters at the diaphragm and intercostal muscles and at the Th9-10 tSCS to exert the same through abdominal muscles. The placement of electrodes for tSCS is an important consideration in order to target specific segments of the spinal cord and promote desired functional outcomes. For promoting upper extremity function, the electrodes were placed at the C3-4, C5, and/or C7-8 segments of the spinal cord [11–15]. These segments are in proximity to the innervation of the upper limb muscles and are targeted to enhance the motor control and responsiveness of the upper extremities. Similarly, for promoting lower extremity function, the electrodes were placed at the Th11-12 and/or L1-2 segments of the spinal cord [16–18]. These segments are relevant for the innervation of the lower limb muscles and were selected to optimize the stimulation effects on the lower extremities.

The principal mechanism of tSCS is a non-invasive activation of inaccessible neuronal networks of the spinal cord likely including the recruitment of afferent fibers (large–medium) in the posterior root in order to elevate spinal network excitability [20,29]. The excitability of spinal interneuronal networks without directly producing action potentials can be readily modulated by changing the networks' physiological state [30]. In addition, the recruitment of cutaneous mechanoreceptors surrounding the electrodes may also contribute to the neuromodulatory effects of tSCS through these polysynaptic connections [31]. In animal models, Guiho et al. [32] observed a potentiation of supraspinal evoked responses with both dorsal epidural SCS and tSCS over the C3-4 and C7–T1 intervertebral spaces in monkeys, but facilitation was stronger with dorsal epidural SCS. It has been reported [10,33] that high-frequency (300-Hz) SCS via a single epidural electrode at the second thoracic spinal level (T2) applied to the dorsal epidural was capable of evoking a physiological recruitment pattern of the inspiratory musculature in canine models of SCI. A reliable approach used to promote functional recovery is to neuromodulate the preserved spinal cord connectome innervating the preserved muscles. The underlying hypothesis for the effect of tSCS is a neuromodulation of spinal sensorimotor networks above, within, and below the lesion toward an elevated functional state that enables and amplifies voluntary motor control. Repeated tSCS over multiple treatment sessions may eventually trigger a cascade of adaptive events, ultimately leading to functional neural reorganization [24], and this can be manifested as chronic (adaptive, learned) functions that can persist for minutes to days. This activity-dependent response is also well suited for respiratory network reorganization and functional recovery following spinal cord injury [24].

Strength and Limitations

This study had a few limitations. (i) We included only individuals with spinal cord injuries with an evolution of more than 3 months. Nevertheless, the IMT group did not show significant changes, whereas tSCS + IMT resulted in significant improvements in both subjective and objective parameters. (ii) We used a subjective scale of breathlessness for the inclusion criteria. Using a subjective scale for the inclusion criteria may indeed have certain limitations, as subjective measures can be influenced by individual perception and interpretation. However, it is encouraging to note that, despite this limitation, the objective assessments still showed significant differences. (iii) Although the time since injury was similar among the participants, individual differences in injury severity or other factors may have influenced the results. (iv) Blinding was not possible, as subjects knew they were receiving electrical stimulation due to the high intensity of tSCS. (v) Long-term follow-up was not possible due to the COVID-19 pandemic and the aim to reduce contact risk. (vi) Finally, the number of participants was small and the stimulation period was short (five days only). Despite these limitations, the study revealed promising results in various respiratory measures.

5. Conclusions

Current evidence supports the potential of tSCS as an adjunctive therapy to accelerate and enhance the rehabilitation process for respiratory impairments following SCI. However, further research is needed to validate these results and establish the long-term benefits of tSCS in this population. Additionally, it is important to explore and understand the optimal parameters of tSCS, including intensity, frequency, and stimulation at different segments of the spinal cord. Tailoring the tSCS intervention to individual needs and optimizing the stimulation parameters may contribute to maximizing the therapeutic benefits and improving clinical outcomes for SCI individuals.

Author Contributions: Conceptualization, H.K.; methodology, H.K.; software, L.G.-A.; validation, H.K. and J.V.; formal analysis, H.K. and S.A.; investigation, L.G.-A., A.R.-A. and H.K.; data curation, L.G.-A., A.R.-A. and H.K.; writing—original draft preparation, H.K.; writing—review and editing, H.K., L.G.-A., S.A., A.R.-A., M.V. and J.V.; visualization, H.K.; supervision, H.K.; project administration, H.K.; funding acquisition, H.K. All authors have read and agreed to the published version of the manuscript.

Funding: This research partially financiated by the "Mike Lane project awarded by Castellers de la Vila de Gràcia" and "REGAIT-PID2021-124111OB-C31 funded by MCIN/AEI/ 10.13039/501100011033 and by ERDF A way of making Europe" to HK.

Institutional Review Board Statement: The study was conducted in accordance with the Declaration of Helsinki, and approved by the Ethics Committee of Institut Guttmann of (protocol code 2019_317 and date of approval: 4 December 2019).

Informed Consent Statement: Informed consent was obtained from all subjects involved in the study.

Data Availability Statement: The data presented in this study are available on request from the corresponding author.

Acknowledgments: We would like to give our thanks to Markus Kofler for his positive feedback about the manuscript.

Conflicts of Interest: The authors declare no conflict of interest.

References

1. Jackson, A.B.; Groomes, T.E. Incidence of respiratory complications following spinal cord injury. *Arch. Phys. Med. Rehabil.* **1994**, *75*, 270–275. [CrossRef] [PubMed]
2. Berlowitz, D.J.; Tamplin, J. Respiratory muscle training for cervical spinal cord injury. *Cochrane Database Syst. Rev.* **2013**, *7*, CD008507. [CrossRef]
3. Arora, S.; Flower, O.; Murray, N.P.; Lee, B.B. Respiratory care of patients with cervical spinal cord injury: A review. *Crit. Care Resusc.* **2012**, *14*, 64–73.
4. Berlowitz, D.J.; Wadsworth, B.; Ross, J. Respiratory problems and management i people with spinal cord injury. *Breathe* **2016**, *12*, 328–340. [CrossRef]
5. Garshick, E.; Kelley, A.; Cohen, S.A.; Garrison, A.; Tun, C.C.G.; Gagnon, D.; Brown, R. A prospective assessment of mortality in chronic spinal cord injury. *Spinal Cord* **2005**, *43*, 408–416. [CrossRef] [PubMed]
6. Waddimba, A.C.; Jain, N.B.; Stolzmann, K.; Gagnon, D.R.; Burgess, J.F., Jr.; Kazis, L.E.; Garshick, E. Predictors of cardiopulmonary hospitalization in chronic spinal cord injury. *Arch. Phys. Med. Rehabil.* **2009**, *90*, 193–200. [CrossRef] [PubMed]
7. Brûlé, J.F.; Leriche, B.; Normand, J.; Khalife, S.; Torabi, D.; Vallée, D. Phrenic stimulation in respiratory paralysis caused by spinal cord injuries. *Neurochirurgie* **1991**, *37*, 127–132.
8. Onders, R.P.; Elmo, M.; Kaplan, C.; Schilz, R.; Katirji, B.; Tinkoff, G. Long-term experience with diaphragm pacing for traumatic spinal cord injury: Early implantation should be considered. *Surgery* **2018**, *164*, 705–711. [CrossRef] [PubMed]
9. Gonzalez-Bermejo, J.; LLontop, C.; Similowski, T.; Morélot-Panzini, C. Respiratory neuromodulation in patients with neurological pathologies: For whom and how? *Ann. Phys. Rehabil. Med.* **2015**, *58*, 238–244. [CrossRef]
10. DiMarco, A.F.; Kowalski, K.E.; Geertman, R.T.; Hromyak, D.R.; Frost, F.S.; Creasey, G.H.; Nemunaitis, G.A. Lower thoracic spinal cord stimulation to restore cough in patients with spinal cord injury: Results of a National Institutes of Health-Sponsored clinical trial. Part II: Clinical outcomes. *Arch. Phys. Med. Rehabil.* **2009**, *90*, 726–732. [CrossRef]
11. García-Alén, L.; Kumru, H.; Castillo-Escario, Y.; Benito-Penalva, J.; Medina-Casanovas, J.; Gerasimenko, Y.P.; Edgerton, V.R.; García-Alías, G.; Vidal, J. Transcutaneous Cervical Spinal Cord Stimulation Combined with Robotic Exoskeleton Rehabilitation for the Upper Limbs in Subjects with Cervical SCI: Clinical Trial. *Biomedicines* **2023**, *11*, 589. [CrossRef] [PubMed]
12. Inanici, F.; Samejima, S.; Gad, P.; Edgerton, V.R.; Hofstetter, C.P.; Moritz, C.T. Transcutaneous electrical spinal stimulation promotes lon-term recovery of upper extremity function in chronic tetraplegia. *IEEE Trans. Neural Syst. Rehabil. Eng.* **2018**, *26*, 1272–1278. [CrossRef] [PubMed]
13. Inanici, F.; Brighton, L.N.; Samejima, S.; Hofstetter, C.P.; Moritz, C.T. Transcutaneous Spinal Cord Stimulation Restores Hand and Arm Function after Spinal Cord Injury. *IEEE Trans. Neural Syst. Rehabil. Eng.* **2021**, *29*, 310–319. [CrossRef] [PubMed]
14. Zhang, F.; Momeni, K.; Ramanujam, A.; Ravi, M.; Carnahan, J.; Kirshblum, S.; Forrest, G.F. Cervical Spinal Cord Transcutaneous Stimulation Improves Upper Extremity and Hand Function in People with Complete Tetraplegia: A Case Study. *IEEE Trans. Neural Syst. Rehabil. Eng.* **2020**, *28*, 3167–3174. [CrossRef] [PubMed]
15. Gad, P.; Lee, S.; Terrafranca, N.; Zhong, H.; Turner, A.; Gerasimenko, Y.P.; Edgerton, V.R. Non-invasive activation of cervical spinal networks after severe paralysis. *J. Neurotrauma* **2018**, *35*, 2145–2158. [CrossRef]
16. Gad, P.; Gerasimenko, Y.; Zdunowski, S.; Turner, A.; Sayenko, D.; Lu, D.C.; Edgerton, V.R. Weight Bearing Over-ground Stepping in an Exoskeleton with Non-invasive Spinal Cord Neuromodulation after Motor Complete Paraplegia. *Front. Neurosci.* **2017**, *11*, 333. [CrossRef]
17. Minassian, K.; Hofstoetter, U.S.; Danner, S.M.; Mayr, W.; Bruce, J.A.; McKay, W.B.; Tansey, K.E. Spinal Rhythm Generation by Step-Induced Feedback and Transcutaneous Posterior Root Stimulation in Complete Spinal Cord–Injured Individuals. *Neurorehabil Neural Repair* **2016**, *30*, 233–243. [CrossRef]
18. Hofstoetter, U.S.; Krenn, M.; Danner, S.M.; Hofer, C.; Kern, H.; McKay, W.B.; Mayr, W.; Minassian, K. Augmentation of Voluntary Locomotor Activity by Transcutaneous Spinal Cord Stimulation in Motor-Incomplete Spinal Cord-Injured Individuals: Augmentation of Locomotion by tSCS in Incomplete SCI. *Artif. Organs* **2015**, *39*, 176–186. [CrossRef]
19. Rath, M.; Vette, A.H.; Ramasubramaniam, S.; Li, K.; Burdick, J.W.; Edgerton, V.R.; Gerasimenko, Y.P.; Sayenko, D.G. Trunk Stability Enabled by Noninvasive Spinal Electrical Stimulation after Spinal Cord Injury. *J. Neurotrauma* **2018**, *35*, 2540–2553. [CrossRef]
20. Sayenko, D.G.; Rath, M.; Ferguson, A.R.; Burdick, J.W.; Hayton, L.A.; Edgerton, V.R.; Gerasimenko, Y.P. Self-Assisted Standing Enabled by Non-Invasive Spinal Stimulation after Spinal Cord Injury. *J. Neurotrauma* **2019**, *36*, 1435–1450. [CrossRef] [PubMed]
21. Gerasimenko, Y.P.; Lu, D.C.; Modaber, M.; Zdunowski, S.; Gad, P.; Sayenko, D.G.; Morikawa, E.; Haakana, P.; Ferguson, A.R.; Roy, R.R.; et al. Noninvasive Reactivation of Motor Descending Control after Paralysis. *J. Neurotrauma* **2015**, *32*, 1968–1980. [CrossRef] [PubMed]

22. Kumru, H.; Flores, A.; Rodríguez-Cañón, M.; Edgerton, V.R.; García, L.; Benito-Penalva, J.; Navarro, X.; Gerasimenko, Y.; García-Alías, G.; Vidal, J. Cervical Electrical Neuromodulation Effectively Enhances Hand Motor Output in Healthy Subjects by Engaging a Use-Dependent Intervention. *J. Clin. Med.* **2021**, *10*, 195. [CrossRef] [PubMed]
23. Kumru, H.; Rodríguez-Cañón, M.; Edgerton, V.R.; García, L.; Soriano, I.; Opisso, E.; Gerasimenko, Y.; Navarro, X.; García-Alías, G.; Vidal, J. Transcutaneous Electrical Neuromodulation of the Cervical Spinal Cord Depends Both on the Stimulation Intensity and the Degree of Voluntary Activity for Training. A Pilot Study. *J. Clin. Med.* **2021**, *10*, 3278. [CrossRef] [PubMed]
24. Gad, P.; Kreydin, E.; Zhong, H.; Edgerton, V.R. Enabling respiratory control after severe chronic tetraplegia: An exploratory case study. *J. Neurophysiol.* **2020**, *124*, 774–780. [CrossRef]
25. Ledsome, J.R.; Sharp, J.M. Pulmonary function in acute cervical cord injury. *Am. Rev. Respir. Dis.* **1981**, *124*, 41–44. [CrossRef] [PubMed]
26. Kirshblum, S.C.; Burns, S.P.; Biering-Sorensen, F.; Donovan, W.; Graves, D.E.; Jha, A.; Johansen, M.; Jones, L.; Krassioukov, A.; Mulcahey, M.J.; et al. International standards for neurological classification of spinal cord injury (Revised 2011). *J. Spinal Cord Med.* **2011**, *34*, 535–546. [CrossRef]
27. Tamplin, J.; Berlowitz, D.J. A systematic review and meta-analysis of the effects of respiratory muscle training on pulmonary function in tetraplegia. *Spinal Cord* **2014**, *52*, 175–180. [CrossRef]
28. Hachmann, J.T.; Grahn, P.J.; Calvert, J.S.; Drubach, D.I.; Lee, K.H.; Lavrov, I.A. Electrical neuromodulation of the respiratory system after spinal cord injury. In *Mayo Clinic Proceedings*; Elsevier: Amsterdam, The Netherlands, 2017; Volume 92, pp. 1401–1414. [CrossRef]
29. Milosevic, M.; Masugi, Y.; Sasaki, A.; Sayenko, D.G.; Nakazawa, K. On the reflex mechanisms of cervical transcutaneous spinal cord stimulation in human subjects. *J. Neurophysiol.* **2019**, *121*, 1672–1679. [CrossRef]
30. Gerasimenko, Y.; Gorodnichev, R.; Moshonkina, T.; Sayenko, D.; Gad, P.; Edgerton, V.R. Transcutaneous electrical spinal-cordstimulation in humans HHS Public Access. *Ann. Phys. Rehabil. Med.* **2015**, *58*, 225–231. [CrossRef]
31. Beekhuizen, K.S.; Field-Fote, E.C. Massed practice versus massed practice with stimulation: Effects on upper extremity function and cortical plasticity in individuals with incomplete cervical spinal cord injury. *Neurorehabilit. Neural Repair* **2005**, *19*, 33–45. [CrossRef]
32. Guiho, T.; Baker, S.N.; Jackson, A. Epidural and transcutaneous spinal cord stimulation facilitates descending inputs to upper-limb motoneurons in monkeys. *J. Neural Eng.* **2021**, *18*, 046011. [CrossRef] [PubMed]
33. DiMarco, A.F.; Kowalski, K.E. Intercostal muscle pacing with high frequency spinal cord stimulation in dogs. *Respir. Physiol. Neurobiol.* **2010**, *171*, 218–224. [CrossRef] [PubMed]

Article

Human Adipose-Derived Stem Cells Reduce Cellular Damage after Experimental Spinal Cord Injury in Rats

Emiliano Neves Vialle [1,*], Letícia Fracaro [2,3], Fabiane Barchiki [2,3], Alejandro Correa Dominguez [4], André de Oliveira Arruda [1], Marcia Olandoski [5], Paulo Roberto Slud Brofman [2,3,†] and Carmen Lúcia Kuniyoshi Rebelatto [2,3,†]

[1] Spine Surgery Group, Cajuru University Hospital, Pontifícia Universidade Católica do Paraná, Curitiba 80215-030, Brazil
[2] Core for Cell Technology, School of Medicine and Life Sciences, Pontifícia Universidade Católica do Paraná, Curitiba 80215-030, Brazil
[3] National Institute of Science and Technology for Regenerative Medicine, INCT-REGENERA, Rio de Janeiro 21941-599, Brazil
[4] Laboratory of Basic Biology of Stem Cells, Carlos Chagas Institute—Fiocruz, Rio de Janeiro 21941-599, Brazil
[5] Department of Biostatistics, School of Medicine, Catholic University of Paraná, Curitiba 80215-030, Brazil
* Correspondence: evialle@hotmail.com; Tel.: +55-41-3223-7860
† These authors contributed equally to this work.

Citation: Vialle, E.N.; Fracaro, L.; Barchiki, F.; Dominguez, A.C.; Arruda, A.d.O.; Olandoski, M.; Brofman, P.R.S.; Kuniyoshi Rebelatto, C.L. Human Adipose-Derived Stem Cells Reduce Cellular Damage after Experimental Spinal Cord Injury in Rats. *Biomedicines* 2023, 11, 1394. https://doi.org/10.3390/biomedicines11051394

Academic Editor: Nicolas Guerout

Received: 28 March 2023
Revised: 28 April 2023
Accepted: 28 April 2023
Published: 8 May 2023

Abstract: Traumatic spinal cord injury (SCI) is a devastating condition without an effective therapy. Cellular therapies are among the promising treatment strategies. Adult stem cells, such as mesenchymal stem cells, are often used clinical research for their immunomodulatory and regenerative potential. This study aimed to evaluate the effect of human adipose tissue-derived stem cells (ADSC) infusion through the cauda equina in rats with SCI. The human ADSC from bariatric surgery was isolated, expanded, and characterized. Wistar rats were subjected to blunt SCI and were divided into four groups. Two experimental groups (EG): EG1 received one ADSC infusion after SCI, and EG2 received two infusions, the first one after SCI and the second infusion seven days after the injury. Control groups (CG1 and CG2) received infusion with a culture medium. In vivo, cell tracking was performed 48 h and seven days after ADSC infusion. The animals were followed up for 40 days after SCI, and immunohistochemical quantification of myelin, neurons, and astrocytes was performed. Cellular tracking showed cell migration towards the injury site. ADSC infusion significantly reduced neuronal loss, although it did not prevent the myelin loss or enhance the area occupied by astrocytes compared to the control group. The results were similar when comparing one or two cell infusions. The injection of ADSC distal to the injured area was shown to be a safe and effective method for cellular administration in spinal cord injury.

Keywords: spinal cord injury; cell therapy; adipose tissue; neurological rehabilitation; experimental surgery

1. Introduction

Traumatic spinal cord injury (SCI) is responsible for high morbidity and mortality rates in our society, affecting a significant portion of the economically active population and causing significant socioeconomic impacts [1,2]. Regarding the pathophysiology of SCI, after the initial triggering factor of the injury, there is a complex and intricate sequence of events that results in neuronal death and functional loss below the affected level, with interruption of the circuits formed by the afferent sensory fibers and efferent motors that communicate between the cerebral cortex and the rest of the body [3,4]. Spinal cord injury is divided into two phases: primary injury—mechanical, acute, with direct injury and compression of neural elements; and secondary—ischemic, related to a reactive biochemical cascade with changes in perfusion and local blood flow [5–7]. Several therapeutic strategies have been proposed to treat or minimize the damage caused by SCI; surgical procedures are not able to act on the biochemical processes of secondary injury or provide stimulus to regeneration [5]. Other options of treatment are corticoids, riluzole, hormones, calcium channel blockers, and specific proteins, which are ineffective in the treatment of SCI [8,9]. The administration of stem cells (SC) presents as a promising strategy since it can potentially interrupt or minimize the cascade of pathophysiological events and provide a favorable environment for specialized tissue regeneration, with possible positive clinical correlation [10–12]. Adult stem cells, such as mesenchymal stem cells (MSC), are often used in clinical research for their immunomodulatory and regenerative potential [9,13]. The adipose tissue is an attractive source of MSC because of the high frequency of adipose-derived stem cells (ADSC); they present few ethical concerns and can be obtained from donors undergoing plastic surgical procedures [14,15]. Considering the complexity of SCI and the different mechanisms of action of the ADSC, the use of a standardized spinal cord contusion model gains scientific strength by controlling several aspects involving the lesion and subsequent evaluations of the possible effects of the administered SC [16]. This paper aims to evaluate the effect of administering human ADSC in the cauda equina of rats submitted to standardized experimental SCI, through histological study of the spinal cord, quantifying cellular changes through immunohistochemistry and histomorphometric analysis; further studying the cellular migration using an in vivo tracking methodology and cell survival rate after transplantation using immunofluorescence.

2. Materials and Methods

2.1. Study Design

The present study is longitudinal, prospective, randomized, and controlled. It was approved by the Institutional Ethics Committee on the Use of Animals (No. 769) and the Ethics Committee in Research with Humans (CAAE 12723713.3.0000.0020—Universidade Católica do Paraná, Curitiba, Brazil). Three biological samples of adipose tissue donors from bariatric surgery were used. Forty-five Wistar rats, *Rattus norvegicus*, young adults with an average of 20 weeks of age, weighing between 300 and 350 g, female, were used. During the study period, all animals received water and food ad libitum and were kept under light control (light-dark cycles of 12 h) and at room temperature (25 ± 1 °C, continuous temperature monitoring). Spinal cord injury was induced in all animals (n = 45). They were divided into experimental groups (EG, n = 20) and control groups (CG, n = 20). The experimental group was divided according to the number of infusions performed: one infusion after SCI induction (EG1, n = 10) or two infusions, the second infusion seven days after the injury (EG2, n = 10). Control groups received culture medium infusion (CG1, n = 10 and CG2, n = 10). The animals were followed up for 40 days after SCI and euthanized for histological analysis (Figure 1).

2.2. Adipose-Derived Stem Cells Isolation and Expansion

Human adipose tissue, obtained through by bariatric surgery, was collected in flasks containing a 2% glucose solution. This tissue was isolated by using enzymatic digestion. Approximately 100 mL of adipose tissue was washed with phosphate buffer saline (PBS) (Gibco™, New York, NY, USA). The tissue was macerated, and enzymatic digestion with collagenase type I (1 mg/mL) (Gibco™, New York, NY, USA) was performed for 30 min at 37 °C, under constant shaking, followed by filtration through 100-μm mesh filter (BD FALCON, BD Biosciences Discovery Labware, Bedford, MA, USA) and centrifugation at $800 \times g$ for 10 min. Contaminating erythrocytes were removed by erythrocyte lysis buffer. The cells were washed with phosphate buffered saline (PBS) (Gibco™, New York, NY, USA), resuspended in Dulbecco's modified Eagle's medium and F12 (DMEM-F12) (Gibco™, New York, NY, USA), supplemented with 10% fetal bovine serum (FBS), penicillin (100 U/mL), and streptomycin (100 μg/mL) (Gibco™, New York, NY, USA), and plated in 75 cm^2 flasks (TPP, Trasadingen, Switzerland), at a density of 1×10^5 cells/cm^2. The first change of medium was performed after 24 h, to remove non-adherent cells, and the other changes were performed twice a week. After 80% confluence, cell dissociation was performed with trypsin/EDTA (Gibco™, New York, NY, USA). The culture medium was discarded, and the cells were washed twice with PBS. Then, trypsin was added to the cells and incubated at 37 °C for 4 min. Next, the trypsin was inactivated with 1% FBS and IMDM (Iscove's Modified Dulbecco's Medium—Gibco™, New York, NY, USA). The cell suspension was centrifuged, the supernatant discarded, and the cells were counted and plated again (first pass—P1). All experiments were performed between the third (P3) and the fifth passage (P5).

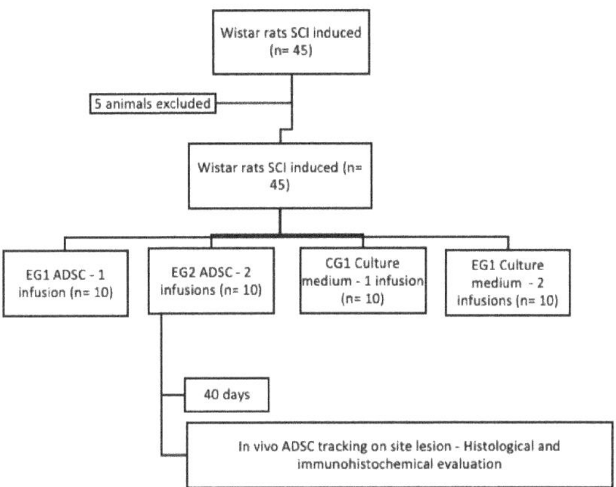

Day 0	Day 2	Day 7	Day 9	Day 14	Day 40
SCI Induction and infusion: ADSC (EG1, EG2) and culture medium (CG1, CG2)	In vivo ADSC tracking (EG1, CG1)	In vivo ADSC tracking (EG1, CG1) Infusion: ADSC (EG2) and culture medium (CG2)	In vivo ADSC tracking (EG2, CG2)	In vivo ADSC tracking (EG2, CG2)	Euthanasia and evaluations

Figure 1. Flow diagram of experimental study. SCI (spinal cord injury); EG (experimental group); ADSC (adipose derived stem cells); CG (control group).

2.3. Adipose-Derived Stem Cells Characterization

The characterization of ADSC was carried out by evaluating cell surface markers by immunophenotyping and by cell differentiation potential. The immunophenotypic analysis followed the criteria defined by the International Society for Cellular Therapy (ISCT) [17]. Approximately 1×10^6 cells were placed in each tube, the cells were washed with PBS (Gibco™, New York, NY, USA) and centrifuged for 5 min at $400 \times g$. After centrifugation, the following fluorochrome-conjugated monoclonal antibodies were added: CD105, CD90, CD73, CD29, CD44, CD14, CD34, CD45, CD31, and CD19 (BD Pharmingen, San Diego, CA, USA). Next, the cells were incubated with the antibodies for 30 min. They were washed with PBS and resuspended in PBS containing 1% paraformaldehyde (Sigma-Aldrich, Saint Louis, MO, USA). Twenty thousand events were collected from each sample, acquired on a BD FACS Calibur flow cytometer (BD Biosciences), and analyzed using FlowJo software version 10.1 (Tree Star, Ashland, OR, USA).

Cell differentiation potential was assessed using commercial media by inducing ADSC differentiation into adipocytes, osteoblasts, and chondrocytes (Lonza, MD, USA). For adipogenic and osteogenic differentiation, 16,000 ADSC were plated in duplicate on glass coverslips (GlassTécnica, Curitiba, Brazil) in 24-well plates (TPP, Trasadingen, Switzerland). The cells were incubated in an incubator (Thermo Fisher Scientific, Carlsbad, CA, USA) at 37 °C, with 5% CO_2 until reaching 80% confluence. Hence, 300 µL of differentiation medium (Lonza, Walkersville, MD, USA) was added and the medium was changed three times a week for 21 days. Control cells were cultured for the same period, in a culture medium without differentiation induction factors. For adipogenic differentiation, cells were fixed with Bouin (Biotec, Pinhais, Basil) for 10 min, washed twice with 70% ethanol (Merck, Damstadt, Germany) and with MiliQ water. The cells were stained with a 0.5% Oil Red O solution (Sigma-Aldrich) for 1 h to visualize the presence of lipid vacuoles. Hematoxylin-Eosin (HE) (Biotec) was used for nuclear staining. For osteogenic differentiation, cells were fixed with 1% paraformaldehyde (Sigma-Aldrich) for 20 min and washed with MiliQ ultra-pure water. Cells were stained with Alizarin Red S (Fluka Chemie, Steinheim, Germany) for 5 min to assess the presence of calcium crystals.

For chondrogenic differentiation, micromass culture was performed. Briefly, 1×10^6 cells were transferred to a conical tube which was centrifuged for 10 min at $400 \times g$ to form a pellet. After centrifugation, 1 mL of differentiation medium was added, and medium exchange was performed three times a week for 21 days. Next, control cells were cultured without differentiation induction factors, for the same period. After 21 days, the cell aggregate was fixed with 10% paraformaldehyde (Sigma-Aldrich) for 1 h, dehydrated with serial dilutions of ethanol (Merck) and placed in a paraffin block. Sections of 4 µm thickness were made and stained with toluidine blue (Sigma-Aldrich). The presence of proteoglycans in the extracellular matrix was evaluated using an optical microscope (Eclipse E200, Nikon, Tokyo, Japan) [18].

2.4. Adipose-Derived Stem Cells Transduction

The cells were transduced with a luciferase gene for the in vivo monitoring of ADSC by bioluminescence. Cell line Human embryonic kidney 293 cells (HEK 293) was transfected with vectors pMD2.G, pCMV_dr8.91, and pMSCV_Luc2_T2A using Lipofectamine 2000 (Invitrogen, Waltham, MA, USA) and maintained for three days in culture. The supernatant containing the viral particles was collected, filtered with a 0.22-µm filter, and ultracentrifuged at 28,000 rpm for 1 h and 30 min. Cells were resuspended in PBS/BSA 1% (Sigma-Aldrich) and stored at −80 °C. ADSC were transduced with the supernatant containing the viral particles and 10 µg/µL of Hexadimethrine Bromide (Polybrene, Sigma Chemical, St. Louis, MO, USA). The medium used for cell transduction was changed every 24 h for three days. After this period, puromycin (Sigma Chemical, St. Louis, MO, USA) was added for cell selection at a final concentration of 10 mM. The IVIS Lumina II (Caliper Life Sciences, Hopkinton, MA, USA) imaging system was used to analyze the bioluminescence emitted by cells expressing luciferase.

2.5. Surgical Procedure

For the induction of standardized experimental SCI, the animals were anesthetized using ketamine (Agener União, São Paulo, Brazil) and 2% xylazine (Agener União, São Paulo, Brazil) intraperitoneally, at a dose of 90 mg/kg and 10 mg/kg weight, respectively. The procedure was started after tests to verify the deep anesthetic plane, absence of corneal reflex, and non-reactivity to caudal compression. Cephalothin 25 mg/kg (Virbac, São Paulo, Brazil) was administered intraperitoneally as preoperative prophylaxis. The animals were positioned in ventral decubitus, on the "Impactor" device (Rutgers University Impactor®, Rutgers, NJ, USA), according to a previously validated method of complete SCI through contusion [19]. Trichotomy was performed, and by palpation of the lower ribs, access to the skin and deep dissection at the level of T10 were accomplished, followed by removal of the posterior vertebral elements (laminectomy), exposing the spinal cord. Under controlled and previously validated adjustments of height and speed, a rod weighing 5 g was dropped on the exposed level, generating a complete lesion proven by the specific software (MASCIS®, Rutgers, NJ, USA) applied to the system for validation of the lesion. Animals that did not have complete SCI were excluded from the study. After SCI induction, hemostasis was performed, and the surgical wound was washed with 0.9% saline solution, and then the skin was sutured with mono nylon 3.0 thread (Shalom, Goiania, Brazil) (Figure 2). All animals remained under anesthesia for approximately two hours in a chamber with a controlled temperature of 30 °C and placed in individual cages with easily accessible food and water. The surgical wound was inspected daily, and the bladder was emptied using the Credé method every 12 h in all animals throughout the postoperative period [20].

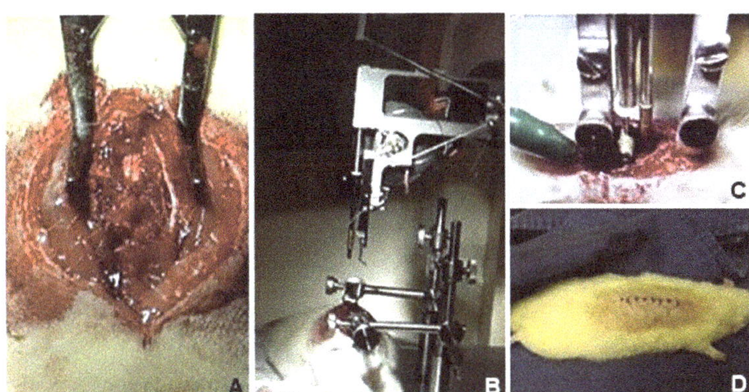

Figure 2. Spinal cord injury procedure. (**A**). Dissection of T9-T10 level followed by laminectomy, exposing the spinal cord. (**B**). Animal positioned on the Impactor device. (**C**). A rod was dropped on the spinal cord, generating a lesion. (**D**). After the procedure, the skin was sutured.

2.6. Cell Transplantation

Under anesthesia, a new surgical incision was made, approximately 1.5 cm distal to the first one, identifying the lumbar levels and the dural sac. Under direct visualization, using a Hamilton syringe (Hamilton®), a solution of 100 µL of DMEM-F12 medium containing 1×10^6 ADSC was infused inside the dural sac of the animals in groups EG1 and EG2. The same procedure was performed with the animals in CG1 and CG2, with 100 µL of culture medium. The incision was sutured with 3.0 mononylon (Bioline, Anápolis, Goias, Brazil). After seven days of injury and first cell infusion, EG2 and CG2 underwent a new infusion of ADSC or culture medium, respectively.

2.7. In Vivo Cell Tracking

The transduced ADSC were infused into three animals from each group. The animals received inhalational anesthesia with isoflurane (Rhodia, Paulínia, Brazil), an intraperitoneal injection of 150 mg/kg of D-luciferin (Perkin Elmer, Waltham, MA, USA) was administered to the animals 5 min before image acquisition. The animals were positioned in the camera in dorsal decubitus and a capture field of 12.7 × 12.7 cm was used (imaging system IVIS LUMINA II, Hopkinton, MA, USA). A series of images were acquired until the maximum peak of the bioluminescence signal was obtained. Images were analyzed using the Living Image software (version 2.50, Xenogen, Caliper Life Sciences, Hopkinton, MA, USA). The signal intensity of the infused ADSC was evaluated 48 h and seven days after the procedure.

2.8. Euthanasia and Tissue Collection

Forty days after SCI induction, the animals were euthanized through high-dose anesthesia, intraperitoneally, with ketamine (240 mg/kg weight) (Agener União, São Paulo, Brazil) and xylazine (30 mg/kg weight) (Agener União), associated with halothane (Tanohalo, Cristália, Brazil) gas in an anesthetic chamber (4–5%) to reduce animal suffering, following euthanasia protocols for experimental studies. First, a thoracotomy was performed to allow intracardiac perfusion with 0.9% saline (JP, Ribeirão Preto, Brazil) until complete exsanguination. Then, through the same route, reaching the right atrium, a 4% paraformaldehyde solution was injected (approximate total volume of 300 mL/animal) (Neon, Suzano, Brazil), allowing the immediate fixation of the nervous tissue and providing ideal conditions for preparation and anatomopathological manipulation.

2.9. Anatomopathological Study—Histology and Immunohistochemistry

After euthanasia, the spinal cord was collected and transferred to a 10% formalin buffered solution. After fixation, histological sections were made, with a thickness of 5 μm from the central area of the lesion. Evident morphological changes were verified macroscopically, evidencing opaque and darker coloration, as well as local thinning, in correspondence with the experimentally induced SCI level (Figure 3).

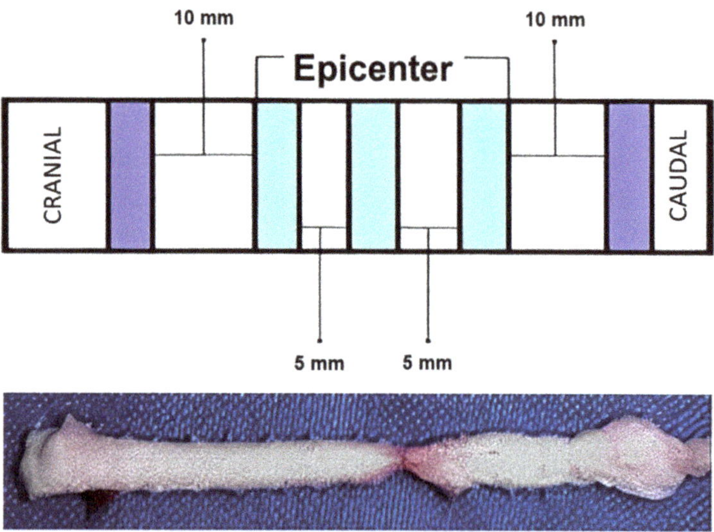

Figure 3. Anatomopathological image of the spinal cord. After removal of the surrounding bone, the tissue was fixed and prepared for histological sections. The central cuts were 5 μm thick, with 5 mm intervals for the central area and 10 mm for the distal cuts. The blue areas had 6 cuts and the violet ones had three cuts.

For histological analysis, samples were stained with hematoxylin-eosin (HE), luxol fast blue (Sigma-Aldrich) and cresyl violet (Sigma-Aldrich). From the analysis of the slides stained with luxol fast blue, a software, Image ProPlus (Media Cybernetics, Silver Spring, MD, USA), based on the color morphometry method, estimated the amount of myelin present in the tissue. Using cresyl violet staining, two independent examiners assessed medullary cellularity, focusing on neuronal presence, by dividing the histological field into equal quadrants (anterior-posterior, right-left, and combinations), using an optical microscope at 200× magnification (Olympus, Tokyo, Japan). Eventual neurons present in medullary sections were counted separately. The software calculated the marked area in units of square micrometers (μm^2). Some values were expressed in the total value of points identified by the program, to better demonstrate the differences between groups [21]. Glial fibrillary acidic protein (GFAP) was evaluated by immunohistochemistry technique to identify activated astrocytes.

2.10. Statistical Analysis

Data analysis was performed using the SPSS Statistics software (v. 20.0). In the cases of histological and immunohistochemical evaluations, the means were compared, and the non-parametric Mann–Whitney test was applied with significance level corrected by Bonferroni, in addition to the one-way analysis of variance (ANOVA, $p = 0.05$). Values $p \leq 0.012$ were considered significant. Student's t test was used for comparison within groups ($p < 0.05$). The data obtained from the in vivo tracking of the cells were not submitted to statistical analysis, being presented in a descriptively way.

3. Results

3.1. Animal Care and Surgical Procedures

The animal model of SCI induction was adequate, with observations as follows. Five animals were excluded from the study: four died, three during the anesthetic period of induction and prolonged maintenance to perform the surgery and evaluate cell tracking in vivo, probably due to idiosyncratic reactions; and one during the late maintenance period, probably related to complications arising from and inherent to the loss of bladder control and neurological function. One animal did not present complete spinal cord injury and was excluded.

3.2. ADSC Characterization

The ADSC surface antigen profile was evaluated, and they were accordingly the criteria defined by ISCT [19]. The cells were positive for CD105, CD90, CD73, CD29, and CD166 markers and negative for CD34, CD45, CD19, CD14, and HLA-DR. Cellular viability before injection into the spinal canal was 98.2% (Figure 4).

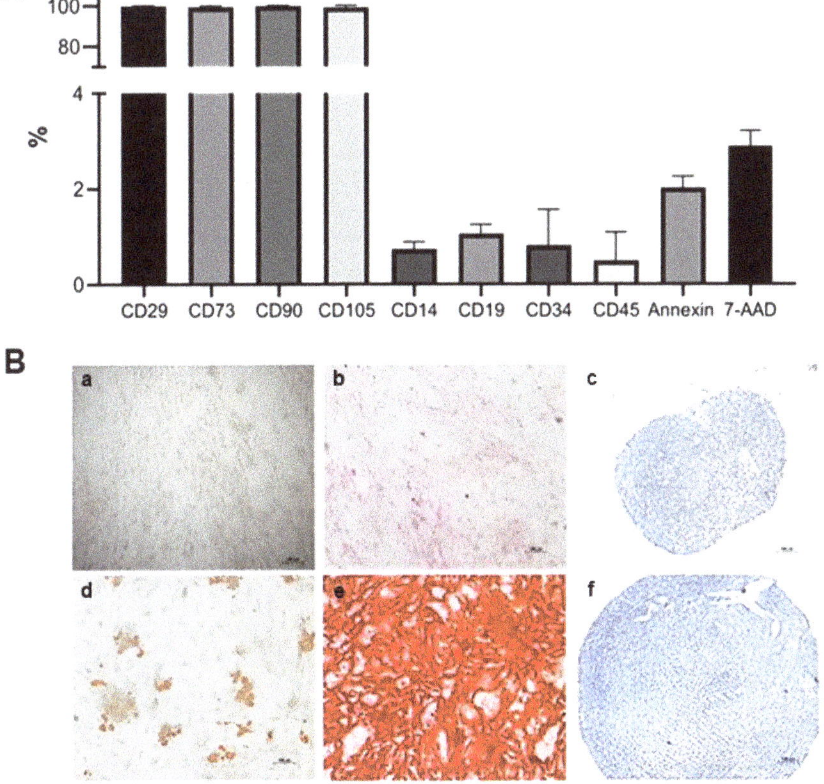

Figure 4. Adipose-derived stem cell characterization. (**A**). Immunophenotypical profile, cell viability, and apoptosis/necrosis. The bars represent the average expression for each marker. (**B**). Adipose-derived stem cell differentiation. Cells were maintained with medium to induce differentiation for 21 days. (**d**). Adipose differentiation was demonstrated with the presence of lipid vacuoles. (**e**). Extracellular matrix mineralization was observed at osteogenic differentiation. (**f**). Chondrocytic differentiation was observed by proteoglycan deposition in the matrix and the presence of lacunae around young chondrocytes. (**a–c**). Control cells were cultured without an induction medium.

After induction of cell differentiation with specific media for each lineage, ADSC showed differentiative potential into adipocytes, osteoblasts, and chondrocytes (Figure 4).

3.3. Histomophometric Analysis of Myelin and Neurons

Adipose-derived stem cell infusion did not change the myelin area significantly in comparison with the control group. In three samples, tissue destruction was extremely intense such that histological analysis was impossible. There was a large variation in the amount of myelin measured within the groups (Figure 5).

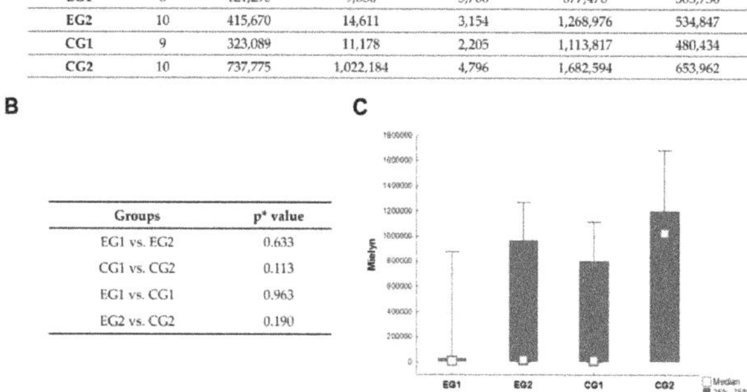

A

		Myelin (μm^2)				
Group	n	Average	Median	Min	Max	SD
EG1	8	121,296	9,830	5,706	877,478	305,736
EG2	10	415,670	14,611	3,154	1,268,976	534,847
CG1	9	323,089	11,178	2,205	1,113,817	480,434
CG2	10	737,775	1,022,184	4,796	1,682,594	653,962

B

Groups	p* value
EG1 vs. EG2	0.633
CG1 vs. CG2	0.113
EG1 vs. CG1	0.963
EG2 vs. CG2	0.190

C

Figure 5. Histomorphometric analysis of myelin area. (**A**,**B**). Comparison of myelin area showed no significant difference between groups. (**C**). Myelin was measured in units of μm^2 and represented by the median values. * Mann-Whitney non-parametric test, $p < 0.012$ (Bonferroni's correction). EG: experimental group; CG: control group; SD: standard deviation.

There was a significant difference regarding number of neurons in the injury area when comparing the experimental and control groups. Groups that received ADSC infusion showed significantly more neurons. There was no difference when comparing one and two ADSC infusions (Figure 6).

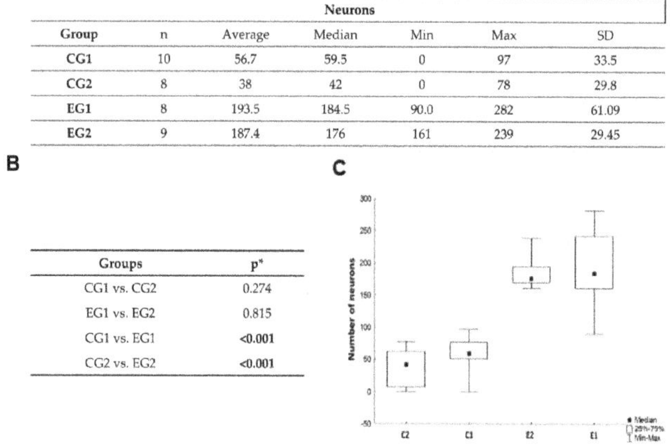

A

		Neurons				
Group	n	Average	Median	Min	Max	SD
CG1	10	56.7	59.5	0	97	33.5
CG2	8	38	42	0	78	29.8
EG1	8	193.5	184.5	90.0	282	61.09
EG2	9	187.4	176	161	239	29.45

B

Groups	p*
CG1 vs. CG2	0.274
EG1 vs. EG2	0.815
CG1 vs. EG1	**<0.001**
CG2 vs. EG2	**<0.001**

C

Figure 6. (**A**) Histomorphometric analysis of neurons. (**B**,**C**) Comparison of neuron numbers between experimental and control groups. Groups that received ADSC infusion, showed a greater number of neurons. * Mann-Whitney's non-parametric test, $p < 0.012$ (Bonferroni's correction). EG: experimental group; CG: control group; SD: standard deviation. Numbers in bold represent statistically significant difference.

3.4. Immunohistochemical Astrocyte Evaluation

The spinal cord was divided in four sections (anterior right and left, posterior right and left). There was no significant difference when comparing experimental and control groups (Tables 1 and 2, Figure 7).

Table 1. Description of astrocytic area in four quadrants. Astrocytes were measured in units of μm^2. Analysis of variance with one factor (ANOVA), $p < 0.05$—EG: experimental group; CG: control group; AR: anterior-right; AL: anterior-left; PR: posterior-right; PL: posterior-left.

Area	Group	N	Average	Median	Min	Max	SD
AR	CG1	9	19,738	18,760	8690	28,785	6500
	CG2	10	16,778	15,287	7870	36,177	8326
	EG1	8	21,547	21,702	15,805	26,511	3597
	EG2	9	20,042	17,650	10,312	34,553	8369
AL	CG1	9	19,882	21,492	7922	27,080	6766
	CG2	10	13,586	12,688	5525	25,587	6325
	EG1	8	21,256	20,114	15,863	29,074	4339
	EG2	9	18,629	18,453	9779	37,710	9531
PR	CG1	9	20,139	20,121	8302	30,488	6559
	CG2	10	18,034	16,628	8737	32,206	7045
	EG1	8	24,795	24,927	15,770	29,270	4107
	EG2	9	19,064	21,752	10,207	31,100	7589
PL	CG1	9	21,411	21,664	10,541	29,412	5929
	CG2	10	15,765	16,124	8161	22,956	4404
	EG1	8	23,672	23,148	16,040	31,780	5776
	EG2	9	19,058	13,830	11,131	38,737	10,120

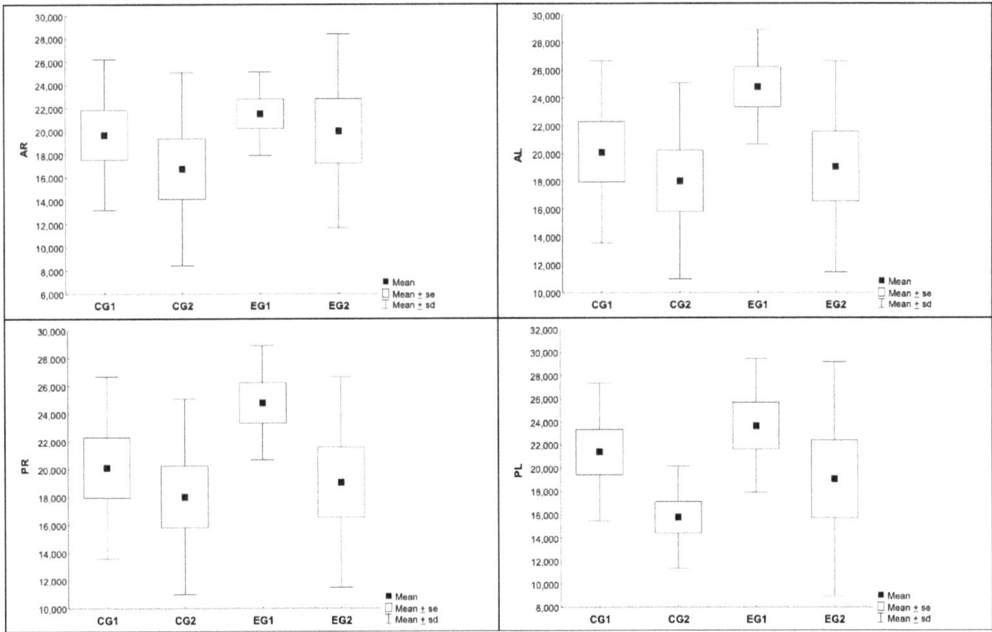

Figure 7. Immunohistochemical evaluation of astrocytic area, divided in four quadrants—EG: experimental group; CG: control group; AR: anterior-right; AL: anterior-left; PR: posterior-right; PL: posterior-left.

Table 2. Comparison of astrocytic area between groups. Student's *t*-test for independent samples, * $p < 0.012$ (Bonferroni's correction); ANOVA, $p < 0.05$—EG: experimental group; CG: control group; AR: anterior-right; AL: anterior-left; PR: posterior-right; PL: posterior-left.

	p * Value			
Groups	**AR**	**AL**	**PR**	**PL**
EG1 × EG2	0.633	0.486	0.077	0.275
CG1 × CG2	0.404	0.051	0.511	0.033
EG1 × CG1	0.497	0.631	0.105	0.439
EG2 × CG2	0.407	0.188	0.763	0.387

3.5. In Vivo Cell Tracking with Bioluminescence Imaging

Bioluminescence evaluation was performed in three animals of each experimental group, at 48 h and seven days after first ADSC infusion (EG1 and EG2). In EG2, the evaluation was performed 48 h and seven days after second cell infusion. At 48 h of evaluation, cells were observed at the injury site and in some animals on the lungs. It was noticed that signal intensity was reduced on the second evaluation for EG1 (day 7) and EG2 group (days 7 and 9) after infusion. In the animals of the groups EG1 and EG2, no cells were observed after seven days of ADSC infusion (Figure 8).

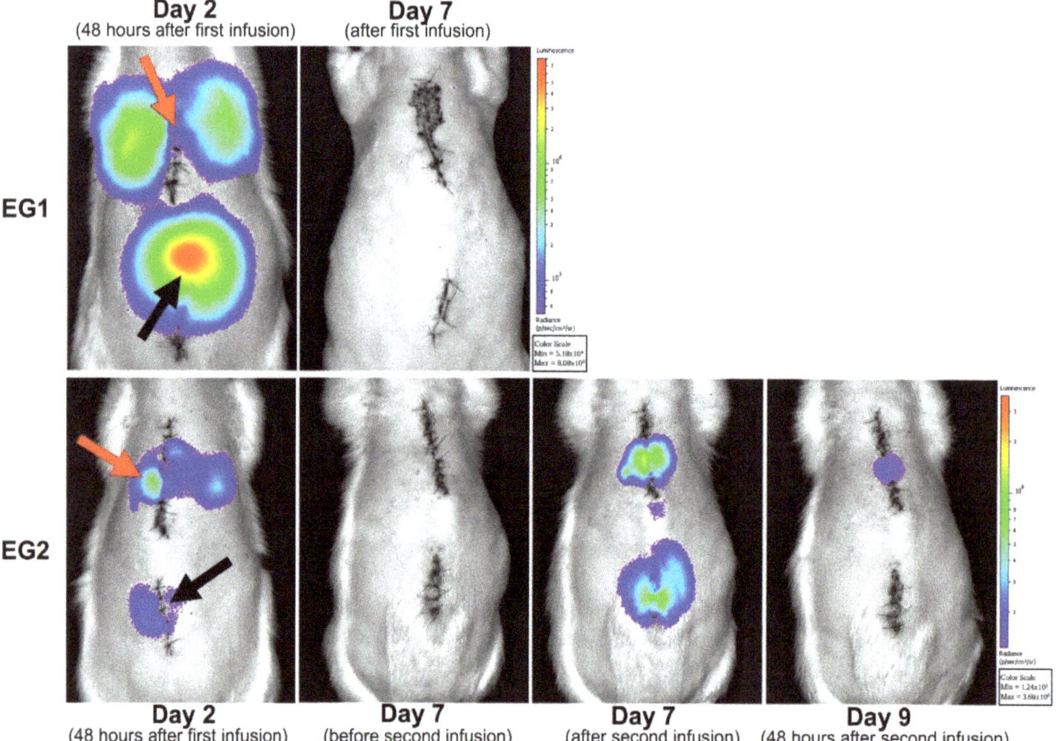

Figure 8. In vivo tracking of ADSC. Representative photos of experimental groups (EG1 and EG2). Cells were observed at the infusion site and in the lungs. The red arrow indicates the injury site, and the black arrow indicates the ADSC infusion site. The images are pseudo-colored and represent the bioluminescent signal emitted during degradation of D-luciferin by the enzyme luciferase present in cells. The color scale represents the intensity of the bioluminescence signal emitted by the cells in photons/second/cm^2/steradian (p/s/cm^2/sr).

4. Discussion

The methodology used in this study, standardized spinal cord contusion, aims to allow for better comparison within our own institution and with other research centers globally. In comparison with clip or balloon compression, the impactor device allows for better similarity in histology and motor recovery and leads to fewer deaths during follow-up [22,23].

The choice of ADSC followed not only our institution's protocol but is also justified by benefits found in other similar studies: good resistance to hypoxic conditions, secretion of neurotrophic factors, and low immunogenicity. It is possible that the combination of hypoxic conditioning and the addition of different cell types will significantly improve the regenerative potential of ADSC [24]. Furthermore, these cells resist hypoxic conditions, secrete neurotrophic factors, and show low immunogenicity [25]. The injection pathway was the main modification in our research strategy, due to unsatisfactory results achieved with direct injection into the injury site, especially in the acute setting. Our previous reports show decreased motricity after direct stem cell injection into the injury site [10]. Although there are reports of cellular migration towards the injury with intravenous or arterial catheterization pathways, cellular concentration is higher when the cells are delivered directly into the nervous system [26]. In the acute setting, the inflammatory process makes it challenging to identify the presence of stem cells, and in the chronic phase, fibrosis limits tissue expansion, limiting the injected volume and presenting evident extravasation from the injury site [27]. As proposed by our protocol, lumbar punction is also accessible for human trials, and understanding the behavior of stem cells injected into this pathway is crucial for developing future human therapeutic strategies.

In vivo, cellular tracking with bioluminescence imaging is a sensitive technique for evaluating administered cell distribution, localization, and migration [28]. Although ADSC have a short life span after infusion [29], this research identified cells at the injury site and on the lungs right after each infusion. Cell tracking with bioluminescence allowed us to observe the migratory capacity of ADSC towards the lesion site; however, the animals that received repeated sedation during the injury recovery period showed a significant mortality rate. Repeated anesthetic procedures can also be a negative factor in evaluating a modulation and cell survival. However, in this study, no significant histological differences were noticed between animals that underwent cell tracking and those that did not. Callera et al. [30] were able to track marked cells using magnetic resonance imaging, confirming migration from the infusion site in the lumbar spine towards the injured area in the thoracic spinal cord. Other methods to track cell migration include magnetic resonance imaging, in vivo microscopy, and micro-computed tomography. However, these methods require expensive equipment, making results reproducibility difficult, especially in low-income countries [31]. The myelin sheath, produced by oligodendrocytes in the central nervous system, renders the structure to the spinal cord and regulates neuronal growth through neurite outgrowth inhibitor (NOGO) molecules secreted by injured or dysfunctional oligodendrocytes. Controlling myelin production and NOGO secretion is critical to spinal cord regeneration research [32,33]. In the present study, ADSC could not effectively modify the structural damage secondary to myelin loss nor activate astrocytes. On the contrary, some studies showed that mesenchymal stromal cell promotes axonal regeneration and myelination [34]. Chen et al. reported that the infusion of ADSC directly to the injury site right after the trauma promoted astrocyte activation and reduced glial scar [25].

In this research, neuronal loss was reduced, which suggests that immunomodulatory factors secreted by the ADSC reduced inflammation, inhibited apoptosis, and protected neurons. These results are in line with other studies showing that the intravenous injection of ADSC had a similar effect on the number of neurons, even with a small concentration of cells migrating towards the injury site on subacute phases of SCI, which suggests a regenerative effect secondary to ADSC administration [35]. Menezes et al. identified laminin deposition on the injury site, indicating neuronal regeneration after cell transplantation [36].

Another key issue in spinal cord regeneration is controlling scar formation. By not controlling or reducing astrocyte migration to the injury area, ADSC could not block the intrinsic process that limits axonal regrowth and neuronal reconnection [37].

The infusion of cells into the dural sac, distal to the spinal cord, allows for a larger injected volume, with low risk of neurologic deterioration, and in vivo cell tracking confirmed cell migration towards the injury site. The downside of repeated anesthetic procedures in the experimental models, would not be a problem on a human trial, since lumbar puncture can be performed on an ambulatory fashion, with local anesthesia.

5. Conclusions

ADSC infusion distal to the injury site was able to reduce neuronal loss significantly, although it did not prevent the myelin loss or enhance the area occupied by astrocytes compared to the control group. The results are similar when comparing one or two cell infusions.

Author Contributions: E.N.V.: study design, experimental procedures, statistical analysis, manuscript preparation; C.L.K.R.: study design, experimental procedures, manuscript preparation; L.F., F.B. and A.C.D.: experimental procedures, manuscript review; A.d.O.A.: experimental procedures, statistical analysis, manuscript preparation; M.O.: statistical analysis; P.R.S.B.: study design, manuscript preparation and review. All authors have read and agreed to the published version of the manuscript.

Funding: This study received a grant from AOSpine Latin America.

Institutional Review Board Statement: This study was approved by the Institutional Ethics Committee on the Use of Animals (No 69) as well as by the Ethics Committee in Research with Humans (CAAE 12723713.3.0000.0020).

Informed Consent Statement: Informed consent was obtained from all subjects involved in the study, allowing for use of adipose tissue from humans undergoing plastic surgery.

Data Availability Statement: Data is available upon request to the main author, according to Institutional policy.

Acknowledgments: The authors thank Crisciele Kuligovski for cell transduction experiments.

Conflicts of Interest: The authors declare no conflict of interest.

References

1. Cripps, R.A.; Lee, B.B.; Wing, P.; Weerts, E.; Mackay, J.; Brown, D. A global map for traumatic spinal cord injury epidemiology: Towards a living data repository for injury prevention. *Spinal Cord* **2011**, *49*, 493–501. [CrossRef] [PubMed]
2. Noonan, V.K.; Fingas, M.; Farry, A.; Baxter, D.; Singh, A.; Fehlings, M.G.; Dvorak, M.F. Incidence and prevalence of spinal cord injury in Canada: A national perspective. *Neuroepidemiology* **2012**, *38*, 219–226. [CrossRef] [PubMed]
3. Choo, A.M.; Liu, J.; Lam, C.K.; Dvorak, M.; Tetzlaff, W.; Oxland, T.R. Contusion, dislocation, and distraction: Primary hemorrhage and membrane permeability in distinct mechanisms of spinal cord injury. *J. Neurosurg. Spine* **2007**, *6*, 255–266. [CrossRef] [PubMed]
4. LaPlaca, M.C.; Simon, C.M.; Prado, G.R.; Cullen, D.K. CNS injury biomechanics and experimental models. *Prog. Brain Res.* **2007**, *161*, 13–26. [CrossRef]
5. Fehlings, M.G.; Perrin, R.G. The role and timing of early decompression for cervical spinal cord injury: Update with a review of recent clinical evidence. *Injury* **2005**, *36* (Suppl. S2), B13–B26. [CrossRef]
6. Hausmann, O.N. Post-traumatic inflammation following spinal cord injury. *Spinal Cord* **2003**, *41*, 369–378. [CrossRef]
7. Kwon, B.K.; Tetzlaff, W.; Grauer, J.N.; Beiner, J.; Vaccaro, A.R. Pathophysiology and pharmacologic treatment of acute spinal cord injury. *Spine J.* **2004**, *4*, 451–464. [CrossRef]
8. Ahuja, C.S.; Martin, A.R.; Fehlings, M. Recent advances in managing a spinal cord injury secondary to trauma. *F1000Research* **2016**, *5*, 1017. [CrossRef]
9. Chen, D.; Zeng, W.; Fu, Y.; Gao, M.; Lv, G. Bone marrow mesenchymal stem cells combined with minocycline improve spinal cord injury in a rat model. *Int. J. Clin. Exp. Pathol.* **2015**, *8*, 11957–11969.
10. Carvalho, K.A.; Cunha, R.C.; Vialle, E.N.; Osiecki, R.; Moreira, G.H.; Simeoni, R.B.; Francisco, J.C.; Guarita-Souza, L.C.; Oliveira, L.; Zocche, L.; et al. Functional outcome of bone marrow stem cells (CD45(+)/CD34(-)) after cell therapy in acute spinal cord injury: In exercise training and in sedentary rats. *Transplant. Proc.* **2008**, *40*, 847–849. [CrossRef]

11. Fonseca, A.F.B.; Scheffer, J.P.; Giraldi-Guimarães, A.; Coelho, B.P.; Medina, R.M.; Oliveira, A.L.A. Comparison among bone marrow mesenchymal stem and mononuclear cells to promote functional recovery after spinal cord injury in rabbits. *Acta Cir. Bras.* **2017**, *32*, 1026–1035. [CrossRef]

12. Lv, C.; Zhang, T.; Li, K.; Gao, K. Bone marrow mesenchymal stem cells improve spinal function of spinal cord injury in rats via TGF-β/Smads signaling pathway. *Exp. Ther. Med.* **2020**, *19*, 3657–3663. [CrossRef]

13. Caplan, A.I. Why are MSCs therapeutic? New data: New insight. *J. Pathol.* **2009**, *217*, 318–324. [CrossRef]

14. Fang, B.; Song, Y.; Zhao, R.C.; Han, Q.; Cao, Y. Treatment of resistant pure red cell aplasia after major ABO-incompatible bone marrow transplantation with human adipose tissue-derived mesenchymal stem cells. *Am. J. Hematol.* **2007**, *82*, 772–773. [CrossRef]

15. Fraser, J.K.; Wulur, I.; Alfonso, Z.; Hedrick, M.H. Fat tissue: An underappreciated source of stem cells for biotechnology. *Trends Biotechnol.* **2006**, *24*, 150–154. [CrossRef]

16. Vialle, E.; Vialle, L.R.G.; Rasera, E.; Cechinel, C.; Leonel, I.; Seyboth, C. Avaliação da recuperação motora em ratos submetidos a lesão medular experimental. *Rev. Bras. Ortop.* **2002**, *37*, 83–91.

17. Dominici, M.; Le Blanc, K.; Mueller, I.; Slaper-Cortenbach, I.; Marini, F.; Krause, D.; Deans, R.; Keating, A.; Prockop, D.; Horwitz, E. Minimal criteria for defining multipotent mesenchymal stromal cells. The International Society for Cellular Therapy position statement. *Cytotherapy* **2006**, *8*, 315–317. [CrossRef]

18. Rebelatto, C.K.; Aguiar, A.M.; Moretão, M.P.; Senegaglia, A.C.; Hansen, P.; Barchiki, F.; Oliveira, J.; Martins, J.; Kuligovski, C.; Mansur, F.; et al. Dissimilar differentiation of mesenchymal stem cells from bone marrow, umbilical cord blood, and adipose tissue. *Exp. Biol. Med.* **2008**, *233*, 901–913. [CrossRef]

19. Vialle, L.R.G.; Fischer, S.; Marcon, J.C.; Vialle, E.; Luzzi, R.; Bleggi-Torres, L.F. Estudo histológico da lesão medular experimental em ratos. *Rev. Bras. Ortop.* **1999**, *34*, 85–89.

20. Meyer, F.; Vialle, L.R.; Vialle, E.N.; Bleggi-Torres, L.F.; Rasera, E.; Leonel, I. Alterações vesicais na lesão medular experimental em ratos. *Acta Cirúrgica Bras.* **2003**, *18*, 203–208. [CrossRef]

21. Vialle, E.N.; Vialle, L.R.; Arruda Ade, O. Histomorphometric analysis of experimental disc degeneration. *Global Spine J.* **2012**, *2*, 129–136. [CrossRef] [PubMed]

22. Talac, R.; Friedman, J.A.; Moore, M.J.; Lu, L.; Jabbari, E.; Windebank, A.J.; Currier, B.L.; Yaszemski, M.J. Animal models of spinal cord injury for evaluation of tissue engineering treatment strategies. *Biomaterials* **2004**, *25*, 1505–1510. [CrossRef] [PubMed]

23. Sharif-Alhoseini, M.; Khormali, M.; Rezaei, M.; Safdarian, M.; Hajighadery, A.; Khalatbari, M.M.; Meknatkhah, S.; Rezvan, M.; Chalangari, M.; Derakhshan, P.; et al. Animal models of spinal cord injury: A systematic review. *Spinal Cord* **2017**, *55*, 714–721. [CrossRef] [PubMed]

24. Villanova Junior, J.A.; Fracaro, L.; Rebelatto, C.L.K.; da Silva, A.J.; Barchiki, F.; Senegaglia, A.C.; Dominguez, A.C.; de Moura, S.A.B.; Pimpão, C.T.; Brofman, P.R.S.; et al. Recovery of motricity and micturition after transplantation of mesenchymal stem cells in rats subjected to spinal cord injury. *Neurosci. Lett.* **2020**, *734*, 135134. [CrossRef]

25. Chen, J.; Wang, L.; Liu, M.; Gao, G.; Zhao, W.; Fu, Q.; Wang, Y. Implantation of adipose-derived mesenchymal stem cell sheets promotes axonal regeneration and restores bladder function after spinal cord injury. *Stem. Cell Res. Ther.* **2022**, *13*, 503. [CrossRef]

26. Dave, S.D.; Patel, C.N.; Vanikar, A.V.; Trivedi, H.L. In vitro differentiation of neural cells from human adipose tissue derived stromal cells. *Neurol. India* **2018**, *66*, 716–721. [CrossRef]

27. Anjum, A.; Yazid, M.D.; Fauzi Daud, M.; Idris, J.; Ng, A.M.H.; Selvi Naicker, A.; Ismail, O.H.R.; Athi Kumar, R.K.; Lokanathan, Y. Spinal Cord Injury: Pathophysiology, Multimolecular Interactions, and Underlying Recovery Mechanisms. *Int. J. Mol. Sci.* **2020**, *21*, 7533. [CrossRef]

28. Kim, T.J.; Tuerkcan, S.; Ceballos, A.; Pratx, G. Modular platform for low-light microscopy. *Biomed Opt. Express.* **2015**, *6*, 4585–4598. [CrossRef]

29. Zaminy, A.; Shokrgozar, M.A.; Sadeghi, Y.; Norouzian, M.; Heidari, M.H.; Piryaei, A. Transplantation of schwann cells differentiated from adipose stem cells improves functional recovery in rat spinal cord injury. *Arch. Iran. Med.* **2013**, *16*, 533–541.

30. Callera, F.; de Melo, C.M. Magnetic resonance tracking of magnetically labeled autologous bone marrow CD34+ cells transplanted into the spinal cord via lumbar puncture technique in patients with chronic spinal cord injury: CD34+ cells' migration into the injured site. *Stem. Cells Dev.* **2007**, *16*, 461–466. [CrossRef]

31. Hubertus, V.; Meyer, L.; Roolfs, L.; Waldmann, L.; Nieminen-Kelhä, M.; Fehlings, M.G.; Vajkoczy, P. In vivo imaging in experimental spinal cord injury–Techniques and trends. *Brain Spine* **2022**, *2*, 100859. [CrossRef]

32. Geoffroy, C.G.; Zheng, B. Myelin-associated inhibitors in axonal growth after CNS injury. *Curr. Opin. Neurobiol.* **2014**, *27*, 31–38. [CrossRef]

33. Lee, J.K.; Geoffroy, C.G.; Chan, A.F.; Tolentino, K.E.; Crawford, M.J.; Leal, M.A.; Kang, B.; Zheng, B. Assessing spinal axon regeneration and sprouting in Nogo-, MAG-, and OMgp-deficient mice. *Neuron* **2010**, *66*, 663–670. [CrossRef]

34. Ashammakhi, N.; Kim, H.J.; Ehsanipour, A.; Bierman, R.D.; Kaarela, O.; Xue, C.; Khademhosseini, A.; Seidlits, S.K. Regenerative Therapies for Spinal Cord Injury. *Tissue Eng. Part B Rev.* **2019**, *25*, 471–491. [CrossRef]

35. Ohta, Y.; Hamaguchi, A.; Ootaki, M.; Watanabe, M.; Takeba, Y.; Iiri, T.; Matsumoto, N.; Takenaga, M. Intravenous infusion of adipose-derived stem/stromal cells improves functional recovery of rats with spinal cord injury. *Cytotherapy* **2017**, *19*, 839–848. [CrossRef]

36. Menezes, K.; Nascimento, M.A.; Gonçalves, J.P.; Cruz, A.S.; Lopes, D.V.; Curzio, B.; Bonamino, M.; de Menezes, J.R.; Borojevic, R.; Rossi, M.I.; et al. Human mesenchymal cells from adipose tissue deposit laminin and promote regeneration of injured spinal cord in rats. *PLoS ONE* **2014**, *9*, e96020. [CrossRef]
37. Faulkner, J.R.; Herrmann, J.E.; Woo, M.J.; Tansey, K.E.; Doan, N.B.; Sofroniew, M.V. Reactive astrocytes protect tissue and preserve function after spinal cord injury. *J. Neurosci.* **2004**, *24*, 2143–2155. [CrossRef]

 *biomedicines*

Article

Microneurotrophin BNN27 Reduces Astrogliosis and Increases Density of Neurons and Implanted Neural Stem Cell-Derived Cells after Spinal Cord Injury

Konstantina Georgelou [1,2], Erasmia-Angeliki Saridaki [1], Kanelina Karali [1,2], Argyri Papagiannaki [1], Ioannis Charalampopoulos [1,2], Achille Gravanis [1,2,†] and Dimitrios S. Tzeranis [2,3,*,†]

[1] Department of Pharmacology, School of Medicine, University of Crete, 71003 Heraklion, Greece
[2] Institute of Molecular Biology and Biotechnology, Foundation for Research and Technology-Hellas, 71003 Heraklion, Greece
[3] Department of Mechanical and Manufacturing Engineering, University of Cyprus, Nicosia 2109, Cyprus
* Correspondence: tzeranis.dimitrios@ucy.ac.cy
† These authors contributed equally to this work.

Abstract: Microneurotrophins, small-molecule mimetics of endogenous neurotrophins, have demonstrated significant therapeutic effects on various animal models of neurological diseases. Nevertheless, their effects on central nervous system injuries remain unknown. Herein, we evaluate the effects of microneurotrophin BNN27, an NGF analog, in the mouse dorsal column crush spinal cord injury (SCI) model. BNN27 was delivered systemically either by itself or combined with neural stem cell (NSC)-seeded collagen-based scaffold grafts, demonstrated recently to improve locomotion performance in the same SCI model. Data validate the ability of NSC-seeded grafts to enhance locomotion recovery, neuronal cell integration with surrounding tissues, axonal elongation and angiogenesis. Our findings also show that systemic administration of BNN27 significantly reduced astrogliosis and increased neuron density in mice SCI lesion sites at 12 weeks post injury. Furthermore, when BNN27 administration was combined with NSC-seeded PCS grafts, BNN27 increased the density of survived implanted NSC-derived cells, possibly addressing a major challenge of NSC-based SCI treatments. In conclusion, this study provides evidence that small-molecule mimetics of endogenous neurotrophins can contribute to effective combinatorial treatments for SCI, by simultaneously regulating key events of SCI and supporting grafted cell therapies in the lesion site.

Keywords: spinal cord injury; neurotrophins; astrogliosis; neuroprotection; neural stem cells; biomaterials

Citation: Georgelou, K.; Saridaki, E.-A.; Karali, K.; Papagiannaki, A.; Charalampopoulos, I.; Gravanis, A.; Tzeranis, D.S. Microneurotrophin BNN27 Reduces Astrogliosis and Increases Density of Neurons and Implanted Neural Stem Cell-Derived Cells after Spinal Cord Injury. *Biomedicines* **2023**, *11*, 1170. https://doi.org/10.3390/biomedicines11041170

Academic Editor: Nicolas Guerout

Received: 4 March 2023
Revised: 2 April 2023
Accepted: 12 April 2023
Published: 13 April 2023

1. Introduction

Spinal cord injury (SCI) leads to partial or complete loss of sensory and/or motor functions. According to the World Health Organization, approximately 400,000 individuals around the world suffer a SCI each year. Despite significant efforts, existing clinical treatments are limited to surgical intervention for spinal cord decompression and the controversial administration of methylprednisolone, a corticosteroid that has serious adverse effects and poor efficacy [1]. The development of effective SCI treatments remains an unmet clinical need.

SCI triggers a complex cascade of interrelated events. Acute primary injury (cell death, hemorrhage, edema and ischemia) proceeds to secondary injury (inflammatory and cytotoxic events) within the next few hours to weeks and leads to chronic injury over the next months to years [2]. The multifactorial nature of SCI calls for the development of combinatorial therapies that will combine the complementary effects of multiple components to simultaneously regulate multiple events.

Significant research focuses on developing combinatorial treatments for SCI that integrate the complementary actions of biomaterials, cell therapies and diffusible growth factors [3]. In such combinatorial treatments, biomaterials contribute structural support, inflammation regulation, delivery, protection and orientation of cells, a substrate for cellular migration and axonal elongation, and localized delivery of diffusible therapeutic molecules (small molecules, biologics) at the lesion [1]. The cellular part of combinatorial treatments usually focuses on replacing neural cells lost during injury and re-establishing connectivity in the lesion site [4]. Among the various kinds of stem cells tested in experimental SCI treatments, neural stem cells (NSCs) have shown promising results in animal models, including the extension of a large number of axons caudally to the lesion and synapse creation with corticospinal tract axons [5,6]. Despite effort to translate NSC-based treatments towards human clinical evaluation [7], much work remains to enhance their efficacy. In this direction, much research focuses on optimizing the biomaterial part of SCI treatments and complementing grafted cells with soluble factors. Recently, we demonstrated that porous collagen scaffold (PCS) grafts can safely deliver NSCs in SCI lesions, enabling their differentiation into neurons and glia. Such NSC-seeded PCS grafts significantly improved locomotion recovery over a period of 12 weeks in a mouse dorsal column crush SCI model [8]. This work focuses on the third part of combinatorial SCI treatments, the supplementation of cell therapies with soluble therapeutic molecules, emphasizing neurotrophic factors.

Much research on SCI treatments has focused on trophic factors' support, due to their role in neuronal survival, axonal guidance and synapse formation [9,10]. Among trophic factors, NGF, the prototypic member of the neurotrophin family, has received significant attention. NGF and its receptors (TrkA, p75^{NTR}) are expressed in the spinal cord [11–13]. The poor regenerative capacity of the CNS compared to the peripheral nervous system (PNS) has been partly attributed to inadequate trophic support, including low levels of NGF following injury. Indeed, following SCI, NGF expression in the spinal cord was detected just in meningeal cells and Schwann cells in nerve roots [9]. SCI does not induce NGF expression in spinal cord oligodendrocytes, in contrast to the PNS axotomy that induces robust NGF expression in Schwann cells. These observations suggest trophic factor supplementation as a potential therapeutic strategy for SCI. Following SCI, NGF administration induced tract-specific neurite growth [14,15], reduced neuronal death [16–18], inhibited SCI-induced autophagy [18], and led to behavioral improvement [16–19]. Significant research has focused on developing means of NGF administration that bypass the poor penetration of NGF via the blood–brain barrier (BBB) [20], including intranasal delivery [13], local BBB disruption using ultrasound [17] and the delivery of NGF-expressing genetically modified cells, such as fibroblasts [14,21,22], Schwann cells [23] or NSCs [18,19], in the lesion site. Overall, despite some promising results, the utilization of NGF in combinatorial research treatments for SCI has been limited, due to unfavorable pharmacokinetics and side effects, including hyperalgesia [24].

Key limitations of endogenous neurotrophins regarding therapeutic applications can be bypassed by microneurotrophins (MNT), 17-carbon small-molecule derivatives of dehydroepiandrosterone that specifically activate neurotrophin receptors and can pass the BBB [25]. BNN27, the seminal MNT, is an NGF analog that can activate both its receptors (TrkA, p75^{NTR}) and lacks the hyperalgesia effects of NGF. BNN27 has demonstrated various effects related to wound healing: BNN27 demonstrated neuroprotective effects on various kinds of neurons (superior cervical ganglion cells, dorsal root ganglion cells, cerebellar granule neurons) mediated via both NGF receptors and downstream pro-survival pathways [26,27], decreased apoptosis of TrkA$^+$ sensory neurons in E13.5 NGF null mice embryos [26], decreased retina cell apoptosis and astrogliosis in a rat model of diabetes-induced retinal damage [28,29], and reduced microgliosis in a mouse cuprizone-induced demyelination model [30]. Nevertheless, the effects of BNN27 have not been evaluated yet in any CNS injury model, either in the form of a monotherapy or as a part of a combinatorial therapy.

The present study evaluates the effects of BNN27 in the mouse dorsal column crush SCI model. BNN27 was delivered systemically either by itself or combined with NSC-seeded PCS grafts demonstrated recently to improve locomotion performance in the same SCI model [8]. Our findings highlight the central role of NSC-seeded PCS grafts in locomotion improvement, the replenishment of neural cells lost due to SCI and the integration of neuronal cells with surrounding CNS tissue 12 weeks post injury (wpi). While systemic administration of BNN27 did not improve locomotion recovery over a 10-week period following SCI, BNN27 significantly reduced astrogliosis and increased neuron density in the lesion site at 12 wpi. Furthermore, BNN27 increased the density of implanted NSC-derived cells in the lesion. Our study provides evidence that small-molecule mimetics of endogenous neurotrophins can contribute to effective combinatorial SCI treatments by regulating key events of SCI and supporting grafted cell therapies in the lesion site.

2. Materials and Methods

2.1. Animals

Animal care and experimentation protocols were performed according to the approval of the Veterinary Directorate of the region of Crete in compliance with Greek Government guidelines, EU guidelines 2010/63/EU, FORTH ethics committee guidelines and in accordance with approved protocols from the Federation of European Laboratory Animal Science Associations (FELASA) on use of laboratory animals. Facility License number: EL91-BIOexp-02; Project title: "Investigating the effects of neurotrophins and their synthetic analogues on the mechanisms of neurodegeneration and neurogenesis in Alzheimer's disease and the effect of neuroimplants based on porous collagen scaffolds in the central nervous system trauma"; approval numbers: 262,272 and 360,667; date of initial approval: 29 October 2018; date of reapproval: 29 November 2021. C57/BL6 [31] and Rosa26-YFP mice (containing a floxed yellow fluorescent protein (YFP) gene) [32] were maintained in climate-controlled conditions (30–50% humidity, 21 ± 2 °C, 12:12 h light/dark cycle) on a 12 h light/dark cycle with ad libitum access to food and water. Rosa26-YFP mice were crossed with CMV-Cre transgenic mice [33] to obtain mice that expressed YFP in all tissues.

2.2. Scaffold Fabrication

PCS sheets were fabricated by lyophilizing a 5 mg/mL microfibrillar collagen I suspension in 50 mM acetic acid as described previously [8]. The resulting 2.5 mm-thick dry sheets were cross-linked via dehydro-thermal treatment (105 °C, 50 mTorr, 24 h). Scaffold structure was verified by scanning electron microscopy.

2.3. Primary Neural Stem Cell Isolation and Culture

NSC isolation was performed from brain tissue (cortex) of mouse embryos (E13.5) of Rosa26-YFP mice crossed with CMV-Cre transgenic mice via the enzymatic digestion of the cortex. Specifically, pregnant mice (gestational day 13.5) were sacrificed via cervical dislocation and embryos were carefully removed. Embryos' cortical hemispheres were washed gently in Hanks' balanced salt solution (Thermo scientific) with 5% penicillin/streptomycin and were mechanically dissociated for NSC isolation in complete NSC medium (Dulbecco's modified Eagle's medium/F12 (Thermo scientific), B27 supplement minus vitamin A (Thermo scientific), 0.6% D-glucose, 50 mg/mL primocin (InvivoGen), 20 μg/mL fibroblast growth factor 2 (R&D), 20 μg/mL epidermal growth factor (R&D)). Then, $2.5 \cdot 10^5$ NSCs were seeded in T25 flasks in 5 mL complete NSC medium and 1 mL of fresh medium was added every other day. Neurospheres formed within 2 days and were dissociated by Accutase (Sigma) at day 4 or 5. Dissociated NSCs were either passaged or used for experiments (passage 3–8). Care was taken in the first passage to pick and keep neurospheres of YFP-expressing embryos.

2.4. Spinal Cord Injury Animal Model

Procedures for SCI and graft implantation followed previously published protocols [8]. Specifically, 3 days prior to injury, $3 \cdot 10^4$ YFP-expressing NSCs were seeded into $1 \times 1 \times 1.5$ mm PCS samples and cultured in complete NSC medium. Prior to implantation, grafts were washed for 30 min in phosphate-buffered saline (PBS). All surfaces, tools, and instruments utilized had been sterilized carefully. C57/BL6 male mice approximately 4 weeks old were anesthetized by 2.5:1 isoflurane:oxygen mix inhalation for 5 min in a scavenger box until breathing slowed down. Before anesthesia, a drop of meloxicam (Metacam) was introduced per os. After verifying the absence of paw reflexes, each mouse was shaved in the level of the humpback and the exposed skin was disinfected. Each mouse was then transferred onto a heat pad, where ophthalmic ointment was applied to avoid eye dryness and 2:1 isoflurane:oxygen mix was applied to maintain deep anesthesia via a mask. In all animal groups ("Uninjured control", "Placebo", "BNN27", "Scaffold NSC + Placebo", "Scaffold NSC + BNN27") a skin incision was made from the base of the humpback until the higher point of the rib cage and exposed muscles were carefully torn. After removing surrounding ligaments, the bone of T10 vertebra was removed and an incision was made in the dura matter above the T13 segment [34]. In four injured animal groups ("Placebo", "BNN27", "Scaffold NSC + Placebo", "Scaffold NSC + BNN27") the dorsal column was crushed by inserting Dumont #5 fine forceps 1 mm deep into the white matter and keeping them closed for 10 sec. This step was repeated once to create a $\approx$1 mm^3 pocket (lesion). In two animal groups ("Scaffold NSC + Placebo", "Scaffold NSC + BNN27") a NSC-seeded graft was placed in the pocket immediately following injury. In all animal groups, the exposed spinal cord tissue was covered by a Geistlich membrane (Geistlich Bio-Gide) and then a hemostatic agent (Lyostypt, B. Braun Company), the muscles and skin were sutured and the skin was disinfected. Then, in animals of the "BNN27" and "Scaffold NSC + BNN27" groups, a small incision was made close to the shoulder and a pellet that contained 18 mg BNN27 (60 day release period; Innovative Research of America #SX999) was implanted under the skin. After inserting the pellet, the skin was sutured and disinfected. Similarly, in animals of the "Placebo" and "Scaffold NSC + Placebo" groups, mice were implanted with a pellet that lacked an active pharmaceutical ingredient (Innovative Research of America #SC111). All animal groups were provided with meloxicam for the next 3 days for analgesia.

2.5. Horizontal Ladder Walking Assay

Locomotion performance of mice over a 10-week period following SCI was assessed via the horizontal ladder walking assay [35], as described previously [8]. Briefly, animals crossed a 1 m-long horizontal ladder consisted of sidewalls made of clear plexiglass connected by aluminum rods (ø2 mm diameter) spaced approximately 1 cm from each other. The ladder was elevated above the ground at the height of two cages. Animals were trained to cross the ladder from a neutral cage to their home cage and always in the same direction.

For handling and acclimation, the week prior to surgeries, mice settled down in the animal facility to become familiarized. For 6 consecutive days prior to testing, animals were handled for 5 min by the examiner and walked through the ladder for 5 min in order to reduce their stress during the testing procedure. Each animal was tested five times per session. Before each session, mice crossed the ladder twice for habituation. For locomotion analysis, each trial was video recorded. Videos of walking mice were analyzed frame-by-frame by two researchers blind to the condition (group, week) of the animal analyzed to identify locomotion errors, defined as steps where a hind limb missed or slipped off or was misplaced on a rod. Each fault was classified using a 7-category scale (0: total miss, 1: deep slip, 2: slight slip, 3: replacement, 4: correction, 5: partial placement, 6: correct placement) as described previously [35]. Locomotion performance was described via the fault rate (fraction of steps classified in categories 0 to 5). The last step before and the first step after a stop were excluded; only consecutive steps were included in the analysis.

While initially each animal group consisted of 9–10 mice, some of them were excluded from our final results in order to ensure consistent injury in all animals. Exclusion criteria were based on mice locomotion performance. Mice of the "Uninjured Control" group were excluded when their mean fault rate over the 10-week monitoring period exceeded 10%. Mice of the remaining four injured groups were excluded when there were strong indications of insufficient injury (e.g., fault rate <20% the day following injury or <15% at

1 wpi) or excessive injury (e.g., fault rate >90% the day following injury or >45% at 1 wpi or >35% at 2 wpi). Mice exclusions were confirmed by histology based on the magnitude of the observed lesion in mice spinal cord sections.

2.6. Histological Evaluation

At 12 wpi, mice were deeply anesthetized using 2.5:1 isoflurane:oxygen mix and transcardially perfused with ice-cold heparinized (10 U/mL) saline followed by 4% paraformaldehyde (PFA) in PBS. Spinal cords were dissected, post-fixed in 4% PFA at 4 °C for 1 h, washed in PBS, and immersed in 30% sucrose solution in 0.1 M phosphate buffer at 4 °C for 24 h. The 1 cm-long part of the spinal cord tissue centered around the lesion was frozen in isopentane at -70 °C and 20 μm-thick parasagittal sections were cut in super-frost slides.

For histological analysis, slides were placed in ice-cold acetone for 5 min, air-dried for 10 min in laminar flow, washed twice in PBS, once in 0.1% Triton X-100 in PBS (PBST) and once in 0.3% PBST, then blocked in 0.1% PBST supplemented with 0.1% bovine serum albumin (BSA) and 10% goat or horse serum at room temperature for 1 h, incubated in primary antibodies (L1: 1:000 rabbit polyclonal antibody (obtained by Dr. Fritz G. Rathjen)); tubulin β3 (Tubb3): 1:1000 Biolegend #801201 (clone Tuj1); glial fibrillary acidic protein (GFAP): 1:2000 Millipore #AB5541; synaptophysin: 1:200 Millipore #MAB368; green fluorescent protein (GFP): 1:200 Minotech #701-1; vesicular glutamate transporter 1 (VGLUT1): 1:400 Synaptic systems #135304; tyrosine hydroxylase (3TH): 1:40 Thermo #MA1-24654; glutamic acid decarboxylase 67 (GAD67): 1:1000 Sigma #MAB5406; α-smooth muscle actin (αSMA): 1:200 Sigma #A5228) diluted in blocking solution at 4 °C overnight, washed twice in 0.1% PBST, incubated with fluorophore-conjugated secondary antibodies (Thermo) diluted 1:1000 in PBS at room temperature for 1 h, washed twice in PBS, counterstained with Hoechst 33,342 and mounted.

Immunostained sections were imaged in a TCS SP8 inverted confocal microscope (Leica Microsystems, Wetzlar, Germany) using a 40× or a 63× oil-immersion objective lens. Images were taken at the epicenter, caudally and rostrally to the lesion. The presence of neurons was quantified by counting the density of Tuj1$^+$ cells in the lesion epicenter. Astrogliosis was quantified by calculating the fraction of pixels that stained positively for GFAP immediately around the lesion boundary, by selecting up to 5 regions (approximately 0.025 mm^2 each) per section. The presence of implanted cells was quantified by counting the density of GFP$^+$ cells in the lesion epicenter, expressed per volume, using images acquired by a 63× objective lens. Synapse formation was quantified by calculating the fraction of pixels that stained positively for synaptophysin in the lesion epicenter. Axonal elongation was quantified by calculating the fraction of pixels that stained positively for L1 in the lesion epicenter as well as in dorsal region rostrally and caudally to the lesion. Angiogenesis was quantified by counting the number of αSMA$^+$ vessels in the lesion epicenter. All imaging analysis was performed in Fiji software. In thresholding operations, care was taken to avoid counting the background of each staining and define positive staining using a common value chosen by an expert user.

2.7. Statistical Analysis

Experimental data are expressed as mean $\pm$ standard error of the mean (SEM). Statistical analysis was performed using the Prism software (Graphpad, La Jolla, CA, USA). Statistically significant effects on locomotion fault rate were assessed by 1-way analysis of variance (ANOVA; for evaluating the overall effects of injury among the groups at postop and 1 wpi) and by 2-way ANOVA (for evaluating the overall effects of BNN27 and NSC-seeded PCS graft treatments) followed by Tukey's post hoc test (for pairwise comparisons between groups at 2–10 wpi). Statistically significant effects in the quantification of GFAP$^+$, synaptophysin$^+$, L1$^+$ pixel fractions, Tuj1$^+$ cell density and αSMA$^+$ vessel number were assessed by 2-way ANOVA (for evaluating the effects of BNN27 and NSC-seeded PCS graft treatments) followed by Tukey's post hoc test (for pairwise comparisons between groups). Statistically significant effects of BNN27 administration on the density and neuronal fate of implanted NSCs were assessed by unpaired two-tailed Student's *t*-tests. All statistical tests assumed a statistical significance level of 0.05.

3. Results

Therapeutic effects of BNN27 after SCI were evaluated using the well-described mouse dorsal column crush SCI model [8]. Immediately after crush, three types of SCI treatments were tested: the first animal group was provided systemic administration of BNN27 via subcutaneous pellet ("BNN27" group; 18mg/pellet released over a 60-day period); the second animal group ("Scaffold NSC + BNN27") was treated with a combinatorial approach consisting of a PCS implant seeded with mouse embryonic E13.5 NSCs grafted inside the lesion site immediately after injury (same graft as the one described in [8]) concurrent with the systemic delivery of BNN27 via subcutaneous pellet; the third animal group ("Scaffold NSC + Placebo") was treated with NSC-seeded PCS implant concurrent with subcutaneous pellet that lacked an active pharmaceutical ingredient. In addition, experimental design included the "Placebo" animal group (injured mice treated with a subcutaneous pellet that lacked an active pharmaceutical ingredient) and the "Uninjured Control" animal group (where, following laminectomy, the exposed spinal cord was not injured). Grafts in the "Scaffold NSC" and "Scaffold NSC + BNN27" groups utilized NSCs isolated from mice obtained by crossing Rosa26-YFP mice containing a floxed YFP gene with CMV-Cre mice and keeping embryos whose NSCs express YFP, enabling their detection via live cell fluorescence imaging or immunocyto/histochemistry (Supplementary Materials: Figure S1).

3.1. BNN27 and NSC Graft Treatment Effects on Locomotion Recovery after SCI

Mice locomotion recovery over a period of 10 weeks following dorsal column crush was evaluated by quantifying their step fault rate in the horizontal ladder walking assay (Figure 1 and Supplementary Materials: Figure S2). During the 10-week period, mice of the "Uninjured Control" group had a time-invariant fault rate ($\mu = 6.9\%$, $\sigma = 2.4\%$). No difference was observed among the four injured groups till the 1st wpi, confirming that injury was consistently performed in all animals (post-op: $P_{1\text{-way-ANOVA}} = 0.94$, $F = 0.12$; week 1: $P_{1\text{-way-ANOVA}} = 0.57$, $F = 0.69$). Between 1 and 10 wpi, all injured groups showed some locomotion improvement. Eventually, differences in fault rate became statistically significant only in mice treated with NSC-seeded PCS grafts compared to the untreated injured group at 9 wpi ("Placebo": $20.9 \pm 0.5\%$, $n = 4$ vs. "Scaffold NSC + Placebo": $13.6 \pm 1.7\%$, $n = 5$, $p = 0.01$; "Placebo" vs. "Scaffold NSC + BNN27": $14.8 \pm 0.7\%$, $n = 5$, $p = 0.04$) and 10 wpi ("Placebo": $22.8 \pm 1.5\%$, $n = 4$ vs. "Scaffold NSC + Placebo": $15.6 \pm 1.2\%$, $n = 5$, $p = 0.04$; "Placebo" vs. "Scaffold NSC + BNN27": $14.8 \pm 2\%$, $n = 5$, $p = 0.02$), replicating the findings of our previous study [8]. While grafting mice with NSC-seeded PCS statistically improved locomotion recovery after 8 weeks (week 9: $P_{2\text{-way-ANOVA}} = 10^{-3}$; week 10: $P_{2\text{-way-ANOVA}} = 10^{-3}$), BNN27 administration did not (week 9: $P_{2\text{-way-ANOVA}} = 0.89$; week 10: $P_{2\text{-way-ANOVA}} = 0.45$). The fault rate in the "BNN27" group was not statistically different from the one of the untreated "Placebo" group (9 weeks: "BNN27": $19.3 \pm 2\%$, $n = 4$, $p = 0.86$; 10 weeks: "BNN27": $21.1 \pm 1.7\%$, $n = 4$, $p = 0.9$). Similarly, BNN27 administration did not further improve locomotion recovery when combined with NSC-seeded PCS grafts since the fault rate in the two mice groups treated with NSC-seeded PCS grafts ("Scaffold NSC + Placebo", "Scaffold NSC + BNN27") was not statistically different over the 10-week period (week 9: $p = 0.91$; week 10: $p = 0.98$).

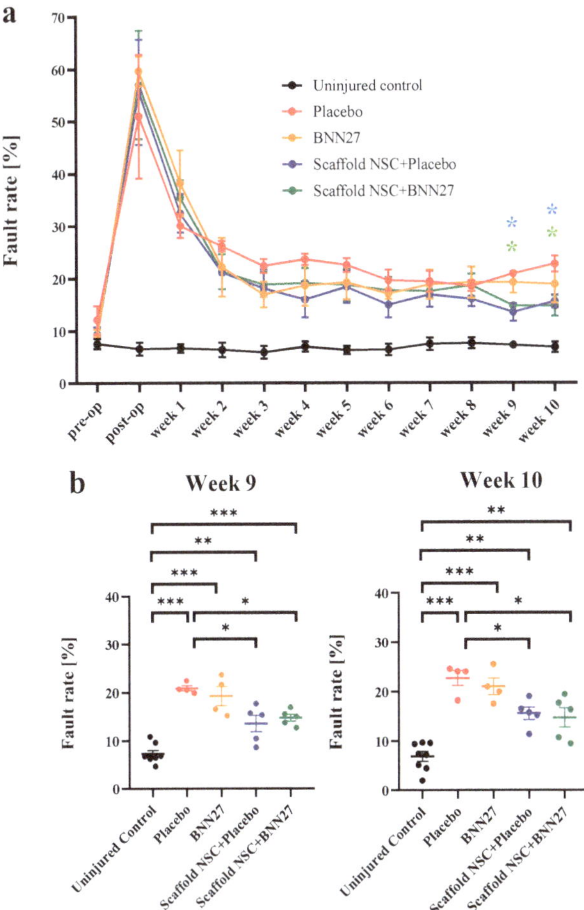

Figure 1. Effects of NSC-seeded PCS grafts and systemic administration of BNN27 on locomotion recovery. (**a**) Quantification of locomotion performance (fault rate) after SCI by the horizontal ladder walking assay over a period of 10 wpi ("Uninjured Control" group: n = 8, "Placebo" and "BNN27" groups: n = 4, "Scaffold NSC" and "Scaffold NSC + BNN27" groups: n = 5). The complete dataset is available in Supplementary Materials: Figure S2. Asterisks indicate the statistically significant difference between the "Placebo" group and the "Scaffold NSC + Placebo" group (blue) or the "Scaffold NSC + BNN27" group (green) at 9 and 10 wpi. (**b**) Dot plot of locomotion fault rate at 9 and 10 wpi. Results are presented as mean ± SEM. * $p < 0.05$, ** $p < 0.01$, *** $p < 0.001$; Tukey's post hoc pairwise test assuming $P_{1\text{-way-ANOVA}} < 0.05$.

3.2. BNN27 and NSC Graft Treatment Effects on Neuron Presence and Astrogliosis at the Lesion Site

The presence of neurons in the lesion epicenter at 12 wpi was evaluated by immunostaining spinal cord parasagittal sections for the Tuj1 neuronal marker and calculating the density of Tuj1$^+$ cells within the lesion epicenter, which was denoted by a GFAP$^+$ boundary region (Figure 2a). While Tuj1$^+$ cells and neurites were present in all four injured groups (Figure 2b), statistical analysis revealed that both NSC-seeded PCS ($P_{2\text{-way-ANOVA}} = 5 \cdot 10^{-4}$) and BNN27 ($P_{2\text{-way-ANOVA}} = 0.01$) had a statistically significant effect on the density of Tuj1$^+$ cells in the lesion epicenter at 12 wpi. A significant increase in Tuj1$^+$ cell density was observed in the presence of BNN27 administration or NSC-seeded PCS grafts with or without BNN27 compared to the untreated "Placebo" group ("Placebo": 397.1 ± 53.2 cells/mm^2, n = 4; "BNN27": 683.8 ± 44.8 cells/mm^2, n = 3, $p = 0.01$; "Scaffold NSC": 776.6 ± 47.2 cells/mm^2, n = 5, $p = 6{\cdot}10^{-4}$; "Scaffold NSC + BNN27": 789.2 ± 50.6 cells/mm^2, n = 4, $p = 7{\cdot}10^{-4}$) (Figure 2d). Simultaneous treatment with BNN27 and NSC-seeded PCS grafts did not have statistically significant synergistic effects on the density of Tuj1$^+$ cells compared to treatment only with NSC-seeded PCS grafts.

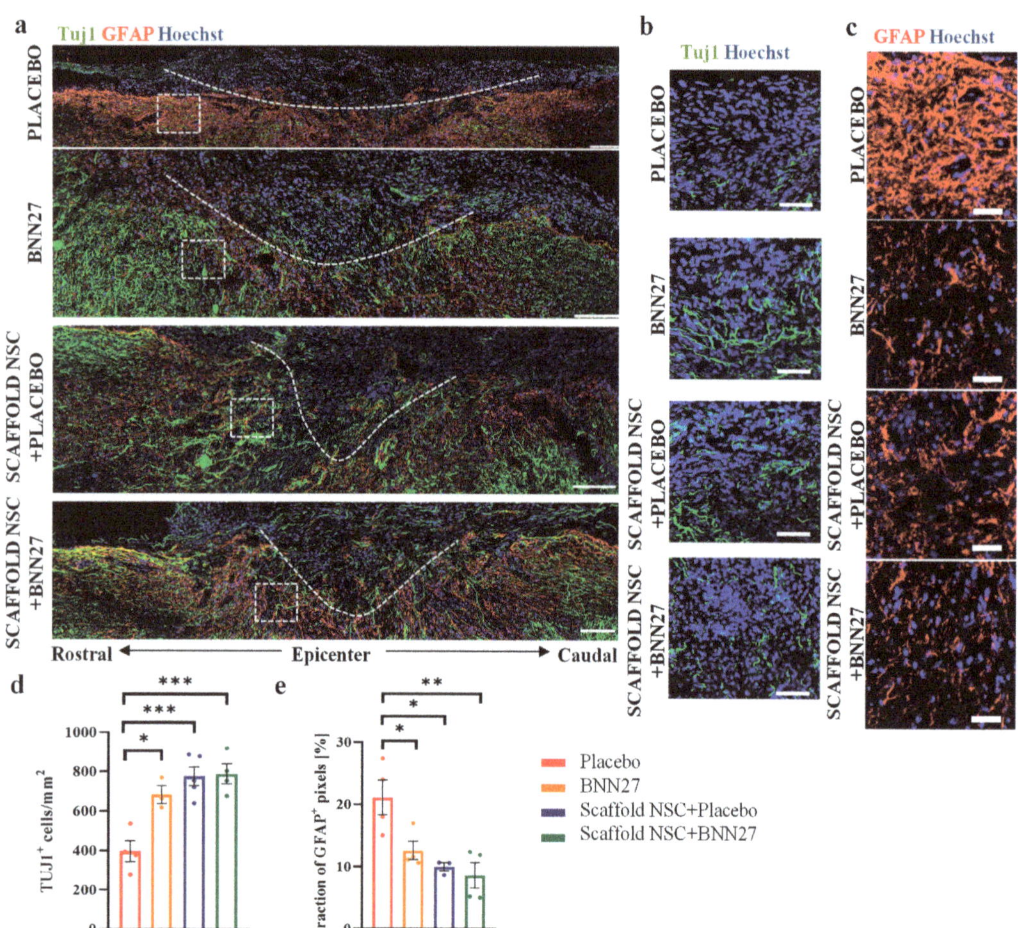

Figure 2. Effects of NSC-seeded PCS grafts and BNN27 administration on neuronal density and astrogliosis. (**a**) Representative fluorescence images of spinal cord parasagittal sections from all four injured groups immunostained for Tuj1 (green) and GFAP (red). The boundaries of lesion site (epicenter) are marked with a dotted line. Scale bars: 100 μm. (**b**) High-magnification fluorescence images in the lesion epicenter of image (**a**) visualizing Tuj1 immunostaining. Scale bars: 30 μm. (**c**) High-magnification fluorescence images of the rectangular ROI of image (**a**) showing GFAP immunostaining at the lesion boundary region. Scale bars: 30 μm. (**d**) Quantification of the density of Tuj1$^+$ cells within the lesion epicenter at 12 wpi in the four injured animal groups ("Placebo": n = 4, "BNN27": n = 3, "Scaffold NSC + Placebo": n = 5, "Scaffold NSC + BNN27" n = 4). (**e**) Quantification of the fraction of GFAP$^+$ pixels (marker of astrogliosis) in the lesion boundary region at 12 wpi in the four injured animal groups ("Placebo": n = 4, "BNN27": n = 4, "Scaffold NSC + Placebo": n = 3, "Scaffold NSC + BNN27" n = 4). Results are presented as mean $\pm$ SEM. * $p < 0.05$, ** $p < 0.01$, *** $p < 0.001$; Tukey's post hoc pairwise test assuming $P_{\text{2-way-ANOVA}} < 0.05$.

Astrogliosis in the SCI lesion site at 12 wpi was evaluated by immunostaining spinal cord parasagittal sections for GFAP (marker of astrocytes) and quantifying the fraction of GFAP$^+$ pixels in the lesion boundary region (Figure 2a,c). Two-way ANOVA revealed that both NSC-seeded PCS ($P_{\text{2-way-ANOVA}} = 4 \cdot 10^{-3}$) and BNN27 ($P_{\text{2-way-ANOVA}} = 0.04$) had a statistically significant effect on the fraction of GFAP$^+$ pixels in the lesion boundary region at 12 wpi. Compared to the untreated "Placebo" group, a significant decrease in GFAP$^+$ pixel fraction was observed in the presence of BNN27 administration or NSC-seeded PCS grafts with or without BNN27 ("Placebo": 21.1 ± 2.8%, n = 4; "BNN27": 12.6 ± 1.5%, n = 4, $p = 0.05$; "Scaffold NSC": 10.0 ± 0.6%, n = 3, $p = 0.02$; "Scaffold NSC + BNN27": 8.7 ± 2.0%, n = 4, $p = 5 \cdot 10^{-3}$) (Figure 2e). BNN27 administration did not lead to statistically significant synergistic effects when combined with NSC-seeded PCS grafts ("Scaffold NSC + Placebo" vs. "Scaffold NSC + BNN27": $p = 0.97$).

3.3. Treatment Effects on Synapse Formation, Axonal Elongation and Angiogenesis at the Lesion Site

The integration of neurons within the lesion area via synapse formation at 12 wpi was evaluated by immunostaining spinal cord parasagittal sections for synaptophysin and then quantifying the fraction of synaptophysin$^+$ pixels within the lesion epicenter (Figure 3a–c). Overall, treatment with NSC-seeded PCS had a statistically significant effect ($P_{\text{2-way-ANOVA}} < 10^{-4}$) on the fraction of synaptophysin$^+$ pixels in the lesion. Both graft-treated groups contained a larger fraction of synaptophysin$^+$ pixels than the untreated "Placebo" group ("Placebo": 4.1 ± 0.7%, n = 3; "Scaffold NSC + Placebo": 8.4 ± 0.7%, n = 3, $p = 2 \cdot 10^{-3}$; "Scaffold NSC + BNN27": 8.7 ± 0.4%, n = 3, $p = 10^{-3}$). On the other hand, BNN27 administration did not increase the fraction of synaptophysin$^+$ pixels ($P_{\text{2-way-ANOVA}} = 0.37$; "Placebo" vs. "BNN27": 4.8 ± 0.2%, n = 3, $p = 0.79$; "Scaffold NSC + Placebo" vs. "Scaffold NSC + BNN27": $p = 0.98$) (Figure 3d).

Ongoing axonal elongation in the lesion site was evaluated by immunostaining spinal cord parasagittal sections for the L1 cell adhesion molecule that is upregulated in elongating axons [36] and quantifying the fraction of L1$^+$ pixels in the lesion epicenter and in the dorsal regions rostrally and caudally to the lesion (Supplementary Materials: Figure S3a). No difference was observed among the four injured groups in the rostral region and in the lesion epicenter. Treatment with NSC-seeded PCS grafts had a statistically significant effect ($P_{\text{2-way-ANOVA}} < 10^{-4}$) on the fraction of L1$^+$ pixels caudally to the lesion (Supplementary Materials: Figure S3b). The two PCS graft-treated groups showed increased axonal elongation in the caudal region compared to the untreated "Placebo" group (caudally: "Placebo": 7.1 ± 0.8%, n = 5; "Scaffold NSC + Placebo": 11.8 ± 0.7%, n = 4, $p < 10^{-3}$; "Scaffold NSC + BNN27": 11 ± 0.4%, n = 5, $p = 2 \cdot 10^{-3}$). BNN27 administration did not affect axonal elongation (caudally: $P_{\text{2-way-ANOVA}} = 0.99$). The fraction of L1$^+$ pixels was not different in the "BNN27" group compared to the untreated "Placebo" group (caudally: "BNN27": 7.9 ± 0.5%, n = 4, $p = 0.83$). Similarly, the combinatorial treatment with NSC-seeded PCS grafts and BNN27 administration did not increase the fraction of L1$^+$ pixels compared to animals treated only with grafts (caudally: "Scaffold NSC + Placebo" vs. "Scaffold NSC + BNN27" $p = 0.83$) (Supplementary Materials: Figure S3b).

The presence of blood vessels within the lesion site was evaluated by immunostaining spinal cord parasagittal sections for αSMA, a marker of vessel-associated pericytes, and quantifying the number of αSMA$^+$ vessels within the lesion epicenter (Figure 3e,f). The results show that treatment with NSC-seeded PCS grafts ($P_{\text{2-way-ANOVA}} = 3 \cdot 10^{-3}$) but not BNN27 administration ($P_{\text{2-way-ANOVA}} = 0.44$) had a statistically significant effect on the number of αSMA$^+$ vessels. The two PCS graft-treated groups contained an increased number of vessels within the lesion epicenter compared to the untreated "Placebo" group ("Placebo": 2.7 ± 0.6%, n = 4; "Scaffold NSC + Placebo": 7.2 ± 1.0%, n = 4, $p = 0.03$; "Scaffold NSC + BNN27": 7.7 ± 0.9%, n = 3, $p = 0.03$). The number of αSMA$^+$ vessels was not different in the "BNN27" group compared to the untreated "Placebo" group ("BNN27": 4.0 ± 1.5%, n = 3, $p = 0.82$). Similarly, combinatorial treatment with NSC-seeded PCS grafts and BNN27 did not increase the number of αSMA$^+$ vessels compared to animals treated only with grafts ("Scaffold NSC + Placebo" vs. "Scaffold NSC + BNN27" $p = 0.99$) (Figure 3g).

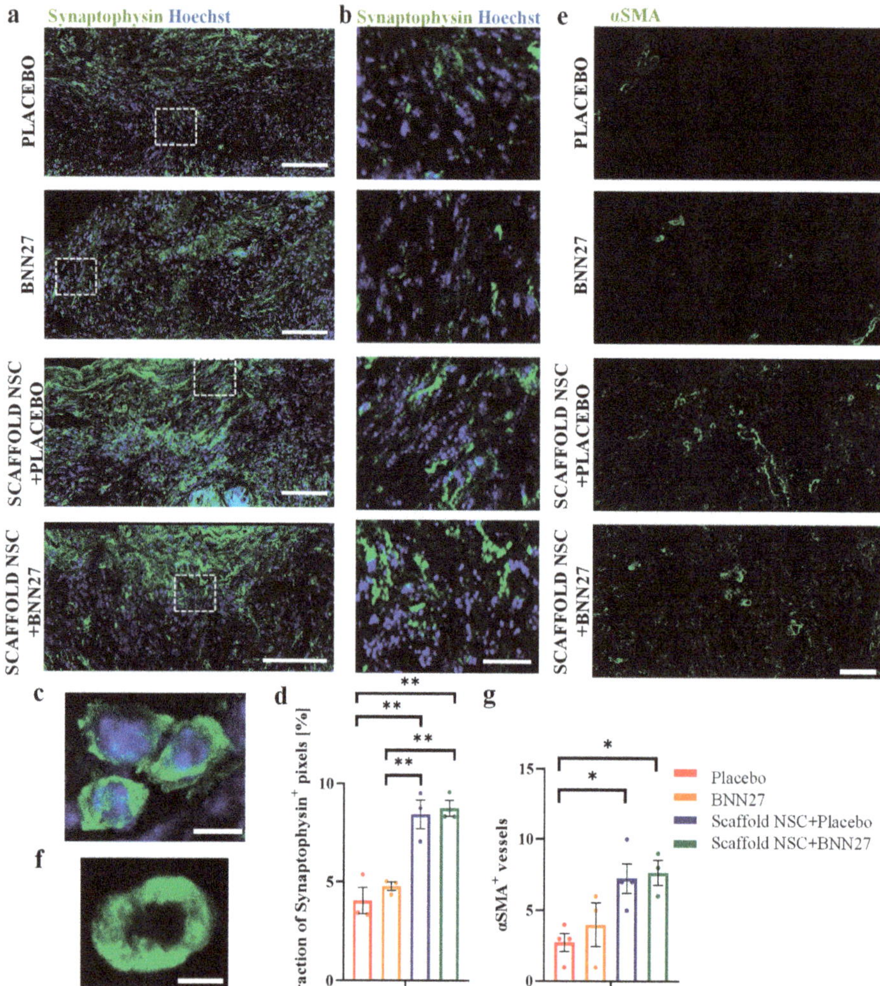

Figure 3. Effects of NSC-seeded PCS grafts on synapse density and angiogenesis. (**a**) Representative fluorescence images of spinal cord parasagittal sections in the lesion epicenter from all four injured groups immunostained for synaptophysin (green). Scale bars: 100 μm. (**b**) High-magnification fluorescence images of the rectangular ROIs of image (**a**) showing synaptophysin immunostaining in the lesion epicenter. Scale bars: 50 μm. (**c**) A representative high-magnification fluorescence image immunostained for synaptophysin. Scale bar: 5 μm. (**d**) Quantification of synaptophysin$^+$ pixel fraction within the lesion epicenter in the four injured groups at 12 wpi (All groups: n = 3). (**e**) Representative fluorescence images of spinal cord parasagittal sections in the lesion epicenter from all four injured groups immunostained for αSMA (green). Scale bar: 50 μm. (**f**) A representative high-magnification fluorescence image of a vessel in the lesion epicenter immunostained for αSMA. Scale bar: 5 μm (**g**) Quantification of the number of αSMA$^+$ vessels within the lesion epicenter in the four injured animal groups at 12 wpi ("Placebo": n = 4, "BNN27": n = 3, "Scaffold NSC + Placebo": n = 4, "Scaffold NSC + BNN27": n = 3). Results are presented as mean ± SEM. * $p < 0.05$, ** $p < 0.01$; Tukey's post hoc pairwise test assuming P$_{2\text{-way-ANOVA}} < 0.05$.

3.4. BNN27 Effects on Implanted NSCs

In our NSC-seeded PCS grafts, we utilized YFP-expressing NSCs so that they could be distinguished from endogenous cells by immunostaining for GFP (Figures 4 and 5a). Immunostaining spinal cord parasagittal sections for GFP (implanted NSC-derived cells), Tuj1 (neurons) and GFAP (astrocytes) revealed an interesting specific spatial pattern on the fate of implanted NSC-derived cells. While Tuj1$^+$GFP$^+$ cells (NSC-derived neurons) remained mostly within the lesion epicenter (Figure 4a), GFAP$^+$GFP$^+$ cells (NSC-derived astrocytes) were localized in the lesion boundary region where they contributed to the formation of glial scar (Figure 4a,b). Double immunostaining for GFP and various neuronal markers revealed that neurons derived from implanted NSCs included inhibitory GAD67$^+$ GABAergic neurons, 3TH$^+$ dopaminergic neurons and VGLUT1$^+$ excitatory glutamatergic neurons (Figure 4c). Interestingly, some GFP$^+$ cells were detected approximately 250 μm outside the lesion, having penetrated the surrounding glial scar (Figure 4d).

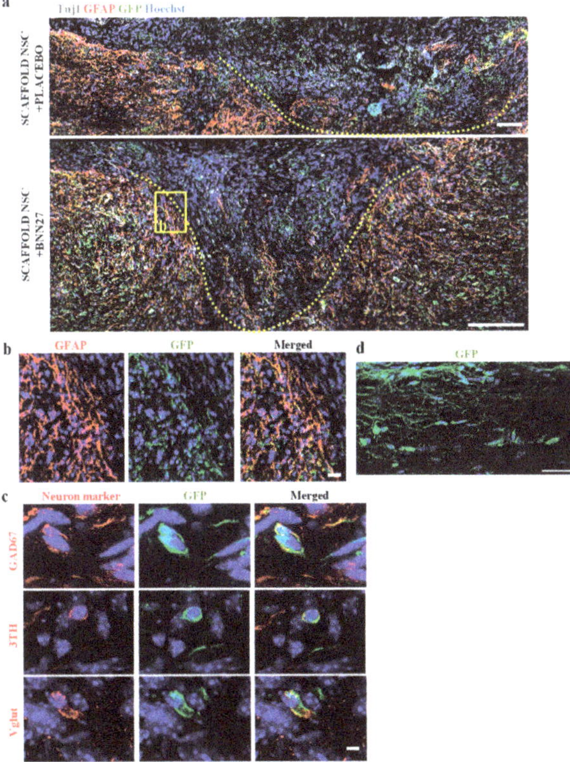

Figure 4. Spatial patterning of implanted NSC fate at the spinal cord lesion site. (**a**) Representative fluorescence images of spinal cord parasagittal sections of SCI lesion immunostained for Tuj1$^+$ neurons (gray), GFAP$^+$ astrocytes (red) and GFP$^+$ implanted NSC-derived cells (green). The boundaries of lesion sites are marked with a dotted line. Scale bars: 100 μm. (**b**) High-magnification fluorescence image of the rectangular insert of image (**a**) showing GFP$^+$GFAP$^+$ implanted NSC-derived astrocytes at the lesion boundary region. Scale bar: 30 μm. (**c**) Representative high-magnification fluorescence images of GFP$^+$ implanted NSC-derived cells double-immunostained with markers of different neuron types: GAD67 (inhibitory GABAergic neurons; upper row), 3TH (dopaminergic neurons; middle row) and VGLUT1 (excitatory glutamatergic neurons; lower row) (red). Scale bar: 3μm. (**d**) Representative fluorescence image of a spinal cord parasagittal section showing GFP$^+$ implanted NSC-derived cells located in the dorsal column, approximately 250 μm caudally to the lesion. Scale bar: 30 μm.

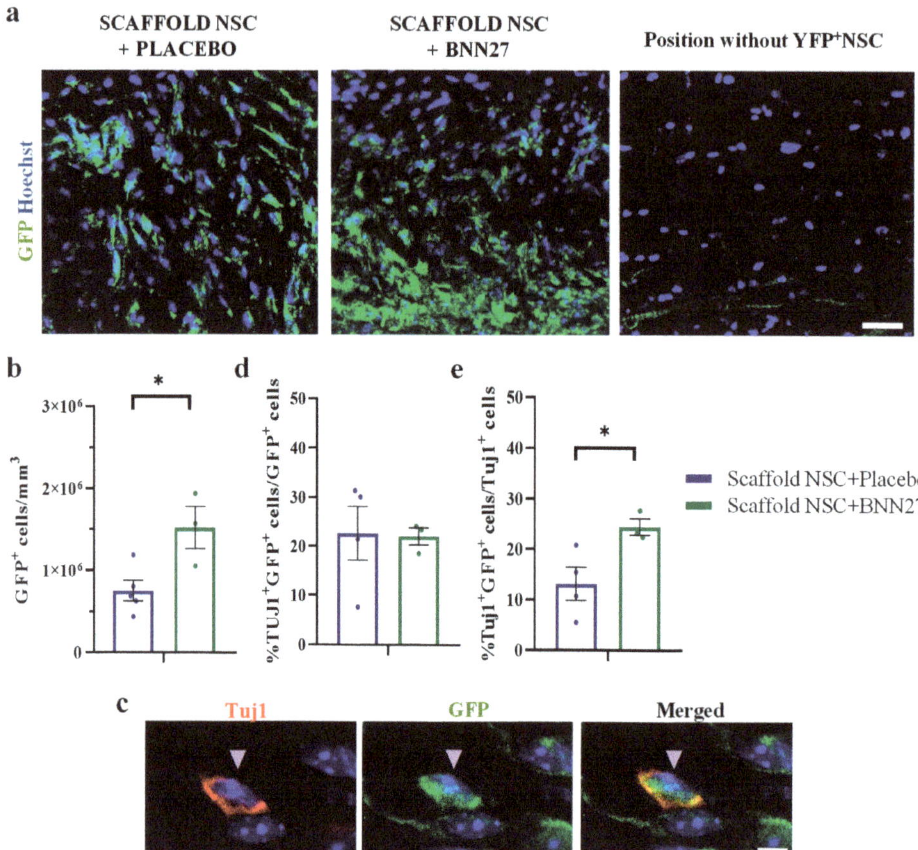

Figure 5. Effects of BNN27 administration on the density and neuronal fate of implanted NSCs. (**a**) Representative high-magnification fluorescence images of GFP$^+$ implanted NSC-derived cells from the two animal groups ("Scaffold NSC + Placebo", "Scaffold NSC + BNN27") treated with NSC-seeded PCS and a region without implanted NSCs. Scale bar: 30 µm. (**b**) Quantification of the density of GFP$^+$ cells in the lesion epicenter at 12 wpi. (**c**) High magnification fluorescence images of cells within the lesion epicenter double-stained for GFP (green; implanted NSC-derived cells) and Tuj1 (red; neurons). Scale bars: 5 µm. (**d**) Quantification of the fraction of GFP$^+$ cells that also stain for the Tuj1 in the lesion epicenter at 12 wpi. (**e**) Quantification of the fraction of Tuj1$^+$ neurons that also stain for GFP in the lesion epicenter at 12 wpi ("Scaffold NSC + Placebo": n = 4, "Scaffold NSC + BNN27": n = 3). Results are presented as mean ± SEM. * $p < 0.05$; unpaired two-tailed Student's *t*-test.

BNN27 effects on the density of implanted NSC-derived cells in the lesion epicenter at 12 wpi were evaluated by counting the number of GFP$^+$ cells (derived from implanted NSCs) in the two animal groups grafted with NSC-seeded PCS grafts. The results show that BNN27 administration increased the density of GFP$^+$ cells in the lesion epicenter ("Scaffold NSC + Placebo": $7.51 \cdot 10^5 \pm 1.25 \cdot 10^5$ cells/mm^3, n = 5 vs. "Scaffold NSC + BNN27": $15.24 \cdot 10^5 \pm 2.57 \cdot 10^5$ cells/mm^3, n = 3, $p = 0.02$) (Figure 5a,b). In order to evaluate the effects of BNN27 administration on the neuronal fate of implanted NSCs, we counted Tuj1$^+$ GFP$^+$ cells in the lesion epicenter (Figure 5c). BNN27 did not affect the fraction of GFP$^+$ cells that had differentiated to Tuj1$^+$ neurons ("Scaffold NSC + Placebo": 22.6 ± 5.4%, n = 4 vs. "Scaffold NSC + BNN27": 22.0 ± 1.7%, n = 3, $p = 0.93$) (Figure 5d). However, BNN27 administration increased the fraction of Tuj1$^+$ neurons that originated from implanted NSCs

("Scaffold NSC + Placebo": 13.1 ± 3.3%, n = 4 vs. "Scaffold NSC + BNN27": 24.4 ± 1.6%, n = 3, *p* = 0.04) (Figure 5e).

4. Discussion

MNTs are considered candidate therapeutic treatments for various neurologic diseases, since they mimic the effects of endogenous neurotrophins, do not cause hyperalgesia and are BBB-permeable, in contrast to endogenous neurotrophins [25]. Despite promising results in several mice neurodegeneration models, so far the therapeutic activity of MNTs had not been evaluated in any CNS injury model. Herein, we build upon our recent work on the effects of NSC-seeded PCS grafts on the dorsal column crush SCI in mice [8] and focus on the effects of BNN27 in SCI either as monotherapy or combined with NSC-seeded PCS grafts. BNN27 was delivered systemically using subcutaneous pellets over a period of approximately 60 days. Subcutaneous systemic administration is expected to provide a relatively steady BNN27 dose with no animal distraction, appropriate for minimizing artifacts in locomotion evaluation. The permeability of BNN27 in rodent CNS was demonstrated in pharmacokinetic studies where BNN27 was detected in rodent brain and retina 30 min after intraperitoneal injection [37,38].

Evaluation of locomotion recovery via the horizontal ladder walking assay showed that grafting SCI lesions with NSC-seeded PCS grafts led to statistically significant loco-motion improvement after dorsal column crush in mice starting on week 9, replicating the results of our previous study [8] and highlighting the central role of NSC-based cell therapies in SCI treatments. On the other hand, systemic administration of BNN27 did not improve locomotion recovery, in agreement with a previous study where local delivery of NGF in rat SCI lesions via genetically modified fibroblasts did not lead to functional im-provement 3 months post-injury [15]. Nevertheless, our histological data provide evidence that BNN27 affected several processes related to SCI wound healing, which are discussed in the following paragraphs.

Following SCI, the formation of glial scar isolates the lesion site to prevent damage spread, yet acts as a major physical barrier that obstructs axonal elongation through the lesion [2]. One way to block the inhibitory effects of a glial scar is to target its extracellular growth inhibitory components. For instance, local administration of chondroitinase ABC (ChABC), a bacterial enzyme, enhances axonal growth in CNS lesion sites by removing inhibitory chondroitin sulfate chains [39,40]. However, the clinical use of ChABC is limited by the safety of its local administration in the lesion site [41]. An alternative approach is to target astrogliosis, the accumulation and activation of astrocytes, thus impeding the major cellular component of the glial scar. Grafting SCI lesions with appropriate biomaterials, such as the PCS utilized in this study, have been reported able to reduce astrogliosis [42,43]. Here, we show that the systemic administration of BNN27 also significantly decreased astrogliosis 12 weeks following SCI (Figure 2a,c,e). This finding agrees with the report that local delivery of NGF (via genetically modified NSCs) in rat SCI lesions reduced astrogliosis [19]. While several studies have demonstrated the ability of BNN27 to re-duce astrogliosis in animal models of CNS degeneration (including cuprizone-induced demyelination in mice [30], the 5xFAD Alzheimer's disease mouse model (Karali et al. in preparation), diabetes-induced retinal damage in rats [28]), here we demonstrate for the first time the anti-astrogliotic effects of BNN27 in an animal model of CNS injury. Com-binatory treatment with NSC-seeded grafts and systemic BNN27 administration further decreased the fraction of GFAP$^+$ pixels compared to the animal groups that received each treatment separately, yet in a non-statistically significant manner (Figure 2e).

Emerging treatments for SCI aim to replace or reduce the loss of neural tissue induced by primary or secondary injury. In this direction, we report that grafting the SCI site with NSC-seeded PCS grafts increased neuron density in the lesion epicenter 12 weeks post SCI (Figure 2a,b,d). In the two groups where mice were grafted with NSC-seeded PCS, the increased neuron density compared to the untreated control could be attributed to multiple possible mechanisms, including the secretion of neurotrophic factors by NSCs [44] and the partial differentiation of NSCs into neurons. Quantification of Tuj1$^+$ neurons in the lesion epicenter shows that 13–25% of neurons were derived from implanted NSCs (Figure 5e), suggesting that neuronal NSC differentiation cannot solely explain the increased neuronal density in the two graft-treated groups. Furthermore, we report that systemic administration of BNN27 increased neuron density in the lesion epicenter at 12 wpi (Figure 2a,b,d). This observation agrees with BNN27 neuroprotection effects reported in various neuron types in several in vitro and in vivo studies [26,27]. It also agrees with published reports on the ability of NGF administration to decrease neuron loss after SCI [16,17]. As previously reported for NGF [9,22], BNN27 effects on neurons following SCI could be mediated either directly (via NGF receptors) or indirectly via effects on other cell types. The combinatorial treatment of NSC-seeded PCS grafts and systemic BNN27 administration did not statistically increase neuronal density compared to the one in the two animal groups that received each treatment separately (Figure 2d), although BNN27 administration increased the fraction of Tuj1$^+$ neurons in the lesion epicenter that originated from NSCs (Figure 5e).

A major motivation of emerging NSC-based therapies for SCI is to deliver a significant population of neural cell precursors in the lesion in order to complement the limited number of endogenous precursor cells and enhance the synthesis of new neural tissue and its connection with the surrounding spared tissue. However, previous studies have highlighted the low post-implantation survival of NSCs in SCI lesions as a major challenge that impedes the long-term efficacy of such NSC-based SCI treatments [45]. For instance, a study showed that just 4.6% of transplanted NSCs survived at 9 wpi [46]. Additionally, implanted NSCs had a low proliferation rate in situ [47]. Overall, NSC survival has been shown to depend strongly on the methodology and timepoint of delivery. NSC survival was lower when NSCs were transplanted in the acute phase of injury (1%) compared to the subacute one (6%) [48,49]. This deficiency could be caused by early immunoreaction events at the lesion site, which are toxic to implanted cells [50,51]. Strategies for protecting implanted NSCs from the harsh SCI lesion environment include delivering NSCs inside biomaterials (such as the PCS utilized in this study) and providing anti-apoptotic support via small molecule compounds or biologics. Indeed, herein we report that BNN27 administration doubled the density of implanted NSC-derived cells within the grafted lesion at 12 wpi from $7.51 \times 10^5 \pm 1.25 \times 10^5$ cells/mm^3 (when injured animals were treated only with NSC-seeded grafts) to $15.24 \times 10^5 \pm 2.57 \times 10^5$ cells/mm^3 (animals treated with both NSC-seeded scaffold grafts and systemic BNN27 administration) (Figure 5a,b). Comparison of these cell density measurements with data from previous studies is usually not straightforward because they utilized different experimental protocols (injury, cell seeding). Furthermore, very few published studies have quantified the number or density of cells derived from implanted NSC grafts within the lesion site. Nevertheless, our measurements on the density of implanted NSC-derived cells in the lesion are of the same order of magnitude as the ones reported in a study that grafted human neural stem/progenitor cells (NSPCs) within a fibrin–thrombin hydrogel [52].

Another major challenge of NSC-based SCI treatments is that NSCs differentiate mostly into glial cells and poorly towards neurons, as the lesion environment is unfavorable for neuronal differentiation [49,51,53–55]. Significant research has attempted to guide the fate of implanted NSCs using biomaterials and growth factors that enhance neuronal survival and differentiation [55]. Herein, our data show that, although BNN27 did not affect the differentiation of implanted NSCs, approximately 20% of implanted NSCs differentiated towards the neuronal fate by 12 wpi (Figure 5d), in agreement with our previous findings [8]. Furthermore, our data show an interesting spatial organization pattern for implanted NSC-derived cells in the presence or absence of BNN27 (Figure 4a). Implanted NSCs that differentiated into astrocytes became part of the glial scar and were localized in the lesion boundary region (Figure 4a,b). On the other hand, implanted NSCs that differentiated towards neurons were localized within the lesion epicenter, replenishing lost neurons. This spatial pattern has also been described by Lien et al. [56]. Finally, we demonstrate that implanted NSCs differentiated to various types of neurons (Figure 4c), including

GABAergic neurons, dopaminergic neurons and glutamatergic neurons in the presence or absence of BNN27. The presence of implanted NSC-derived GAD67$^+$ GABAergic neurons, VGLUT1$^+$ glutamatergic and 3TH$^+$ dopaminergic neurons has been reported in rat SCI models [43,57].

The observed BNN27 effects on the density and fate of implanted NSC-derived cells could originate in various mechanisms, including direct BNN27 effects on NSCs, NSC-derived neural cells and intermediate precursors or indirect effects via other cells that participate in SCI wound healing. Regarding BNN27 effects on NSCs, various types of embryonic rodent NSCs have been reported to express both p75^{NTR} and TrkA NGF receptors [58,59]. Moreover, there are reports that NGF can increase proliferation in NSCs isolated from embryonic E14 rat brain or adult rat spinal cord [60,61]. BNN27 has demonstrated neuroprotective effects on various types of neurons mediated via NGF receptors [26,27], although no prior study has demonstrated direct effects on spinal cord neurons or NSC-derived neurons. Finally, BNN27 neuroprotective effects could also be mediated by indirect effects in SCI lesions. For instance, the ability of BNN27 to decrease astrogliosis, described above, could facilitate NSC survival and enhance the integration of NSC-derived neurons.

Apart from quantifying the effects of BNN27 in SCI, the present study further highlights the ability of NSC-seeded PCS grafts to enhance important SCI processes, which were not affected by BNN27 administration. First, grafting SCI lesions with NSC-seeded PCS grafts resulted in significantly larger fraction of synaptophysin$^+$ pixels (Figure 3a–d), suggesting enhanced neuronal integration and communication in the lesion epicenter. Indeed, specific NSPC grafts have been shown able to organize, become synaptically active and interact with host axons [62]. Second, grafting SCI lesions with NSC-seeded PCS enhanced axonal elongation caudally to the epicenter (Supplementary Materials: Figure S3), in agreement with previous measurements of axonal elongation at 6 and 9 wpi [8]. Several studies have shown that NSPC implantation can lead to extensive axonal regeneration of both host axons into the graft and axons of NSPC-derived neurons caudally to the lesion over long distances [5,57]. Third, our work shows that GFP$^+$ implanted NSC-derived cells crossed the astrogliotic boundary and migrated a few hundred micrometers caudally to the lesion, but not much further to create ectopic colonies, a key limitation of NSC delivery in suspension [8]. Finally, our data show that NSC-seeded PCS enhanced angiogenesis, a critical process related to enhanced functional recovery, since endogenous angiogenesis following SCI is not sufficient to replace lost blood vessels [63,64]. Herein, we show that NSC-seeded PCS grafts increased the number of αSMA$^+$ vessels within the lesion site (Figure 3e–g), in agreement with the reported ability of transplanted embryonic stem cell-derived neural progenitors to enhance angiogenesis in mouse SCI lesions [65,66].

5. Conclusions

This study presents the first demonstration of therapeutic effects of MNTs in an animal model of CNS injury. It shows that systemic administration of BNN27 had significant effects on astrogliosis and neuronal density following dorsal column SCI in mice. BNN27 administration also significantly increased the density of cells derived from implanted NSCs, possibly addressing a major challenge of emerging NSC-based cell therapies. Considering a wider perspective, our study suggests that MNTs can contribute to combinatorial treatments for CNS injuries by simultaneously enhancing cell therapies and modulating the loss of neuronal tissue. MNT contribution in such treatments could be further enhanced by optimizing MNT dosing, exploiting novel MNT designs that target specific neurotrophin receptors [67,68] or utilizing targeted methods to control their spatiotemporal delivery within CNS lesions.

Supplementary Materials: The following supporting information can be downloaded at: https://www.mdpi.com/article/10.3390/biomedicines11041170/s1, Figure S1: YFP-expressing NSCs; Figure S2: Raw data of the fault rate response in the horizontal ladder walking assay per experimental group; Figure S3: Effects of NSC-seeded PCS on axonal elongation.

Author Contributions: Conceptualization, D.S.T., A.G., I.C. and K.G.; methodology, K.G., D.S.T., A.G., I.C. and K.K.; validation, K.G., D.S.T. and K.K.; formal analysis, K.G., D.S.T., E.-A.S. and A.P.; investigation, K.G. and E.-A.S.; resources, D.S.T., A.G., I.C. and K.K.; data curation, K.G. and D.S.T.; writing—original draft preparation, K.G. and D.S.T.; writing—review and editing, K.G., D.S.T., A.G., I.C. and K.K.; visualization, K.G. and D.S.T.; supervision, D.S.T., A.G. and I.C.; project administration, D.S.T., A.G., I.C. and K.K; funding acquisition, D.S.T., K.K. and K.G. All authors have read and agreed to the published version of the manuscript.

Funding: This research was funded by the Hellenic Foundation for Research and Innovation (HFRI) and the General Secretariat for Research and Technology (GSRT) under grant agreement No 1635 to D.T. and K.K., and a State Scholarships Foundation (IKY) scholarship to K.G.

Institutional Review Board Statement: Animal care and experimentation protocols were performed according to the approval of the Veterinary Directorate of the region of Crete in compliance with Greek Government guidelines, EU guidelines 2010/63/EU, FORTH ethics committee guidelines and in accordance with approved protocols from the Federation of European Laboratory Animal Science Associations (FELASA) on use of laboratory animals. Facility License number: EL91-BIOexp-02, Project title: "Investigating the effects of neurotrophins and their synthetic analogues on the mechanisms of neurodegeneration and neurogenesis in Alzheimer's disease and the effect of neuroimplants based on porous collagen scaffolds in the central nervous system trauma", Approval numbers: 262272 and 360667, date of initial approval: 29 October 2018. Date of reapproval: 29 November 2021.

Informed Consent Statement: Not applicable.

Data Availability Statement: Data are available within the article or in Supplementary Materials. Raw data are available upon reasonable request.

Acknowledgments: We thank D. Karagogeos (University of Crete) for the kind gift of L1 antibody and F. Moschogiannaki for assistance in scaffold fabrication.

Conflicts of Interest: The authors declare no conflict of interest.

Abbreviations

ANOVA: analysis of variance, BBB: blood–brain barrier, BSA: bovine serum albumin, ChABC: chondroitinase ABC, CNS: central nervous system, GAD67: glutamic acid decarboxylase 67, GFAP: glial fibrillary acidic protein, GFP: green fluorescent protein, MNT: microneurotrophin, NSC: neural stem cell, NSPC: neural stem/progenitor cell, PBS: phosphate-buffered saline, PBST: Triton X-100 in PBS, PCS: porous collagen scaffold, PFA: paraformaldehyde, PNS: peripheral nervous system, ROI: region of interest, SCI: spinal cord injury, SEM: standard error of the mean, VGLUT1: vesicular glutamate transporter 1, wpi: week post injury, YFP: yellow fluorescent protein, αSMA: α-smooth muscle actin, 3TH: tyrosine hydroxylase.

References

1. Zimmermann, R.; Alves, Y.V.; Sperling, L.E.; Pranke, P. Nanotechnology for the Treatment of Spinal Cord Injury. *Tissue Eng. Part B Rev.* **2021**, *27*, 353–365. [CrossRef]
2. Silva, N.A.; Sousa, N.; Reis, R.L.; Salgado, A.J. From basics to clinical: A comprehensive review on spinal cord injury. *Prog. Neurobiol.* **2014**, *114*, 25–57. [CrossRef] [PubMed]
3. Salgado, A.J.; Gomes, E.D.; Silva, N.A. Combinatorial therapies for spinal cord injury: Strategies to induce regeneration. *Neural Regen. Res.* **2019**, *14*, 69–71. [CrossRef] [PubMed]
4. Assinck, P.; Duncan, G.J.; Hilton, B.J.; Plemel, J.R.; Tetzlaff, W. Cell transplantation therapy for spinal cord injury. *Nat. Neurosci.* **2017**, *20*, 637–647. [CrossRef] [PubMed]

5. Kadoya, K.; Lu, P.; Nguyen, K.; Lee-Kubli, C.; Kumamaru, H.; Yao, L.; Knackert, J.; Poplawski, G.; Dulin, J.N.; Strobl, H.; et al. Spinal cord reconstitution with homologous neural grafts enables robust corticospinal regeneration. *Nat. Med.* **2016**, *22*, 479–487. [CrossRef]

6. Lu, P.; Gomes-Leal, W.; Anil, S.; Dobkins, G.; Huie, J.R.; Ferguson, A.; Graham, L.; Tuszynski, M. Origins of Neural Progenitor Cell-Derived Axons Projecting Caudally after Spinal Cord Injury. *Stem Cell Rep.* **2019**, *13*, 105–114. [CrossRef] [PubMed]

7. Curtis, E.; Martin, J.R.; Gabel, B.; Sidhu, N.; Rzesiewicz, T.K.; Mandeville, R.; Van Gorp, S.; Leerink, M.; Tadokoro, T.; Marsala, S.; et al. A First-in-Human, Phase I Study of Neural Stem Cell Transplantation for Chronic Spinal Cord Injury. *Cell Stem Cell* **2018**, *22*, 941–950.e6. [CrossRef] [PubMed]

8. Kourgiantaki, A.; Tzeranis, D.S.; Karali, K.; Georgelou, K.; Bampoula, E.; Psilodimitrakopoulos, S.; Yannas, I.V.; Stratakis, E.; Sidiropoulou, K.; Charalampopoulos, I.; et al. Neural stem cell delivery via porous collagen scaffolds promotes neuronal differentiation and locomotion recovery in spinal cord injury. *NPJ Regen. Med.* **2020**, *5*, 12. [CrossRef]

9. Widenfalk, J.; Lundströmer, K.; Jubran, M.; Brené, S.; Olson, L. Neurotrophic factors and receptors in the immature and adult spinal cord after mechanical injury or kainic acid. *J. Neurosci.* **2001**, *21*, 3457–3475. [CrossRef]

10. Griffin, J.M.; Bradke, F. Therapeutic repair for spinal cord injury: Combinatory approaches to address a multifaceted problem. *EMBO Mol. Med.* **2020**, *12*, e11505. [CrossRef]

11. Krenz, N.R.; Weaver, L.C. Nerve Growth Factor in Glia and Inflammatory Cells of the Injured Rat Spinal Cord. *J. Neurochem.* **2001**, *74*, 730–739. [CrossRef] [PubMed]

12. Davis-López De Carrizosa, M.A.; Morado-Díaz, C.J.; Morcuende, S.; De La Cruz, R.R.; Pastor, Á.M. Nerve growth factor regulates the firing patterns and synaptic composition of motoneurons. *J. Neurosci.* **2010**, *30*, 8308–8319. [CrossRef]

13. Aloe, L.; Bianchi, P.; De Bellis, A.; Soligo, M.; Rocco, M.L. Intranasal nerve growth factor bypasses the blood-brain barrier and affects spinal cord neurons in spinal cord injury. *Neural Regen. Res.* **2014**, *9*, 1025–1030. [CrossRef] [PubMed]

14. Grill, R.; Blesch, A.; Tuszynski, M. Robust Growth of Chronically Injured Spinal Cord Axons Induced by Grafts of Genetically Modified NGF-Secreting Cells. *Exp. Neurol.* **1997**, *148*, 444–452. [CrossRef] [PubMed]

15. Tuszynski, M.H.; Murai, K.; Blesch, A.; Grill, R.; Miller, I. Functional Characterization of Ngf-Secreting Cell Grafts to the Acutely Injured Spinal Cord. *Cell Transplant.* **1997**, *6*, 361–368. [CrossRef] [PubMed]

16. Zhang, H.; Wu, F.; Kong, X.; Yang, J.; Chen, H.; Deng, L.; Cheng, Y.; Ye, L.; Zhu, S.; Zhang, X.; et al. Nerve growth factor improves functional recovery by inhibiting endoplasmic reticulum stress-induced neuronal apoptosis in rats with spinal cord injury. *J. Transl. Med.* **2014**, *12*, 130. [CrossRef]

17. Song, Z.; Wang, Z.; Shen, J.; Xu, S.; Hu, Z. Nerve growth factor delivery by ultrasound-mediated nanobubble destruction as a treatment for acute spinal cord injury in rats. *Int. J. Nanomed.* **2017**, *12*, 1717–1729. [CrossRef] [PubMed]

18. Wu, Q.; Xiang, Z.; Ying, Y.; Huang, Z.; Tu, Y.; Chen, M.; Ye, J.; Dou, H.; Sheng, S.; Li, X.; et al. Nerve growth factor (NGF) with hypoxia response elements loaded by adeno-associated virus (AAV) combined with neural stem cells improve the spinal cord injury recovery. *Cell Death Discov.* **2021**, *7*, 301. [CrossRef]

19. Wang, L.; Gu, S.; Gan, J.; Tian, Y.; Zhang, F.; Zhao, H.; Lei, D. Neural Stem Cells Overexpressing Nerve Growth Factor Improve Functional Recovery in Rats Following Spinal Cord Injury via Modulating Microenvironment and Enhancing Endogenous Neurogenesis. *Front. Cell. Neurosci.* **2021**, *15*, 773375. [CrossRef]

20. Faustino, C.; Rijo, P.; Reis, C.P. Nanotechnological strategies for nerve growth factor delivery: Therapeutic implications in Alzheimer's disease. *Pharmacol. Res.* **2017**, *120*, 68–87. [CrossRef]

21. Tuszynski, M.H.; Peterson, D.A.; Ray, J.; Baird, A.; Nakahara, Y.; Gages, F.H. Fibroblasts genetically modified to produce Nerve Growth Factor Induce Robust Neuritic Ingrowth after Grafting to the Spinal Cord. *Exp. Neurol.* **1994**, *126*, 1–14. [CrossRef]

22. Tuszynski, M.H.; Gabriel, K.; Gage, F.H.; Suhr, S.; Meyer, S.; Rosetti, A. Nerve growth factor delivery by gene transfer induces dif-ferential outgrowth of sensory, motor, and noradrenergic neurites after adult spinal cord injury. *Exp. Neurol.* **1996**, *137*, 157–173. [CrossRef] [PubMed]

23. Feng, S.-Q.; Kong, X.-H.; Liu, Y.; Ban, D.-X.; Ning, G.-Z.; Chen, J.-T.; Guo, S.-F.; Wang, P. Regeneration of spinal cord with cell and gene therapy. *Orthop Surg.* **2009**, *1*, 153–163. [CrossRef] [PubMed]

24. Keefe, K.M.; Sheikh, I.S.; Smith, G.M. Targeting Neurotrophins to Specific Populations of Neurons: NGF, BDNF, and NT-3 and Their Relevance for Treatment of Spinal Cord Injury. *Int. J. Mol. Sci.* **2017**, *18*, 548. [CrossRef]

25. Gravanis, A.; Pediaditakis, I.; Charalampopoulos, I. Synthetic microneurotrophins in therapeutics of neurodegeneration. *Oncotarget* **2017**, *8*, 9005–9006. [CrossRef]

26. Pediaditakis, I.; Efstathopoulos, P.; Prousis, K.C.; Zervou, M.; Arévalo, J.C.; Alexaki, V.I.; Nikoletopoulou, V.; Karagianni, E.; Potamitis, C.; Tavernarakis, N.; et al. Selective and differential interactions of BNN27, a novel C17-spiroepoxy steroid derivative, with TrkA receptors, regulating neuronal survival and differentiation. *Neuropharmacology* **2016**, *111*, 266–282. [CrossRef] [PubMed]

27. Pediaditakis, I.; Kourgiantaki, A.; Prousis, K.C.; Potamitis, C.; Xanthopoulos, K.P.; Zervou, M.; Calogeropoulou, T.; Charalampopoulos, I.; Gravanis, A. BNN27, a 17-Spiroepoxy Steroid Derivative, Interacts With and Activates p75 Neurotrophin Receptor, Rescuing Cerebellar Granule Neurons from Apoptosis. *Front. Pharmacol.* **2016**, *7*, 512. [CrossRef]

28. Ibán-Arias, R.; Lisa, S.; Mastrodimou, N.; Kokona, D.; Koulakis, E.; Iordanidou, P.; Kouvarakis, A.; Fothiadaki, M.; Papadogkonaki, S.; Sotiriou, A.; et al. The Synthetic Microneurotrophin BNN27 Affects Retinal Function in Rats With Streptozotocin-Induced Diabetes. *Diabetes* **2017**, *67*, 321–333. [CrossRef]

29. Ibán-Arias, R.; Lisa, S.; Poulaki, S.; Mastrodimou, N.; Charalampopoulos, I.; Gravanis, A.; Thermos, K. Effect of topical administration of the microneurotrophin BNN27 in the diabetic rat retina. *Graefe's Arch. Clin. Exp. Ophthalmol.* **2019**, *257*, 2429–2436. [CrossRef] [PubMed]

30. Bonetto, G.; Charalampopoulos, I.; Gravanis, A.; Karagogeos, D. The novel synthetic microneurotrophin BNN27 protects mature oligodendrocytes against cuprizone-induced death, through the NGF receptor TrkA. *Glia* **2017**, *65*, 1376–1394. [CrossRef]

31. The Jackson Laboratory. C57BL/6J. Available online: https://www.jax.org/strain/000664 (accessed on 18 November 2022).

32. Mouse Genome Informatics. 129X1/SvJ. Available online: http://www.informatics.jax.org/strain/MGI:3044210 (accessed on 18 November 2022).

33. The Jackson Laboratory. B6.C-Tg(CMV-cre)1Cgn/Je. Available online: https://www.jax.org/strain/0060543044210 (accessed on 18 November 2022).

34. Harrison, M.; O'Brien, A.; Adams, L.; Cowin, G.; Ruitenberg, M.J.; Sengul, G.; Watson, C. Vertebral landmarks for the identification of spinal cord segments in the mouse. *Neuroimage* **2013**, *68*, 22–29. [CrossRef]

35. Farr, T.D.; Liu, L.; Colwell, K.L.; Whishaw, I.Q.; Metz, G.A. Bilateral alteration in stepping pattern after unilateral motor cortex injury: A new test strategy for analysis of skilled limb movements in neurological mouse models. *J. Neurosci. Methods* **2006**, *153*, 104–113. [CrossRef] [PubMed]

36. Savvaki, M.; Kafetzis, G.; Kaplanis, S.; Ktena, N.; Theodorakis, K.; Karagogeos, D. Neuronal, but not glial, Contactin 2 negatively regulates axon regeneration in the injured adult optic nerve. *Eur. J. Neurosci.* **2021**, *53*, 1705–1721. [CrossRef]

37. Bennett, J.P.; O'Brien, L.C.; Brohawn, D.G. Pharmacological properties of microneurotrophin drugs developed for treatment of amyotrophic lateral sclerosis. *Biochem. Pharmacol.* **2016**, *117*, 68–77. [CrossRef] [PubMed]

38. Tsika, C.; Tzatzarakis, M.N.; Antimisiaris, S.G.; Tsoka, P.; Efstathopoulos, P.; Charalampopoulos, I.; Gravanis, A.; Tsilimbaris, M.K. Quantification of BNN27, a novel neuroprotective 17-spiroepoxy dehydroepiandrosterone derivative in the blood and retina of rodents, after single intraperitoneal administration. *Pharmacol. Res. Perspect.* **2021**, *9*, e00724. [CrossRef]

39. Bradbury, E.J.; Moon, L.D.F.; Popat, R.J.; King, V.R.; Bennett, G.S.; Patel, P.N.; Fawcett, J.W.; McMahon, S.B. Chondroitinase ABC promotes functional recovery after spinal cord injury. *Nature* **2002**, *416*, 636–640. [CrossRef] [PubMed]

40. Ferraro, G.B.; Alabed, Y.; Fournier, A. Molecular Targets to Promote Central Nervous System Regeneration. *Curr. Neurovascular Res.* **2004**, *1*, 61–75. [CrossRef]

41. Gaudet, A.D.; Fonken, L.K. Glial Cells Shape Pathology and Repair After Spinal Cord Injury. *Neurotherapeutics* **2018**, *15*, 554–577. [CrossRef]

42. Spilker, M.H.; Yannas, I.V.; Kostyk, S.K.; Norregaard, T.V.; Hsu, H.P.; Spector, M. The Effects of Tubulation on Healing and Scar Formation after Transection of the Adult Rat Spinal Cord. *Restor. Neurol. Neurosci.* **2021**, *18*, 23–38.

43. Li, X.; Liu, S.; Zhao, Y.; Li, J.; Ding, W.; Han, S.; Chen, B.; Xiao, Z.; Dai, J. Training Neural Stem Cells on Functional Collagen Scaffolds for Severe Spinal Cord Injury Repair. *Adv. Funct. Mater.* **2016**, *26*, 5835–5847. [CrossRef]

44. Lu, P.; Jones, L.; Snyder, E.; Tuszynski, M. Neural stem cells constitutively secrete neurotrophic factors and promote extensive host axonal growth after spinal cord injury. *Exp. Neurol.* **2003**, *181*, 115–129. [CrossRef]

45. Lepore, A.; Fischer, I. Lineage-restricted neural precursors survive, migrate, and differentiate following transplantation into the injured adult spinal cord. *Exp. Neurol.* **2005**, *194*, 230–242. [CrossRef]

46. Hwang, D.H.; Shin, H.Y.; Kwon, M.J.; Choi, J.Y.; Ryu, B.-Y.; Kim, B.G. Survival of Neural Stem Cell Grafts in the Lesioned Spinal Cord Is Enhanced by a Combination of Treadmill Locomotor Training via Insulin-Like Growth Factor-1 Signaling. *J. Neurosci.* **2014**, *34*, 12788–12800. [CrossRef] [PubMed]

47. Karimi-Abdolrezaee, S.; Eftekharpour, E.; Wang, J.; Morshead, C.M.; Fehlings, M.G. Delayed Transplantation of Adult Neural Precursor Cells Promotes Remyelination and Functional Neurological Recovery after Spinal Cord Injury. *J. Neurosci.* **2006**, *26*, 3377–3389. [CrossRef] [PubMed]

48. Parr, A.M.; Kulbatski, I.; Tator, C.H. Transplantation of Adult Rat Spinal Cord Stem/Progenitor Cells for Spinal Cord Injury. *J. Neurotrauma* **2007**, *24*, 835–845. [CrossRef] [PubMed]

49. Liu, S.; Chen, Z. Employing Endogenous NSCs to Promote Recovery of Spinal Cord Injury. *Stem Cells Int.* **2019**, *2019*, 1958631. [CrossRef] [PubMed]

50. Wilcox, J.T.; Satkunendrarajah, K.; Zuccato, J.A.; Nassiri, F.; Fehlings, M.G. Neural Precursor Cell Transplantation Enhances Functional Recovery and Reduces Astrogliosis in Bilateral Compressive/Contusive Cervical Spinal Cord Injury. *STEM CELLS Transl. Med.* **2014**, *3*, 1148–1159. [CrossRef]

51. Younsi, A.; Zheng, G.; Riemann, L.; Scherer, M.; Zhang, H.; Tail, M.; Hatami, M.; Skutella, T.; Unterberg, A.; Zweckberger, K. Long-Term Effects of Neural Precursor Cell Transplantation on Secondary Injury Processes and Functional Recovery after Severe Cervical Contusion-Compression Spinal Cord Injury. *Int. J. Mol. Sci.* **2021**, *22*, 13106. [CrossRef]

52. Rosenzweig, E.S.; Brock, J.H.; Lu, P.; Kumamaru, H.; Salegio, E.A.; Kadoya, K.; Weber, J.L.; Liang, J.J.; Moseanko, R.; Hawbecker, S.; et al. Restorative effects of human neural stem cell grafts on the primate spinal cord. *Nat. Med.* **2018**, *24*, 484–490. [CrossRef]

53. Cao, Q.L.; Zhang, Y.P.; Howard, R.M.; Walters, W.M.; Tsoulfas, P.; Whittemore, S.R. Pluripotent stem cells engrafted into the normal or lesioned adult rat spinal cord are restricted to a glial lineage. *Exp. Neurol.* **2001**, *167*, 48–58. [CrossRef]

54. Zhu, Y.; Uezono, N.; Yasui, T.; Nakashima, K. Neural stem cell therapy aiming at better functional recovery after spinal cord injury. *Dev. Dyn.* **2017**, *247*, 75–84. [CrossRef] [PubMed]

55. Xue, W.; Fan, C.; Chen, B.; Zhao, Y.; Xiao, Z.; Dai, J. Direct neuronal differentiation of neural stem cells for spinal cord injury repair. *STEM CELLS* **2021**, *39*, 1025–1032. [CrossRef] [PubMed]

56. Lien, B.V.; Tuszynski, M.H.; Lu, P. Astrocytes migrate from human neural stem cell grafts and functionally integrate into the injured rat spinal cord. *Exp. Neurol.* **2019**, *314*, 46–57. [CrossRef] [PubMed]

57. Lu, P.; Wang, Y.; Graham, L.; McHale, K.; Gao, M.; Wu, D.; Brock, J.; Blesch, A.; Rosenzweig, E.S.; Havton, L.A.; et al. Long-Distance Growth and Connectivity of Neural Stem Cells after Severe Spinal Cord Injury. *Cell* **2012**, *150*, 1264–1273. [CrossRef] [PubMed]

58. Lachyankar, M.B.; Condon, P.J.; Quesenberry, P.J.; Litofsky, N.; Recht, L.D.; Ross, A.H. Embryonic Precursor Cells That Express Trk Receptors: Induction of Different Cell Fates by NGF, BDNF, NT-3, and CNTF. *Exp. Neurol.* **1997**, *144*, 350–360. [CrossRef]

59. Kumar, V.; Gupta, A.K.; Shukla, R.K.; Tripathi, V.K.; Jahan, S.; Pandey, A.; Srivastava, A.; Agrawal, M.; Yadav, S.; Khanna, V.K.; et al. Molecular Mechanism of Switching of TrkA/p75NTR Signaling in Monocrotophos Induced Neurotoxicity. *Sci. Rep.* **2015**, *5*, srep14038. [CrossRef]

60. Oliveira, S.L.; Trujillo, C.A.; Negraes, P.D.; Ulrich, H. Effects of ATP and NGF on Proliferation and Migration of Neural Precursor Cells. *Neurochem. Res.* **2015**, *40*, 1849–1857. [CrossRef] [PubMed]

61. Han, Y.; Kim, K.T. Neural growth factor stimulates proliferation of spinal cord derived-neural precursor/stem cells. *J. Korean Neurosurg. Soc.* **2016**, *59*, 437–441. [CrossRef]

62. Ceto, S.; Sekiguchi, K.J.; Takashima, Y.; Nimmerjahn, A.; Tuszynski, M.H. Neural Stem Cell Grafts Form Extensive Synaptic Networks that Integrate with Host Circuits after Spinal Cord Injury. *Cell Stem Cell* **2020**, *27*, 430–440.e5. [CrossRef] [PubMed]

63. Yao, C.; Cao, X.; Yu, B. Revascularization After Traumatic Spinal Cord Injury. *Front. Physiol.* **2021**, *12*, 631500. [CrossRef] [PubMed]

64. Tsivelekas, K.K.; Evangelopoulos, D.S.; Pallis, D.; Benetos, I.S.; Papadakis, S.A.; Vlamis, J.; Pneumaticos, S.G. Angiogenesis in Spinal Cord Injury: Progress and Treatment. *Cureus* **2022**, *14*, e25475. [CrossRef]

65. Kumagai, G.; Okada, Y.; Yamane, J.; Nagoshi, N.; Kitamura, K.; Mukaino, M.; Tsuji, O.; Fujiyoshi, K.; Katoh, H.; Okada, S.; et al. Roles of ES Cell-Derived Gliogenic Neural Stem/Progenitor Cells in Functional Recovery after Spinal Cord Injury. *PLoS ONE* **2009**, *4*, e7706. [CrossRef] [PubMed]

66. Nori, S.; Okada, Y.; Yasuda, A.; Tsuji, O.; Takahashi, Y.; Kobayashi, Y.; Fujiyoshi, K.; Koike, M.; Uchiyama, Y.; Ikeda, E.; et al. Grafted human-induced pluripotent stem-cell-derived neurospheres promote motor functional recovery after spinal cord injury in mice. *Proc. Natl. Acad. Sci. USA* **2011**, *108*, 16825–16830. [CrossRef] [PubMed]

67. Rogdakis, T.; Charou, D.; Latorrata, A.; Papadimitriou, E.; Tsengenes, A.; Athanasiou, C.; Papadopoulou, M.; Chalikiopoulou, C.; Katsila, T.; Ramos, I.; et al. Development and Biological Characterization of a Novel Selective TrkA Agonist with Neuroprotective Properties against Amyloid Toxicity. *Biomedicines* **2022**, *10*, 614. [CrossRef] [PubMed]

68. Yilmaz, C.; Rogdakis, T.; Latorrata, A.; Thanou, E.; Karadima, E.; Papadimitriou, E.; Siapi, E.; Li, K.W.; Katsila, T.; Calogeropoulou, T.; et al. ENT-A010, a Novel Steroid Derivative, Displays Neuroprotective Functions and Modulates Microglial Responses. *Biomolecules* **2022**, *12*, 424. [CrossRef] [PubMed]

Article

The Role of Transcranial Magnetic Stimulation, Peripheral Electrotherapy, and Neurophysiology Tests for Managing Incomplete Spinal Cord Injury

Katarzyna Leszczyńska [1,2,†] and Juliusz Huber [1,*,†]

[1] Department of Pathophysiology of Locomotor Organs, Poznan University of Medical Sciences, 28 Czerwca 1956 no 135/147, 60-545 Poznań, Poland; kleszczynska@ump.edu.pl
[2] Department of Neurosurgery, Wroclaw Medical University, Borowska 213, 50-556 Wroclaw, Poland
[*] Correspondence: juliusz.huber@ump.edu.pl; Tel.: +48-504-041-843
[†] These authors contributed equally to this work.

Abstract: Efforts to find therapeutic methods that support spinal cord functional regeneration continue to be desirable. Natural recovery is limited, so high hopes are being placed on neuromodulation methods which promote neuroplasticity, such as repetitive transcranial magnetic stimulation (rTMS) and electrical stimulation used as treatment options for managing incomplete spinal cord injury (iSCI) apart from kinesiotherapy. However, there is still no agreement on the methodology and algorithms for treatment with these methods. The search for effective therapy is also hampered by the use of different, often subjective in nature, evaluation methods and difficulties in assessing the actual results of the therapy versus the phenomenon of spontaneous spinal cord regeneration. In this study, an analysis was performed on the database of five trials, and the cumulative data are presented. Participants (iSCI patients) were divided into five groups on the basis of the treatment they had received: rTMS and kinesiotherapy (N = 36), peripheral electrotherapy and kinesiotherapy (N = 65), kinesiotherapy alone (N = 55), rTMS only (N = 34), and peripheral electrotherapy mainly (N = 53). We present changes in amplitudes and frequencies of the motor units' action potentials recorded by surface electromyography (sEMG) from the tibialis anterior—the index muscle for the lower extremity and the percentage of improvement in sEMG results before and after the applied therapies. The increase in values in sEMG parameters represents the better ability of motor units to recruit and, thus, improvement of neural efferent transmission. Our results indicate that peripheral electrotherapy provides a higher percentage of neurophysiological improvement than rTMS; however, the use of any of these additional stimulation methods (rTMS or peripheral electrotherapy) provided better results than the use of kinesiotherapy alone. The best improvement of tibialis anterior motor units' activity in iSCI patients provided the application of electrotherapy conjoined with kinesiotherapy and rTMS conjoined with kinesiotherapy. We also undertook a review of the current literature to identify and summarise available works which address the use of rTMS or peripheral electrotherapy as neuromodulation treatment options in patients after iSCI. Our goal is to encourage other clinicians to implement both types of stimulation into the neurorehabilitation program for subjects after iSCI and evaluate their effectiveness with neurophysiological tests such as sEMG so further results and algorithms can be compared across studies. Facilitating the motor rehabilitation process by combining two rehabilitation procedures together was confirmed.

Keywords: incomplete spinal cord injury; neuromodulation methods; nerve electrostimulation; repetitive transcranial magnetic stimulation; kinesiotherapy; electromyography

Citation: Leszczyńska, K.; Huber, J. The Role of Transcranial Magnetic Stimulation, Peripheral Electrotherapy, and Neurophysiology Tests for Managing Incomplete Spinal Cord Injury. *Biomedicines* **2023**, *11*, 1035. https://doi.org/10.3390/biomedicines11041035

Academic Editor: Nicolas Guerout

Received: 23 February 2023
Revised: 20 March 2023
Accepted: 24 March 2023
Published: 27 March 2023

1. Introduction

Incomplete spinal cord injury (iSCI) handicaps neuronal impulse transmission between the supraspinal centres and muscles, leading to varying degrees of paralysis and sensory impairment that greatly impact functional ability. According to the data presented by the WHO, between 250,000 and 500,000 people suffer a spinal cord injury (SCI) every year, and millions of people are currently living with this highly disabling injury [1]. Speaking of causes of SCI, car accidents rank first, with a score of 37.5%, followed by falls with a score of 31.3% in the U.S. population. Thanks to medical advances, the length of survival after spinal cord injuries has been significantly extended [2]. These are people living with disabilities that affect not only them but their entire environment and are very costly to treat and severely hampered in their daily lives. Many of these people do not return to work and are sentenced to the lifetime costs of rehabilitation and long-term care expenses [2]. Approximately 50% of subjects affected by spinal cord injury present a lesion in the cervical and thoracic regions, so there is a high percentage of this population that present severe limitations in the execution of basic activities of daily life [3]. Regardless of the mechanism of injury and highly heterogeneous pathophysiology [4], most often, patients have varying degrees of loss to the sensory and motor functions below the level of the injury, the presence of pain, and a wide range of autonomic dysfunctions. In addition, people with iSCI have many problems with mental health, and their condition very often affects the entire community [5]. Over the past decades, neurosurgical and rehabilitation advances have opened new horizons for spinal cord regeneration. Today, efforts to find methods that support cortical and spinal cord plasticity and functional recovery after iSCI continue to be even more desirable and still remain controversial [6]. Although many different treatment options have been suggested for the management of SCI, including surgery, kinesiotherapy, physical therapy, cell-based therapy, or a combination of these approaches [5–9], there is still no agreement on which method provides the best chance of clinical improvement [10]. Moreover, the more presented clinical interventions are often difficult to compare due to the lack of standardised objective evaluation methods.

Natural, spontaneous spinal cord recovery, which occurs in most individuals with SCI, is limited [7,10], so high hopes are being placed on neuromodulation and stimulation methods such as repetitive transcranial magnetic stimulation (rTMS) and peripheral electrical stimulation used as treatment options for managing complete and incomplete spinal cord injury (iSCI) [10–14]. Thus, we have focused on providing evidence of two treatment options' effectiveness that may have the potential to support and enhance functional recovery in patients with iSCI. In so doing, we have conducted studies on repetitive transcranial magnetic stimulation (rTMS) combined with kinesiotherapy [15], as well as peripheral nerve electrostimulation combined with kinesiotherapy [16]. Both methods provided better results in clinical neurophysiology tests compared to the group treated with kinesiotherapy alone. In this study, another analysis was performed in order to compare these two neuromodulation methods and kinesiotherapy. Our goal was to assess which of these treatment options provides the best chance of clinical improvement for iSCI patients. We also reviewed the current most frequently applied protocols of rTMS and peripheral electrotherapy and presented a review of the available in-literature algorithms regarding both methods used for the management of iSCI patients. Moreover, we want to provide more evidence concerning the application of both types of stimulations in the spinal cord injury rehabilitation process. As proof steadily grows in favour of electrical and magnetic stimulation as therapeutic tools, there is a need to develop standardised algorithm protocols for their application and evaluation. It is our hope and goal to provide the foundation and motivation for other clinicians to investigate and implement both types of stimulation and assess their effectiveness with neurophysiological tests so further results can be compared across studies. This likely will lead to evidence-based neuromodulation and neurorehabilitation programs for people after incomplete spinal cord injury [17] and will, as a result, improve the quality of life of these patients.

2. Materials and Methods

For the purposes of this work, we combined and upgraded the data from two previous studies, created a new set of data, and added results of the treatment in two groups of iSCI patients who received either rTMS or electrotherapy treatment, in order to try to answer the question of whether one of the neuromodulation and treatment methods is significantly better. We upgraded about 30% of the data in each patient group. We chose the data for patients by considering the spinal injury levels, similar ASIA C and D incidences in each group, averaged time from injury, and average before-after treatment observation time, as well as the BMI characteristics and gender. In addition, we once again gathered and recalculated the tibialis anterior muscle surface electromyography (sEMG) recording results for patients treated with kinesiotherapy alone, and this time, we presented the results for a total group of 55 patients.

2.1. Patients and Healthy Subjects Included in the Study

The study on the effects of peripheral electrostimulation only (Electro group) included 53 patients, while peripheral electrostimulation conjoined with kinesiotherapy (Electro + K group) included 65 patients with confirmed C4 to Th12 spinal cord injury. The study on the application of rTMS only (rTMS group) included 34 patients, and rTMS conjoined with kinesiotherapy (rTMS + K group) included 35 patients with confirmed C4-Th12 spinal cord injury. The group that received only kinesiotherapy consisted of 55 patients with C4-Th12 spinal cord injury (K group). The control group of healthy volunteers (N = 45, 15 women and 30 men) was examined to obtain the reference values of control neurophysiological recordings. The age of the healthy group was 37.9 ± 4.0 (range from 24 to 51), their height was 161–180 cm with a mean of 168.1 ± 3.6 cm, and their weight was 51–81 with a mean of 60.7 ± 5.3. The patients and healthy volunteers did not differ significantly in terms of age, height, and weight ($p = 0.7$). We did not find differences in age, height, and weight between groups of patients with $p < 0.05$.

The general practitioner, neurosurgeon, neurologist, and neurophysiologist evaluated the health status of both controls and patients.

The majority of our patients suffered an incomplete spinal cord injury that occurred in a traffic accident, which is confirmed by data from the literature that clearly indicate traffic crashes as the most common cause [18]. All patients were treated neurosurgically (spinal stabilisation), and their general health was stabilised before they started rehabilitation. Clinical studies were conducted according to the American Spinal Injury Association (ASIA) impairment scale, and the tests revealed AIS C and D in patients in all five studied groups prior to rehabilitation in almost equal proportions (Table 1). Both grades stand for the preservation of sensation to a greater extent and incomplete motor impairment, but the difference is that at least half of the key muscles below the level of injury are not strong enough to move against gravity (a muscle grade less than 3) in the case of the ASIA C, and at least half of the key muscles below the level of injury are strong enough to move against gravity (a muscle grade is 3 or more) in the case of ASIA D.

The preservation of the spinal cord structure in 1/3 to 1/4 verified by MRI was the main inclusion criterion. It was also important to us that patients enrolled in the study were no less than 4 months but no more than a year and a half after injury (Table 1). The average observation time before–after therapy in all groups of patients treated with the different modalities which were similar to the duration of the applied therapy lasted about 8.6 months in rTMS + K, K, rTMS, and Electro and about 9.9 months in Electro + K. The main exclusion criteria from the study were contraindications to the use of stimulation, i.e., pregnancy; severe head injury; episodes of epilepsy; severe disorders of the cardiovascular system; electronic implants and devices such as a pacemaker, an insulin pump, a baclofen pump, or a cochlear implant; stroke; episodes of plexopathies during treatment; inflammatory diseases; and myelopathies before or diagnosed after the incident. Patients who were ineligible for stimulation treatments but agreed to participate in the study were included in the kinesiotherapy-only treatment group. All the patients were informed and understood the potential for no benefit and the risk of all the procedures. The study was conducted according to the guidelines of the Declaration of Helsinki and was approved by the Bioethics Committee of the Medical University (decision no. 942/2021).

Table 1. Demographic, anthropometric, and traumas characteristics as well as the average duration of the applied therapies (before–after observation time) in the iSCI patients treated with different modalities. Minimum and maximum values, mean values, and standard deviations are presented.

Variable Patients Group	Age (Years)	Height (cm)	Weight (kg)	Averaged Time from Injury (Months)	Averaged Observation Time (Before–After) (Months)	ASIA AIS Score	Injury Level
rTMS + K 12♀, 23♂ N = 35	23–51 38.1 ± 3.5	157–178 170.1 ± 2.9	50–86 60.7 ± 4.1	6.7 ± 1.4	8.9	C = 26 D = 9	C4–C8 = 18 Th1–Th12 = 17
Electro + K 23♀, 42♂ N = 65	25–56 37.2 ± 4.1	161–184 172.1 ± 3.6	50–85 62.1 ± 4.3	7.4 ± 1.9	9.9	C = 54 D = 11	C4–C8 = 29 Th1–Th12 = 36
K 18♀, 37♂ N = 55	24–53 37.4 ± 3.6	159–183 173.2 ± 3.6	49–84 63.5 ± 3.7	7.1 ± 1.8	8.7	C = 46 D = 9	C4–C8 = 27 Th1–Th12 = 28
rTMS 12♀, 22♂ N = 34	25–58 38.1 ± 3.3	163–180 174.2 ± 3.2	51–84 64.6 ± 3.7	6.9 ± 1.6	8.4	C = 24 D = 10	C4–C8 = 15 Th1–Th12 = 19
Electro 19♀, 34♂ N = 53	24–60 38.1 ± 4.0	158–179 175.4 ± 3.8	50—80 62.8 ± 3.6	6.7± 1.2	8.7	C = 45 D = 8	C4–C8 = 24 Th1–Th12 = 29
p-value (difference)							
rTMS + K vs. Electro + K	NS	NS	0.05	0.05	NS	NA	NA
Electro + K vs. K	NS	NS	NS	NS	0.05	NA	NA
rTMS + K vs. K	NS	0.05	0.05	0.05	NS	NA	NA
rTMS + K vs. rTMS	NS	0.05	0.05	NS	NS	NA	NA
rTMS vs. Electro	NS	NS	0.05	NS	NS	NA	NA
rTMS vs. K	NS	NS	NS	NS	NS	NA	NA
Electro vs. K	NS	0.05	NS	NS	NS	NA	NA

Abbreviations: rTMS + K—a group of patients treated with rTMS and kinesiotherapy, Electro + K—a group of patients treated with electrotherapy and kinesiotherapy, K—patients treated with kinesiotherapy only, rTMS—patients treated with rTMS only, Electro—patients treated with electrotherapy mainly, NS—non-significant, NA—non-applicable.

2.2. Kinesiotherapy

There is no universal, recommended approach to the rehabilitation of patients with spinal cord injury in the literature [19]. This is mainly due to the different levels of damage, the severity of neurological dysfunction, and the different clinical condition of patients both before and after the injury. Intensive and regular exercise supervised by physiotherapists is the cornerstone of patient rehabilitation, but a kinesiotherapy program must be tailored individually. The stimulation methods (rTMS and electrical stimulation) described in this paper that can support neuroplasticity can only be an adjunct to kinesiotherapy but should never replace it.

The spinal cord injury patients from our study underwent a rehabilitation program created and developed over the years in the Neurorehabilitation Center for Treatment of Spinal Cord Injuries AKSON, Wroclaw, Poland. It is one of the few such places in the country dedicated specifically to patients who have suffered spinal cord injuries, being staffed by physiotherapists who constantly consult with neurosurgeons, neurologists, and neurophysiologists. There, exercises are individually tailored to patients by qualified physiotherapists, but there are a few universal rules—patients exercise for three months for several hours a day (4–6 h per day, five days a week), and these are very intense workouts in which a patient is an active participant. The programme includes postural stabilisation exercises, exercises without and with loadings (Figure 1D), stretching, and learning daily activities. The effectiveness of rehabilitation is not only assessed clinically but also by neurophysiological tests, for which are able to show even subclinical changes in voluntary motor function (e.g., clinically, the movement in the patient's extremity is not detected, but there is observed an increase in amplitudes in sEMG recording, which indicates better improvement in recruitment of the muscle motor units). When asked to contract a muscle, patients can see a change in the sEMG recording on the recorder's screen. Such biofeedback helps patients to see their progress and motivate them to work further.

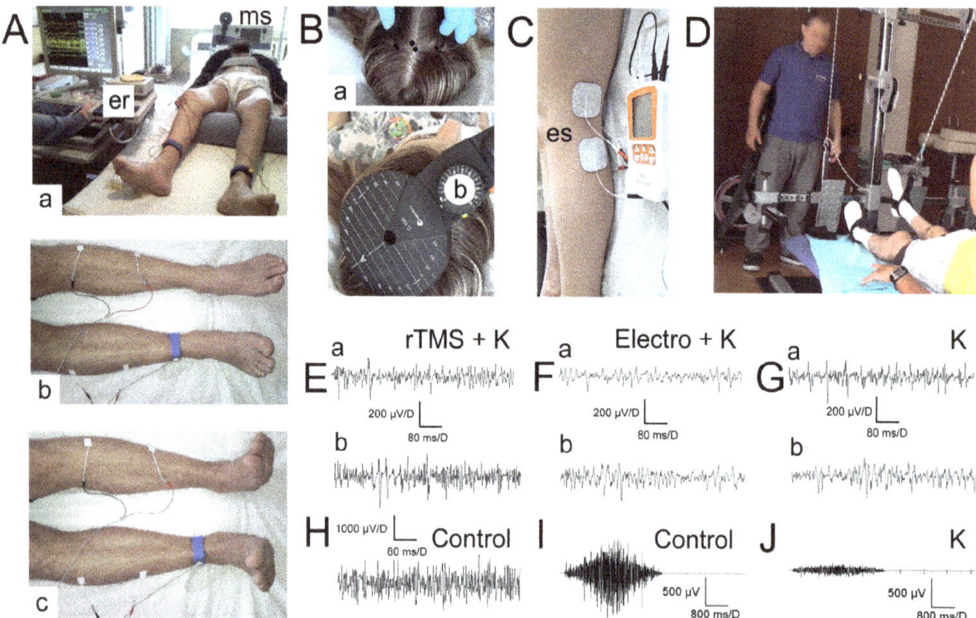

Figure 1. Photographs illustrating methodological principles of sEMG recordings (**A**): (a) electromyography recorder "er" and the magnetic stimulation device "ms"; (b) bilateral location of sEMG bipolar surface electrodes during recordings from tibialis anterior muscles at rest; (c) sEMG recording during the attempt of 5 s lasting muscles maximal contraction and applied therapies in five groups of iSCI patients. (**B**) rTMS or rTMS + K—groups of patients treated with rTMS and additionally with kinesiotherapy: (a) marking of the "hot spots" for repetitive transcranial magnetic stimulations; (b) location of coil during releasing the trains of stimuli from the magnetic stimulation device "ms". (**C**) Electro or Electro + K—groups of patients treated with electrotherapy and additionally with kinesiotherapy ("es")—bipolar electrode placement and the device for functional electrical stimulation during nerves electrotherapy. (**D**) K—patients treated with kinesiotherapy only (system for lower extremities strengthening exercises with loadings is shown). (**E–G**) Examples of sEMG recordings from anterior tibialis muscles during the attempt of maximal contraction in patients belonging to three studied groups before (a) and after (b) therapies. sEMG recording from one of the healthy volunteers is shown in (**H**) for comparison. (**I,J**) Examples of sEMG recordings presenting 5 s lasting maximal contraction with a slower time base in one of the subjects from the control group and a patient from the (**K**) group, respectively. Calibration bars for different amplifications and time bases are presented.

Most of the rehabilitation specialists agree that patients with iSCI should start physical therapy as early as post-injury day one, and it is recommended to provide at least a once-daily session with the target goal being >20 min of maximum-tolerated activity. Thus, we are well aware that the very intensive rehabilitation program that our patients undergo [14,15,20] is much more than the standard and average kinesiotherapy algorithm [20].

2.3. Transcranial Magnetic Stimulation (TMS)

Transcranial magnetic stimulation (TMS) is a method of non-invasive stimulation through the intact scalp. The magnetic field evokes an electrical response in the cortical neurons [10,15]. This method is very commonly used in various psychiatric disorders.

TMS is also used in diagnostics to produce motor-evoked potentials (MEP) to test the functional integrity of the corticospinal tract [15]. By stimulating the motor centre, it is verified if the impulse is conducted from the brain through the spinal cord to the peripheral nerves and effectors. MEP parameters such as amplitudes, latencies, and response morphology are very useful diagnostic tools that help to reveal abnormalities in efferent transmission [21]. However, to our knowledge, we are one of the very few research teams in Poland (represented by clinicians from Poznań and Wrocław) that uses and analyses rTMS (repetitive transcranial magnetic stimulation) in the therapeutic (not only diagnostic) process of patients with spinal cord injuries. TMS is restricted in those people having magnetic implants, events of epilepsy, or recent adverse neurological or cardiac events [22]. Using such exclusion criteria, it is in vain to find reports in the literature of serious side effects of this therapy, which is considered safe and non-invasive.

Non-invasive, repetitive transcranial magnetic stimulation increases descending activity so isolated neuronal circuits below the injury can receive supraspinal input. It may induce cortical and spinal cord plasticity and restore connectivity between cells [23]. The remaining connections are strengthened by rTMS, which helps the nervous system to use these surviving connections and rewire itself. Magnetic stimulation, unlike electrical stimulation, is painless, provided the appropriate parameters are used. Electrical stimulation requires large electric currents, which may cause very painful and unpleasant contractions of scalp muscles and sensory reception activation [21]. This is why magnetic stimulation seems to be more suitable for clinical use. This stimulation of the motor cortex and elicitation of a response in the muscle in this way is painless; non-invasive; and, in our experience, well tolerated by iSCI patients. It also does not require a surgical procedure like direct epidural spinal cord stimulation. It is, therefore, simpler and less expensive to apply [5,19].

In the literature, we can find various papers evaluating the use of TMS on small groups of patients and with different algorithms. On the basis of the available literature, we created our own stimulation algorithm, which differs primarily in the applied intensity of stimulation.

The rTMS study tested the effect of treatment using transcranial magnetic stimulation in patients after iSCI. Therapy lasted from 2 to 14 months, with an average of 8 months, and consisted of 3–5 sessions per month with a maximum of 15 sessions in the whole treatment period. The differences in length and frequency were mostly due to the social reasons of the patients. The devices we use for both rTMS and diagnostic MEPs are the MagPro R30 and MagPro X100 magnetic stimulator with MagOption (Medtronic A/S, Skøvlunde, Denmark) (Figure 1B). The motor cortex area with greater representation for the lower limbs has been stimulated with a 15–25 Hz frequency and pulse strength that has been 70–80% of a patient's resting motor threshold (RMT), usually not more than 45% of the maximal device's output. Patients receive 800 pulses per hemisphere. Trains last 2 s and consist of 40 pulses. Trains are separated by 28 s intervals. These are the fixed parameters of our algorithm. However, patients with poorer neurophysiological test results (e.g., lower amplitudes in sEMG recordings) have higher frequency values applied. An individual RMT is determined by the diagnostic MEPs performed prior to the stimulation. Every patient may have a different RMT because it is the minimal stimulation intensity needed to evoke a reliable motor response seen as a twitch in a tested muscle or, if the muscle contraction is unnoticeable, >50 μV peak-to-peak amplitude above the background electromyographic activity. On the basis of the RMT determined in this way, the stimulation strength is then set, which in the case of our algorithm, does not exceed 70–80% of the individual RMT. For stimulation, we used a circular coil (C-100, 12 cm in diameter) (Figure 1B(b)). We applied it over the scalp at the expected location of the motor cortex, previously confirmed by performing MEP (Figure 1 B(a)). It is the anatomical origin of the neurons and fibres of origin of the descending corticospinal pathway.

2.4. Peripheral Electrotherapy

The electrical stimulation applied to the nerves is more commonly used in the rehabilitation of patients. It is more readily available, less expensive, and more common in clinical use to aid neuroplasticity processes [10], but, for some reason, is still rarely used in patients after spinal cord injuries. There is a poor understanding of the mechanisms underlying the improvement of neurological function, but it is theorised that peripheral stimulation may prevent increased levels of neurotrophic factors, stimulate neuronal outgrowth, and increase the excitability of neuronal networks below the lesion; thus, muscle atrophy can be prevented [10,24,25]. This seems to be a very important safeguard against secondary pathological changes in SCI patients that appear in motor fibres of lower extremities despite no direct damage to the lumbosacral region of the spine [16]; however, there are a lack of data on this issue in the literature because most of the studies are conducted on animals, rarely on humans, and using neurophysiological tests as the feedback [26–30].

We performed an individually adjusted, home-based electrostimulation dedicated to the peroneal and tibialis nerves. The device is a personal, mobile, four-channel stimulator. Patients were given a portable device to take home, on which the stimulation algorithm was programmed (NeuroTrac® Sports XL, UK). They were also given precise instructions on how to use the device and how to apply the two pairs of self-adhesive stimulating electrodes (Axelgaard Ultrastim Wire Neurostimulation Electrodes with MultiStick Gel, 5×5 cm, UK), which were placed bilaterally in the area of the popliteal fossa on the anatomical passage of tibial and peroneal nerves, with the anode placed more proximal to the cathode (Figure 1C). Bipolar, rectangular electric pulses were applied in series. Stimulus strength was the only parameter that was able to be adjusted by patients. Patients were instructed that they could increase the strength of the stimulus on their own until they could see the toe move but not yet feel pain. All the rest parameters, such as frequency (20–70 Hz), single stimulus duration (11–22 ms), train duration (2–3 s), the interval between trains (2–3 s), and session duration (15–20 min) were programmed by the neurophysiologists and tailored to the patient on the basis of their electromyography and electroneurography recordings conducted prior to the treatment. Patients were unable to change these parameters themselves. Patients with more severe neurogenic changes had a longer single stimulus duration and lower frequency set. The stimulator made it possible to read the actual stimulation time used by the patient. This was an additional incentive for patients to follow the recommendations because they knew that whether they followed the plan or not was verified. Patients followed the stimulation regime, as evidenced by the very similar actual stimulation times read from the stimulators to those we expected from the patients.

2.5. Neurophysiological Feedback

We compared the effectiveness of both stimulation methods using bilateral surface electromyography (sEMG) to evaluate motor unit recruitment in the lower extremities (Figure 1A(a–c)). We took the tibialis anterior (TA) as the index muscle. sEMG was performed before and after the applied treatment using the KeyPoint Diagnostic System (Medtronic A/S, Skøvlunde, Denmark), and all patients were lying in the supine position. We applied standard, disposable Ag/AgCl surface recording electrodes (5 mm^2 of an active surface) with an active electrode placed on the muscle belly, a reference electrode placed on the distal tendon of the same muscle, and a ground electrode placed on the distal part of the leg—according to the Guidelines of the International Federation of Clinical Neurophysiology—European Chapter [15,16]. We set the upper 10 kHz and the lower 20 Hz filters in the recorder. The sEMG examination consisted of two stages. First, we asked the patient to lie down comfortably and try to relax the muscles of the lower limbs as much as possible (Figure 1A(b)). Then, the patients were instructed to contract the muscles under examination, and, on the command of the examiner, the patient made the strongest possible contraction of the muscles in the lower limbs and maintained it for 5 s. They performed 3 attempts each time, separated by 1 min rest. The examiner selected the best attempt for analysis—the one with the highest mean amplitude (in µV)—peak-to-peak with reference to the isoelectric line. The output measures were the parameters of the amplitude measured in μV and the frequency of muscle motor unit action potential recruitment measured in Hz. Frequency index (3–0) was scored on the basis of the calculations of the motor unit action potential recruitment during maximal contraction in sEMG recording: 3 = 95–70 Hz—normal; 2 = 65–40 Hz—moderate abnormality; 1 = 35–10 Hz—severe abnormality; 0 = no contraction. sEMG recordings were performed at the base time of 80 ms/D and the amplification of 20–1000 μV/D (Figure 1E–H).

2.6. Statistical Analysis

Data were analysed with Statistica, version 13.1 (StatSoft, Kraków, Poland). Descriptive statistics included mean values, standard deviations (SD), and minimum (min) and maximum (max) values for measurable variables. The normality distribution and homogeneity of variances were conducted with Shapiro–Wilk tests and with Levene's tests in some cases. In patients of five groups with incomplete spinal cord injuries at the cervical and thoracic levels, the mean values of parameters from sEMG tests were compared before and after therapy using Student's t-test, Welch's test, the Mann–Whitney test, or Wilcoxon's test. It was assumed that a comparison of values at $p \leq 0.05$ determined significant statistical differences. Changes in the outcome measures from sEMG recordings (parameters of amplitudes and frequency indexes) before and after treatment were also expressed in percentages. The preliminary statistical analysis determined the required sample size using the primary outcome variables from sEMG recordings from anterior tibialis muscles before and after treatment with a power of 80% and a significance level of 0.05 (two-tailed). The mean and standard deviation (SD) were calculated using the data from the first ten subjects. The sample size software estimated that at least 25 subjects from each of the studied groups were needed.

3. Results

None of our patients reported side effects from stimulation, whether rTMS or peripheral electrotherapy.

Subjects belonging to the five groups of iSCI patients treated with different methods did not differ greatly in terms of demographic and anthropometric characteristics, as well as the severity and extent of spinal cord injuries (Table 1). Although the number of patients in each group was different, all subjects represented similar injury levels and ASIA scores. The differences in weight, height, time from injury, and observation time were either not significant or at the limit of the statistical significance ($p = 0.05$). It is not likely that these variables and factors influenced the differences found in the comparison of neurophysiological results in the studied groups of patients (Table 2).

Table 2 shows a summary of the sEMG parameters recorded from the anterior tibialis muscle in iSCI patients belonging to five therapeutic groups before and after the applied treatment. All of them differed from those recorded in a control group of healthy volunteers (Table 3), both with references to the recorded values of amplitudes and the frequencies of firings of motor units' action potentials (Figure 1E–H). After the therapies were applied, a significant improvement of the mean amplitudes in sEMG recordings from anterior tibialis muscles were found in patients from the rTMS + K group (at $p = 0.008$) by 32.1% and Electro + K group (at $p = 0.009$) by 44.4% in comparison to the values recorded before therapy. A significant increase at $p = 0.03$ of the sEMG amplitude parameter after therapies was observed in patients belonging to the Electro (27.5%) and rTMS (26.4%) groups. A significant improvement (at $p = 0.04$) in the frequency index of sEMG recordings after the therapy was observed in patients from the Electro + K group by a 22.2% difference, although by 16.6% was also ascertained in patients treated with electrotherapy only, and by 11.7% in rTMS + K and rTMS groups. A slight difference in the improvement of the amplitude parameter by 14.8%, without a significant improvement in the frequency index, was found in the group of patients treated only with kinesiotherapy (K).

Table 2. A summary of the sEMG parameters measured from the anterior tibialis muscles (cumulative data for the right and left sides) in the iSCI patients belonging to five therapeutic groups before and after the applied treatment. p—significant differences are marked in bold. %—the percentage of change.

Groups of Patients	rTMS + K		Electro + K		K		rTMS		Electro	
Parameter	Before	After	Before	After	Before	After	Before	After	Before	After
TA sEMG										
Amplitude (µV)	50–2100 232.7 ± 94.5 1–3	50–2300 343.0 ± 79.2 1–3	50–2000 191.4 ± 94.7 1–3	50–2400 344.2 ± 88.3 1–3	50–1900 221.0 ± 87.1 1–3	50–2000 259.3 ± 93.8 1–3	50–2000 229.3 ± 83.1 1–3	50–2300 311.9 ± 81.6 1–3	50–2000 210.5 ± 85.6 1–3	50–2300 290.6 ± 82.1 1–3
Frequency index	1.5 ± 0.6	1.7 ± 0.5	1.4 ± 0.5	1.8 ± 0.5	1.4 ± 0.5	1.5 ± 0.6	1.5 ± 0.4	1.7 ± 0.6	1.5 ± 0.5	1.8 ± 0.5
Neurophysiological improvement (before vs. after)										
Amplitude (µV)	**0.008**	32.1%	**0.009**	44.4%	**0.05**	14.8%	**0.03**	26.4%	**0.03**	27.5%
Frequency index	**0.05**	11.7%	**0.04**	22.2%	NS	6.6%	**0.05**	11.7%	**0.05**	16.6%

Abbreviations: rTMS + K—a group of patients treated with rTMS and kinesiotherapy, Electro + K—a group of patients treated with electrotherapy and kinesiotherapy, K—patients treated with kinesiotherapy only, rTMS—patients treated with rTMS only, Electro—patients treated with electrotherapy mainly, Frequency index (3–0)—frequency of motor unit action potentials recruitment during maximal contraction sEMG recording: 3 = 95–70 Hz—normal; 2 = 65–40 Hz—moderate abnormality; 1 = 35–10 Hz—severe abnormality; 0 = no contraction; NS—non-significant.

Table 3. Reference values of surface electromyography (sEMG) parameters that were recorded during an attempt of maximal contraction in a group of healthy volunteers (N = 45).

Recording	Parameter	Healthy Volunteers (Control)	
TA sEMG	Amplitude (μV)	600–2450	785.3 ± 100.2
	Frequency index	3–3	3.0 ± 0.5

Abbreviations: TA—anterior tibialis muscle; sEMG—surface electromyography. Frequency index (3–0)—frequency of motor unit action potentials recruitment during maximal contraction sEMG recording: 3 = 95–70 Hz—normal; 2 = 65–40 Hz—moderate abnormality; 1 = 35–10 Hz—severe abnormality; 0 = no contraction.

In all the groups of patients treated with five types of therapies, there were no observed improvements in the parameters of the amplitude and frequency of sEMG recording from the anterior tibialis muscle during the attempt of maximal contraction comparable to those recorded in healthy volunteers from the control group (Table 3).

4. Discussion

This study compared the effectiveness of three different treatment options, namely, combinations of rTMS, peripheral electrotherapy, and kinesiotherapy, evaluated with clinical neurophysiology measurements in patients with iSCI. Our data show that better improvement of the average values of sEMG amplitudes more than firing frequencies of motor units action potentials recorded from the anterior tibialis muscle and better general neurophysiological improvement were found in iSCI patients treated with peripheral electrotherapy combined with kinesiotherapy and rTMS combined with kinesiotherapy. Patients treated with kinesiotherapy alone achieved worse results.

Tables 4 and 5, with the data on the algorithms of rTMS and peripheral electrotherapy described in the literature, show how the parameters and methods used to evaluate therapy varied in patients after iSCI.

Table 4. Data on parameters of transcranial magnetic stimulation in patients with spinal cord injury were gathered from available literature and from observations where effects confirmed its efficiency.

Reference	No of Patients Who Received Stimulation	Stimulation Algorithm	Interval between Trains of Stimuli	Intensity of Stimulation, Maximal % Stimulus Output	Number of Sessions Per Day, Week, Month	Evaluation Method of Effectiveness
Belci et al., 2004 [11]	4	10 Hz 720 pulses in total	100 ms	90%	5 sessions of 1 h, one per day	ASIA scale 5 days of sham rTMS delivered over the occipital cortex
Kumru et al., 2010 [12]	15	20 Hz 40 pulses in a train 1600 pulses in total	28 s	90%	15 sessions, 5 per week, 3 weeks of observation in total	Clinical evaluation of spasticity, EMG Sham stimulation with a coil disconnected from the main stimulator unit
Kuppuswamy et al., 2011 [13]	15	5 Hz 900 pulses in total	8 s	80%	5 days, 1 per day (sham or active, randomly)	ASIA scale, upper limb functional tests, MEP, autonomic measures; sham stimulation with only 5% of real stimulator output

Table 4. *Cont.*

Reference	No of Patients Who Received Stimulation	Stimulation Algorithm	Interval between Trains of Stimuli	Intensity of Stimulation, Maximal % Stimulus Output	Number of Sessions Per Day, Week, Month	Evaluation Method of Effectiveness
Benito et al., 2012 [14]	7	20 Hz 40 pulses in a train, 1800 pulses in total, 20 trains	28 s	90%	15 days, 1 per day (sham or active)	Functional gait assessment and clinical spasticity assessment Double-blind, sham-controlled trial; sham stimulation with a coil disconnected from the main stimulator unit
Nardone et al., 2014 [31]	9	20 Hz 1600 pulses in total	28 s	100%	4 days, 1 session per day	Clinical evaluation of spasticity, EMG Sham stimulation with a coil disconnected from the main stimulator unit
De Araujo et al., 2017 [32]	10	5 Hz 50 pulses in a train 12 trains	10 s	100%	10 sessions over 2 weeks	ASIA scale, Fugl–Meyer scale, Modified Ashworth Scale, socio-demographic questionnaire, Mini-Mental State Examination, EMG, EEG 5 sessions active and 5 blind for each person, with a two-week washout period
Wincek et al., 2021 [32]	26	15–25 Hz 40 pulses in a train, 20 trains 1600 pulses in total	28 s	38–40%	3–5 sessions per month for 5 months	sEMG, MEP; results were compared to patients treated with kinesiotherapy alone

Table 5. Data on parameters of peripheral electrostimulation in patients with spinal cord injury gathered from the available literature and from observations where effects confirmed its efficiency.

Reference	No of Patients	Place of Stimulation	Stimulation Frequency (Hz) and Train Duration (s)	Single Stimulus Duration and Strength (mA), Interval between Trains of Stimuli	Observation Time (Months), Numbers of Session	Evaluation Method of Effectiveness	Additional Comments
Baldi et al., 1998 [25]	26	hip extensors, knee extensors and knee flexors	60 Hz	0.375 ms 140 mA	6 months 3 times per week 30 min per day	Total body lean body mass, lower limb lean body mass, and gluteal lean body mass assessed by using dual-energy X-ray absorptiometry	All the parameters were fixed and the same for all the patients

Table 5. *Cont.*

Reference	No of Patients	Place of Stimulation	Stimulation Frequency (Hz) and Train Duration (s)	Single Stimulus Duration and Strength (mA), Interval between Trains of Stimuli	Observation Time (Months), Numbers of Session	Evaluation Method of Effectiveness	Additional Comments
Kern et al., 2010 [27]	22	leg muscles	-	12–150 ms 250 mA	2 years 5 times a week 30 min per day	EMG, knee torque measurement every 12 weeks, muscle cross-sectional area measured by computed tomography, muscle biopsies	Home-based therapy
Pieber et al., 2015 [30]	24	tibialis anterior muscle	1 Hz	200 ms 500 ms 8 mA	4 sessions, 1–2 min each	Medical Research Council Scale Intensity was recorded in absolute values in mA and relatively provided in percentage to the first stimulation of each patient.	-
Lee et al., 2015 [29]	22	median and common peroneal nerves	100 Hz 2 s	0.4 ms 12–30 mA	6 weeks, 5 days a week 30 min per day	EMG ENG non-stimulated leg used as a control	All the parameters were fixed and the same for all the patients
Zheng et al., 2020 [17]	5	knee extensors	30 Hz	30–99 mA	12 weeks of therapy, sessions twice a week	Quadriceps femoris torque, blood lipid profilin, c-reactive protein, spasticity evaluation, quality of life index	-
Huber et al., 2021 [16]	42	peroneal and tibial nerves	20–70 Hz 2–3 s	11–22 ms 18–45 mA 2–3 s	6–14 months, 5 days a week, 15–20 min per day	EMG ENG Results were compared to patients treated with kinesiotherapy alone	Home-based therapy Stimulus strength was the only parameter adjusted by patients

In the literature, there are many examples of rTMS applications in the treatment of psychiatric disorders [33], as well as the use of TMS as a diagnostic tool that, by means of MEP, reflects the integrity of the corticospinal tract and the impulse conduction from the motor cortex to the muscles [34,35]. However, there are still very few scientific reports using rTMS as an adjunct to treatment in patients with spinal cord injury [11–14,31,32]. The effect of rTMS stimulation on the cortical structure function depends mainly on the frequency of pulses during stimulation. The high frequency used in rTMS (>10 Hz) is known as excitatory. It increases activity in the neurons of the corticospinal pathway and also reduces corticospinal inhibition. Therefore, we use frequency values above 20 Hz in therapy, as opposed to low frequencies, which would have the effect of reducing cortical excitability [15]. Compared to the therapy algorithms used in the literature (see Table 4), we used a much lower stimulation strength—70–80% of the individually selected RMT, which is no more than 0.8–1 Tesla (usually 38–45% of the maximum device's output and not close to 90–100% as in other studies). In our opinion, such a low stimulus strength is still effective; it stimulates structures up to 3–5 cm deep into the cortex, but at the same time, it is less painful for the patient, which has an impact on engagement in therapy. Stimulus strength of more than 100% RMT (80% and more of the maximum device's output) when performing diagnostic MEPs is painful for patients. Not to mention such high stimulus values applied in series and repeated hundreds of times during rTMS. Such painful therapy does not have a positive effect on patient compliance. In addition to other studies, we used a significantly longer observation period.

Perez et al. [35] presented their study on healthy humans, which indicates the presence of short-term plasticity within spinal inhibitory circuits under the influence of electrical nerve stimulation. They hypothesised that the pattern of electrical input might be a crucial factor for inducing changes in the spinal circuit and that such stimulation may improve walking ability in patients with spinal cord injury [35]. However, in the case of currently used peripheral electrotherapy, it is very common to use the same stimulation parameters without neurophysiological-based individual adjustment [29]. Moreover, these algorithms differ greatly in their values and in the way they are assessed. Single stimulus duration may vary from 0.1 to 500 ms, frequency 1 to 100 Hz, and intensity is set up even to 250 mA [36]. Our algorithm has parameter ranges that are individually tailored to the patient on the basis of their neurophysiological test results (see Table 5). As with rTMS, here, too, we performed a much longer observation period.

A very interesting summary of the literature on the use of TMS as a therapeutic method was used by Brihmat et al. [37]. He and his team provide a very detailed analysis of various stimulation algorithms and their evaluation methods. The conclusions are similar to ours—there are still many gaps in the rTMS algorithms presentation and, in general, in SCI research. This summary also confirms that TMS is a minimally invasive, safe, and easy-to-use method that may provide various positive clinical and neurophysiological results and should be considered in the rehabilitation of the SCI population [37].

The activity of the anterior tibial muscle during the maximal contraction used as the marker of motor improvement following the application of different rehabilitation procedures in iSCI patients was chosen on the basis of the suggestions of other researchers [38] as well as previous experiences [9,15,16]. During electrotherapy procedures, we stimulated peroneal nerves innervating tibial muscles, for which sEMG were recorded [16]. We concluded in the previous study that the activity of thigh muscles (e.g., rectus femoris muscles) was usually better in iSCI patients because they are innervated by femoral nerves whose trunks are larger and include a greater spectrum of axons than peroneal nerves [9]. The secondary, extensive degenerating processes in motor fibres and muscle atrophy are clearly detected in distal than proximal parts of nerves of the lower extremities [16,25]. Therefore, in cases of patients after iSCI one year after the incident, it is recommended to choose sEMG diagnostic recordings from the tibial muscle instead of the foot extensors, although they are both innervated from peroneal nerves [19,38].

The increased amplitude (in μV), more than the frequency index (in Hz) parameters, in the sEMG recording after the applied treatment (see Table 1) indicates better recruitment of motor units of the examined muscles after therapy. Such neurophysiological improvement is not necessarily always simultaneous with clinical improvement, but it certainly indicates positive changes in the muscles of the lower extremities in the patients treated with neuromodulation methods. Such changes may also correlate with the improvement of the transmission of neural impulses. The question arises, is any of these stimulation methods better? Our data show that better average sEMG amplitudes for the anterior tibialis muscle were recorded by patients treated with peripheral electrotherapy and kinesiotherapy. However, it should be noted that each of these stimulation methods has a different mechanism for improving patients' neurophysiological status and neuroplasticity, so it may be beneficial to use both methods, if possible. However, this hypothesis requires further research. Nevertheless, our results clearly indicate that the use of even one of the methods of stimulation, regardless of the supraspinal and spinal or peripheral mechanism, provides better results than kinesiotherapy alone. Due to the length of the therapy (several months) and the strenuousness of the repeated stimulation and neurophysiology tests, we took great care to ensure that these procedures were as painless as possible for patients and as little traumatising as possible, having in mind patients' comfort and well-being. Hence, the algorithms with a much lower electrical and magnetic stimulation intensity than those used by other researchers (see Tables 4 and 5) and the use of surface electromyography (sEMG) instead of invasive needle electromyography for evaluation of the treatment results were considered. For diagnostic purposes, sEMG is contemporarily considered a functional test precise and specific enough to evaluate the entire efferent transmission from the supraspinal level to the muscles in iSCI patients treated with the different classical conservative methods as well as the newest based on the neuromodulation [38,39]. Nevertheless, a key difference in the use of the rTMS and electrical stimulation of nerves is their availability. Patients qualified for rTMS had to attend the treatment facility regularly for stimulation application. This requires more time and financial investment in terms of travel. Patients qualified for electrotherapy applied stimulation independently at home. This, too, may have affected the patient's involvement in the therapy process.

In looking for ways to improve the functional status of people after an incomplete spinal cord injury, it is important to remember that the dysfunction is not only due to primary injury, local synaptic and neuronal loss, inflammation, and the later formation of scar tissue, which are most often treated with pharmacology and surgery. The main problem for these people becomes a change in the flow of information in ascending and descending pathways. Over time, the neuronal projections become weaker, and this may lead to secondary degenerative changes in the subcortical and cortical structures and brain reorganisation [3,5,40]. Therefore, it seems reasonable to look for stimulation methods that enhance activity-dependent neuroplasticity and residual signals and alter neuronal activity [10,41]. Such a method may support neurorehabilitation by fulfilling the premise of Hebb's theory—neurons that fire together wire together [42,43]. Coordinated activity of a presynaptic terminal and a postsynaptic cell would lead to the strengthening of that synaptic connection between them, and conversely, uncoordinated activity, which is observed in iSCI patients, between synaptic partners would weaken their synaptic connections. It speaks powerfully about the consequences of coordinated activity for not only synaptic strength but also the long-term consequences that require the building up of new synaptic connections. Currently, many different methods of stimulation are being tested, and hopes are being held that a proper stimulation algorithm can modulate excitability; thus, we may be able to establish new synaptic connections, improve neural transmission, and inhibit and reverse the degenerative process in axons. Stimulation methods such as transcranial magnetic stimulation and peripheral electrotherapy should be included in the therapy in the first year after the injury because the first six months to a year appears to be a crucial time period for spinal cord regeneration; structural and functional reorganisation of the spine; and the brain and, thus, functional recovery [9,11,44–46]. However, we still need more evidence because the mechanisms underlying stimulus-induced recovery are not well understood.

Incomplete spinal cord injury is a pathological complex syndrome in which the treatment can be disturbed with many factors such as chronic non-bacterial osteomyelitis [47], and hence many modern surgical [48] and non-surgical [49] therapeutic methods are incorporated. The main problem in the management and treatment of patients with spinal cord injuries is the use of different algorithms and methods of evaluating these patients and the lack of consistency between observed effects [9]. The stimulation algorithms presented in Tables 4 and 5 are difficult to compare because the authors use different scales, such as the ASIA scale, the visual analogue scale of pain (VAS), the Ashworth scale, and different clinical assessment tools that are subjective in nature, thus making it especially challenging to report the results of applied therapy, interpret them in a consistent manner, and compare them across studies. In this work, we would like to emphasise the important role of neurophysiological tests such as electromyography, eletroneurography, and motor-evoked potential examinations. These methods allow clinicians to assess the progress of patients' physical recovery objectively; therefore, in our opinion, they should be an obligatory part of the assessment of patients after spinal cord injury. Such methods of assessing the effects of treatment will make it possible to compare further results, which will contribute to providing more evidence of the effectiveness of applied treatment algorithms. This will make discussions between groups of researchers and the comparison of research results much easier. Moreover, such evaluation will also make it possible to assess subclinical changes that are impossible to capture by the scales commonly used in the neurological evaluation of patients and standard physical examination. Capturing changes (even subclinical), such as an increase in amplitudes in sEMG or in MEP responses, is evidence of improvement and can be a signal that the treatment algorithm used is heading in the right direction, which, especially for patients, is a follow-up and great motivation for further work and rehabilitation. This is of utmost importance in patients for whom spinal cord injury is a life-altering trauma; thus, they look for any improvement in their health status. The use of this neurophysiological test not only allows for the comparison of the effectiveness of different treatment options (such as the comparison presented in this study) but also allows for the modification and personalisation of the process, bringing us closer to creating an evidence-based rehabilitation algorithm.

The challenge and the main limitation of all research related to spinal cord injury is the highly heterogeneous pathophysiology and different dynamic of the spontaneous regeneration process [5]. Most cases involve incomplete injury, but even in this group, the severity depends on the level of damage and which parts of the spinal cord are affected. Such a diversity of cases and pre- and post-injury health status makes it very difficult to compare patients. One must always take this into account in studies and try, as in our work [15,16], to include patients with damage at a similar level and with similar neurophysiological status. Unlike other authors [12–14,30,33], for our study, we did not use sham stimulation. Moreover, as it appears in the data presented in Table 5, the protocols for electrical stimulation parameters are different in different studies regarding muscle and nerve electrotherapy. Therefore, they cannot be sufficiently compared and justified. We believe that the small number of spinal cord injury patients who are eligible for stimulation and are willing to participate in several months of therapy deserve to be given every possible chance for treatment. Such people, seeking every possible chance of therapy and functional recovery, would find it difficult to face a possible placebo approach. For this reason, we compared patients treated only with kinesiotherapy who were unwilling or ineligible for additional therapy with patients receiving additional stimulation treatment. We are aware of the scientific downside of such an approach, but ethical considerations prevail here to ensure that our patients receive the best possible treatment algorithm, especially in a tragic event such as a spinal cord injury, which is life-changing not only for the patient but also for those around them.

We hope that the evidence gathered here on the effective application of transcranial magnetic stimulation or peripheral electrostimulation as methods facilitating functional recovery in patients with spinal cord injuries will encourage clinicians to use these methods, evaluate their effectiveness with neurophysiological tests, and present further results.

5. Summary of Recommendation

Rehabilitation should be started as soon as possible—it should be intensive and individually tailored to a patient's needs. It is a foundational and indispensable element of every treatment program for any patient after iSCI, regardless of the level of damage. Kinesiotherapy effects alone are not satisfactory enough. The treatment should include repetitive transcranial magnetic stimulation or peripheral electrotherapy or both in the first year after injury—the first six months to a year appears to be a crucial time period for spinal cord regeneration and functional recovery. The possible algorithms are proposed in Tables 4 and 5. The effects of therapy should be evaluated not only with the clinical methods but using objective clinical neurophysiology tests such as electromyography, electroneurography, and motor-evoked potentials examinations. Observation of even the subclinical changes expressed in improvements in clinical neurophysiology tests can be an additional motivation for patients to continue the rehabilitation programme.

6. Conclusions

The concept of neuromodulation, aiming to improve the activity in spinal centres of iSCI patients, assumes a combination of afferent pulsation (provided by electrotherapy or to some degree by different procedures of kinesiotherapy) and efferent (provided by rTMS) to obtain the facilitated response in general exemplified in improving motor function according to Hebb's theory [42]. Since the results of treatment in two more groups of patients only with rTMS and only with nerve electrotherapy were presented, the effectiveness of these technologies for the treatment of spinal cord injury consequences can be better understood. Transcranial magnetic stimulation and peripheral electrostimulation appear to be methods that support and facilitate the recovery process in iSCI patients. Both simulation methods should be complementary to standard kinesiotherapy since they improve patients' neurophysiological status and provide better sEMG results than kinesiotherapy alone. In addition, the entire neurorehabilitation process should be evaluated by objective methods of clinical neurophysiology, such as electromyography. Such a method of assessing the effects of treatment will make it possible to compare further results, which will contribute to providing more evidence of the effectiveness of applied treatment algorithms. Moreover, such evaluation will also make it possible to assess subclinical changes that are impossible to capture by the scales commonly used in the neurological evaluation of patients and standard physical examination. All of these allow us to modify and personalise the management of spinal cord injury patients and may bring us all closer to the main task ahead, which is establishing an evidence-based neurorehabilitation program for subjects after iSCI.

Author Contributions: Conceptualisation, K.L. and J.H.; methodology, K.L. and J.H.; software, J.H.; validation, K.L. and J.H., formal analysis, K.L. and J.H.; investigation, K.L. and J.H; resources, J.H.; data curation, K.L. and J.H.; writing—original draft preparation, K.L.; writing—review and editing, K.L. and J.H.; visualisation—J.H., supervision, J.H.; project administration, K.L. All authors have read and agreed to the published version of the manuscript.

Funding: This research received no external funding.

Institutional Review Board Statement: The study was conducted in accordance with the Declaration of Helsinki and approved by the Bioethics Committee of Poznan University of Medical Science, decision no. 942/21 dated on 13 January 2022.

Informed Consent Statement: Informed consent was obtained from all subjects involved in the study.

Data Availability Statement: All the data generated or analyzed during this study are included in this published article.

Conflicts of Interest: The authors declare no conflict of interest.

References

1. Bickenbach, J.; Officer, A.; Shakespeare, T.; von Groote, P.; World Health Organization; The International Spinal Cord Society. *International Perspectives on Spinal Cord Injury/Edited by Jerome Bickenbach*; World Health Organization: Geneva, Switzerland, 2013. Available online: https://apps.who.int/iris/handle/10665/94190 (accessed on 6 December 2022).
2. National Spinal Cord Injury Statistical Center. *2021 Annual Report*; National Spinal Cord Injury Statistical Center: Birmingham, AL, USA, 2021.
3. Laliberte, A.M.; Goltash, S.; Lalonde, N.; Bui, T.V. Propriospinal Neurons: Essential Elements of Locomotor Control in the Intact and Possibly the Injured Spinal Cord. *Front. Cell. Neurosci.* **2019**, *12*, 512. [CrossRef]
4. Alizadeh, A.; Dyck, S.M.; Karimi-Abdolrezaee, S. Traumatic Spinal Cord Injury: An Overview of Pathophysiology, Models and Acute Injury Mechanisms. *Front. Neurol.* **2019**, *10*, 282. [CrossRef] [PubMed]
5. Leemhuis, E.; Favieri, F.; Forte, G.; Pazzaglia, M. Integrated Neuroregenerative Techniques for Plasticity of the Injured Spinal Cord. *Biomedicines* **2022**, *10*, 2563. [CrossRef] [PubMed]
6. Davletshin, E.; Sabirov, D.; Rizvanov, A.; Mukhamedshina, Y. Combined Approaches Leading to Synergistic Therapeutic Effects in Spinal Cord Injury: State of the Art. *Front. Biosci.* **2022**, *27*, 334. [CrossRef] [PubMed]

7. Fawcett, J.W.; Curt, A.; Steeves, J.D.; Coleman, W.P.; Tuszynski, M.H.; Lammertse, D.; Bartlett, P.F.; Blight, A.R.; Dietz, V.; Ditunno, J.; et al. Guidelines for the Conduct of Clinical Trials for Spinal Cord Injury as Developed by the ICCP Panel: Spontaneous Recovery after Spinal Cord Injury and Statistical Power Needed for Therapeutic Clinical Trials. *Spinal Cord* **2007**, *45*, 190–205. [CrossRef] [PubMed]
8. Wang, T.Y.; Park, C.; Zhang, H.; Rahimpour, S.; Murphy, K.R.; Goodwin, C.R.; Karikari, I.O.; Than, K.D.; Shaffrey, C.I.; Foster, N.; et al. Management of Acute Traumatic Spinal Cord Injury: A Review of the Literature. *Front. Surg.* **2021**, *8*, 698736. [CrossRef] [PubMed]
9. Tabakow, P.; Jarmundowicz, W.; Czapiga, B.; Fortuna, W.; Miedzybrodzki, R.; Czyz, M.; Huber, J.; Szarek, D.; Okurowski, S.; Szewczyk, P.; et al. Transplantation of Autologous Olfactory Ensheathing Cells in Complete Human Spinal Cord Injury. *Cell Transplant.* **2013**, *22*, 1591–1612. [CrossRef]
10. James, N.D.; McMahon, S.B.; Field-Fote, E.C.; Bradbury, E.J. Neuromodulation in the Restoration of Function after Spinal Cord Injury. *Lancet Neurol.* **2018**, *17*, 905–917. [CrossRef]
11. Belci, M.; Catley, M.; Husain, M.; Frankel, H.L.; Davey, N.J. Magnetic brain stimulation can improve clinical outcome in incomplete spinal cord injured patients. *Spinal Cord* **2004**, *42*, 417–419. [CrossRef]
12. Kumru, H.; Murillo, N.; Vidal Samso, J.; Valls-Sole, J.; Edwards, D.; Pelayo, R.; Valero-Cabre, A.; Tormos, J.M.; Pascual-Leone, A. Reduction of Spasticity with Repetitive Transcranial Magnetic Stimulation in Patients with Spinal Cord Injury. *Neurorehabilit. Neural Repair* **2010**, *24*, 435–441. [CrossRef]
13. Kuppuswamy, A.; Balasubramaniam, A.V.; Maksimovic, R.; Mathiasm, C.J.; Gall, A.; Craggs, M.D.; Ellaway, P.H. Action of 5 Hz repetitive transcranial magnetic stimulation on sensory, motor and autonomic function in human spinal cord injury. *Clin. Neurophysiol.* **2011**, *122*, 2452–2461. [CrossRef] [PubMed]
14. Benito, J.; Kumru, H.; Murillo, N.; Costa, U.; Medina, J.; Tormos, J.; Pascual-Leone, A.; Vidal, J. Motor and Gait Improvement in Patients With Incomplete Spinal Cord Injury Induced by High-Frequency Repetitive Transcranial Magnetic Stimulation. *Top. Spinal Cord Inj. Rehabil.* **2012**, *18*, 106–112. [CrossRef]
15. Wincek, A.; Huber, J.; Leszczyńska, K.; Fortuna, W.; Okurowski, S.; Chmielak, K.; Tabakow, P. The Long-Term Effect of Treatment Using the Transcranial Magnetic Stimulation RTMS in Patients after Incomplete Cervical or Thoracic Spinal Cord Injury. *JCM* **2021**, *10*, 2975. [CrossRef] [PubMed]
16. Huber, J.; Leszczyńska, K.; Wincek, A.; Szymankiewicz-Szukała, A.; Fortuna, W.; Okurowski, S.; Tabakow, P. The Role of Peripheral Nerve Electrotherapy in Functional Recovery of Muscle Motor Units in Patients after Incomplete Spinal Cord Injury. *Appl. Sci.* **2021**, *11*, 9764. [CrossRef]
17. Zheng, Y.; Mao, Y.R.; Yuan, T.F.; Xu, D.S.; Cheng, L.M. Multimodal treatment for spinal cord injury: A sword of neuroregeneration upon neuromodulation. *Neural Regen. Res.* **2020**, *15*, 1437–1450. [PubMed]
18. Vasanthan, L.T.; Nehrujee, A.; Solomon, J.; Tilak, M. Electrical Stimulation for People with Spinal Cord Injury. *Cochrane Database Syst. Rev.* **2019**, *2019*, CD013481. [CrossRef]
19. Wincek, A.; Huber, J.; Leszczyńska, K.; Fortuna, W.; Okurowski, S.; Tabakow, P. The Results of a Long Term Uniform System of Neurorehabilitation in Patients with Incomplete Thoracic Spinal Cord Injury. *Ann. Agric. Environ. Med.* **2021**, *29*, 94–102. [CrossRef]
20. Harvey, L.A. Physiotherapy rehabilitation for people with spinal cord injuries. *J. Physiother.* **2016**, *62*, 4–11. [CrossRef]
21. Awad, B.I.; Carmody, M.A.; Zhang, X.; Lin, V.W.; Steinmetz, M.P. Transcranial Magnetic Stimulation After Spinal Cord Injury. *World Neurosurg.* **2015**, *83*, 232–235. [CrossRef]
22. Rossi, S.; Antal, A.; Bestmann, S.; Bikson, M.; Brewer, C.; Brockmöller, J.; Carpenter, L.L.; Cincotta, M.; Chen, R.; Daskalakis, J.D.; et al. Safety and Recommendations for TMS Use in Healthy Subjects and Patient Populations, with Updates on Training, Ethical and Regulatory Issues: Expert Guidelines. *Clin. Neurophysiol.* **2021**, *132*, 269–306. [CrossRef]
23. Ellaway, P. Induction of Central Nervous System Plasticity by Repetitive Transcranial Magnetic Stimulation to Promote Sensorimotor Recovery in Incomplete Spinal Cord Injury. *Front. Integr. Neurosci.* **2014**, *8*, 42. [CrossRef]
24. Gajewska-Woźniak, O.; Skup, M.; Kasicki, S.; Ziemlińska, E.; Czarkowska-Bauch, J. Enhancing Proprioceptive Input to Motoneurons Differentially Affects Expression of Neurotrophin 3 and Brain-Derived Neurotrophic Factor in Rat Hoffmann-Reflex Circuitry. *PLoS ONE* **2013**, *8*, e65937. [CrossRef] [PubMed]
25. Baldi, J.C.; Jackson, R.D.; Moraille, R.; Mysiw, W.J. Muscle Atrophy Is Prevented in Patients with Acute Spinal Cord Injury Using Functional Electrical Stimulation. *Spinal Cord* **1998**, *36*, 463–469. [CrossRef] [PubMed]
26. Chan, K.M.; Curran, M.W.T.; Gordon, T. The Use of Brief Post-Surgical Low Frequency Electrical Stimulation to Enhance Nerve Regeneration in Clinical Practice: Use of Brief Post-Surgical Low Frequency Electrical Stimulation. *J. Physiol.* **2016**, *594*, 3553–3559. [CrossRef]
27. Kern, H.; Carraro, U.; Adami, N.; Biral, D.; Hofer, C.; Forstner, C.; Mödlin, M.; Vogelauer, M.; Pond, A.; Boncompagni, S.; et al. Home-Based Functional Electrical Stimulation Rescues Permanently Denervated Muscles in Paraplegic Patients with Complete Lower Motor Neuron Lesion. *Neurorehabilit. Neural Repair* **2010**, *24*, 709–721. [CrossRef]
28. Günter, C.; Delbeke, J.; Ortiz-Catalan, M. Correction to: Safety of Long-Term Electrical Peripheral Nerve Stimulation: Review of the State of the Art. *J. NeuroEng. Rehabil.* **2020**, *17*, 77. [CrossRef]
29. Lee, M.; Kiernan, M.C.; Macefield, V.G.; Lee, B.B.; Lin, C.S.-Y. Short-Term Peripheral Nerve Stimulation Ameliorates Axonal Dysfunction after Spinal Cord Injury. *J. Neurophysiol.* **2015**, *113*, 3209–3218. [CrossRef] [PubMed]

30. Pieber, K.; Herceg, M.; Paternostro-Sluga, T.; Schuhfried, O. Optimizing Stimulation Parameters in Functional Electrical Stimulation of Denervated Muscles: A Cross-Sectional Study. *J. NeuroEng. Rehabil.* **2015**, *12*, 51. [CrossRef] [PubMed]

31. Nardone, R.; Höller, Y.; Thomschewski, A.; Brigo, F.; Orioli, A.; Höller, P.; Golaszewski, S.; Trinka, E. RTMS Modulates Reciprocal Inhibition in Patients with Traumatic Spinal Cord Injury. *Spinal Cord* **2014**, *52*, 831–835. [CrossRef] [PubMed]

32. de Araújo, A.V.L.; Barbosa, V.R.N.; Galdino, G.S.; Fregni, F.; Massetti, T.; Fontes, S.L.; de Oliveira Silva, D.; da Silva, T.D.; Monteiro, C.B.D.M.; Tonks, J.; et al. Effects of High-Frequency Transcranial Magnetic Stimulation on Functional Performance in Individuals with Incomplete Spinal Cord Injury: Study Protocol for a Randomized Controlled Trial. *Trials* **2017**, *18*, 522. [CrossRef]

33. Chail, A.; Saini, R.; Bhat, P.; Srivastava, K.; Chauhan, V. Transcranial Magnetic Stimulation: A Review of Its Evolution and Current Applications. *Ind. Psychiatry J.* **2018**, *27*, 172. [PubMed]

34. Arora, T.; Desai, N.; Kirshblum, S.; Chen, R. Utility of Transcranial Magnetic Stimulation in the Assessment of Spinal Cord Injury: Current Status and Future Directions. *Front. Rehabil. Sci.* **2022**, *3*, 1005–1111. [CrossRef] [PubMed]

35. Perez, M.A.; Field-Fote, E.C.; Floeter, M.K. Patterned Sensory Stimulation Induces Plasticity in Reciprocal Ia Inhibition in Humans. *J. Neurosci.* **2003**, *23*, 2014–2018. [CrossRef]

36. Wincek, A.; Wietrzak, P.; Mielewczyk, Z.; Kędzia, G.; Mieczyński, K.; Leszczyńska, K. Review of effective nerve electrostimulation parameters applied for treatment of patients after incomplete spinal cord injuries. *Issue Rehabil. Orthop. Neurophysiol. Sport Promot.* **2020**, *31*, 25–33.

37. Brihmat, N.; Allexandre, D.; Saleh, S.; Zhong, J.; Yue, G.H.; Forrest, G.F. Stimulation Parameters Used During Repetitive Transcranial Magnetic Stimulation for Motor Recovery and Corticospinal Excitability Modulation in SCI: A Scoping Review. *Front. Hum. Neurosci.* **2022**, *16*, 800349. [CrossRef] [PubMed]

38. Balbinot, G. Surface EMG in Subacute and Chronic Care after Traumatic Spinal Cord Injuries. *Trauma Care* **2022**, *2*, 381–391. [CrossRef]

39. Korupolu, R.; Stampas, A.; Singh, M.; Zhou, P.; Francisco, G. Electrophysiological Outcome Measures in Spinal Cord Injury Clinical Trials: A Systematic Review. *Top. Spinal Cord Inj. Rehabil.* **2019**, *25*, 340–354. [CrossRef]

40. Zholudeva, L.V.; Qiang, L.; Marchenko, V.; Dougherty, K.J.; Sakiyama-Elbert, S.E.; Lane, M.A. The Neuroplastic and Therapeutic Potential of Spinal Interneurons in the Injured Spinal Cord. *Trends Neurosci.* **2018**, *41*, 625–639. [CrossRef]

41. Gordon, T. Electrical Stimulation to Enhance Axon Regeneration After Peripheral Nerve Injuries in Animal Models and Humans. *Neurotherapeutics* **2016**, *13*, 295–310. [CrossRef]

42. Grau, J.W. Learning from the Spinal Cord: How the Study of Spinal Cord Plasticity Informs Our View of Learning. *Neurobiol. Learn. Mem.* **2014**, *108*, 155–171. [CrossRef]

43. Wise, Y. Electrical Stimulation and Motor Recovery. *Cell Transplant.* **2015**, *24*, 429–446.

44. Curt, A.; Schwab, M.E.; Dietz, V. Providing the Clinical Basis for New Interventional Therapies: Refined Diagnosis and Assessment of Recovery after Spinal Cord Injury. *Spinal Cord* **2004**, *42*, 1–6. [CrossRef] [PubMed]

45. Curt, A.; Van Hedel, H.J.A.; Klaus, D.; Dietz, V.; EM-SCI Study Group. Recovery from a Spinal Cord Injury: Significance of Compensation, Neural Plasticity, and Repair. *J. Neurotrauma* **2008**, *25*, 677–685. [CrossRef] [PubMed]

46. Hou, J.; Xiang, Z.; Yan, R.; Zhao, M.; Wu, Y.; Zhong, J.; Guo, L.; Li, H.; Wang, J.; Wu, J.; et al. Motor Recovery at 6 Months after Admission Is Related to Structural and Functional Reorganization of the Spine and Brain in Patients with Spinal Cord Injury: Neural Correlates of Motor Recovery After SCI. *Hum. Brain Mapp.* **2016**, *37*, 2195–2209. [CrossRef]

47. Kubaszewski, Ł.; Wojdasiewicz, P.; Rożek, M.; Słowińska, I.E.; Romanowska-Próchnicka, K.; Słowiński, R.; Poniatowski, Ł.A.; Gasik, R. Syndromes with chronic non-bacterial osteomyelitis in the spine. *Reumatologia* **2015**, *53*, 328–336. [CrossRef]

48. Tykocki, T.; Poniatowski, Ł.A.; Czyz, M.; Wynne-Jones, G. Oblique corpectomy in the cervical spine. *Spinal Cord* **2018**, *56*, 426–435. [CrossRef]

49. De Miguel-Rubio, A.; Muñoz-Pérez, L.; Alba-Rueda, A.; Arias-Avila, M.; Rodrigues-de-Souza, D.P. A Therapeutic Approach Using the Combined Application of Virtual Reality with Robotics for the Treatment of Patients with Spinal Cord Injury: A Systematic Review. *Int. J. Environ. Res. Public Health* **2022**, *19*, 8772. [CrossRef]

Article

Combined Transcutaneous Electrical Spinal Cord Stimulation and Task-Specific Rehabilitation Improves Trunk and Sitting Functions in People with Chronic Tetraplegia

Niraj Singh Tharu [1], Monzurul Alam [1,*], Yan To Ling [1], Arnold YL Wong [2,3] and Yong-Ping Zheng [1,3,*]

[1] Department of Biomedical Engineering, The Hong Kong Polytechnic University, Hong Kong SAR, China
[2] Department of Rehabilitation Sciences, The Hong Kong Polytechnic University, Hong Kong SAR, China
[3] Research Institute for Smart Ageing, The Hong Kong Polytechnic University, Hong Kong SAR, China
* Correspondence: md.malam@connect.polyu.hk (M.A); yongping.zheng@polyu.edu.hk (Y.-P.Z.)

Abstract: The aim of this study was to examine the effects of transcutaneous electrical spinal cord stimulation (TSCS) and conventional task-specific rehabilitation (TSR) on trunk control and sitting stability in people with chronic tetraplegia secondary to a spinal cord injury (SCI). Five individuals with complete cervical (C4–C7) cord injury participated in 24-week therapy that combined TSCS and TSR in the first 12 weeks, followed by TSR alone for another 12 weeks. The TSCS was delivered simultaneously at T11 and L1 spinal levels, at a frequency ranging from 20–30 Hz with 0.1–1.0 ms. pulse width biphasically. Although the neurological prognosis did not manifest after either treatment, the results show that there were significant increases in forward reach distance (10.3 ± 4.5 cm), right lateral reach distance (3.7 ± 1.8 cm), and left lateral reach distance (3.0 ± 0.9 cm) after the combinational treatment (TSCS+TSR). The stimulation also significantly improved the participants' trunk control and function in sitting. Additionally, the trunk range of motion and the electromyographic response of the trunk muscles were significantly elevated after TSCS+TSR. The TSCS+TSR intervention improved independent trunk control with significantly increased static and dynamic sitting balance, which were maintained throughout the TSR period and the follow-up period, indicating long-term sustainable recovery.

Keywords: transcutaneous electrical spinal cord stimulation; trunk control; sitting balance; tetraplegia; spinal cord injury

Citation: Tharu, N.S.; Alam, M.; Ling, Y.T.; Wong, A.Y.; Zheng, Y.-P. Combined Transcutaneous Electrical Spinal Cord Stimulation and Task-Specific Rehabilitation Improves Trunk and Sitting Functions in People with Chronic Tetraplegia. *Biomedicines* **2023**, *11*, 34. https://doi.org/10.3390/biomedicines11010034

Academic Editor: Nicolas Guerout

Received: 5 December 2022
Revised: 19 December 2022
Accepted: 20 December 2022
Published: 23 December 2022

1. Introduction

Spinal cord injury (SCI) often results in a permanent loss of motor, sensory, or autonomic functions below the level of injury [1,2]. The worldwide annual incidence and prevalence of SCI range from 8 to 246 and from 236 to 1298 per million people, respectively [3]. Depending on the level of neurological damage, SCI may cause tetraplegia or paraplegia [4]. One-third of people with SCI are found to have tetraplegia, and 50% of them have a complete lesion [5], with C5 (cervical vertebrae) being the most commonly affected level [6]. People with tetraplegia experience severe trunk muscle weakness that inhibits sitting and trunk functions, thus restricting their overall functional activity [7]. In addition, individuals with complete or incomplete tetraplegia have greater trunk impairment, which results in more trunk instability and sitting imbalance than people with paraplegia [8,9], and often struggle to perform the daily functions of sitting [10,11]. They are also more prone to postural instability and fall-related injuries [12]. For people with tetraplegia, trunk stability is more important than trunk mobility [13]. Individuals with tetraplegia report significantly more limitations and restrictions in their activities of daily living (ADL) than people with paraplegia [14]. Similarly, more than 60% of SCI survivors consider trunk function to be an important factor for mobility and performing ADL [15]. Around 70–80%

of people with SCI are wheelchair-bound, and consider trunk control an essential factor for those with sitting difficulties [16].

Traditionally, SCI management focused on teaching compensatory skills. Recently, this approach has gradually shifted toward neural restoration through the use of innovative, intensive therapy strategies that successfully address physiological changes and regeneration concepts [17,18]. Of various approaches (activity-based therapy, impairment-based training, etc.), task-specific rehabilitation (TSR) has been shown to be more beneficial than conventional rehabilitation for motor recovery in people with chronic SCI [19,20]. "The TSR is a therapeutic approach that involves intensive practice of actions or functional tasks" [21], which is based on the idea that motor output can be shaped and retrained in response to specific sensory inputs [22]. It focuses on regaining muscular strength and improving functional abilities through targeted and repeated specific exercises [23]. Motor or sensory recovery is often restricted to minimal success in people with tetraplegia [24,25]. In addition, TSR has been observed to be advantageous for improving upper limb functions [26]. The locomotion ability was reported to be enhanced in individuals with incomplete SCI using TSR strategies [27,28]. Although TSR has shown some benefits over conventional rehabilitation, more studies are needed to confirm its efficacy [29,30].

It used to be believed that people with complete SCI would not experience motor recovery below the lesion [31]. Yet prior research has indicated that individuals with complete or incomplete SCI showed motor and sensory improvement with neuromodulation treatment [32]. In addition, transcutaneous electrical spinal cord stimulation (TSCS) has recently emerged as a potential treatment for SCI, among other neuromodulation techniques [33,34]. It modulates the activity of neural pathways through spinal stimulation to generate therapeutic benefits [35]. The TSCS produces an electric field that triggers the neural connectome via sensory pathways in the dorsal roots, which provides subthreshold excitation to interneurons and motor neurons distal to the injury. Motor neurons near the threshold are then more rapidly triggered and produce volitional movement [36]. Previous studies indicated that TSCS may decrease spasticity [37], modify neuronal connections [38], assist in locomotion [39] and stepping [40], and initiate volitional movement [41]. Moreover, TSCS produced immediate trunk self-control in people with chronic SCI with a more steady and upright sitting position [12], and quickly restored the ability to sit up straight [42]. However, these studies on trunk control have only been conducted on people with paraplegia or incomplete cervical SCI [11,43]. More research is warranted to investigate trunk recovery in people with complete tetraplegia [42].

The current study aimed to investigate the effects of adding TSCS to TSR treatment on trunk control and sitting function in people with tetraplegia following complete cervical SCI. To achieve this aim, two specific objectives were set to investigate the (1) efficacy of TSCS for improving trunk control and sitting function with TSR in people with tetraplegia; and (2) effect of TSCS on motor and sensory functions of the study participants. We hypothesized that the combined treatment (TSCS+TSR) could improve trunk and sitting functions in people with complete chronic tetraplegia. To our knowledge, this is the first study to report the use of TSCS for trunk control in people with complete cervical SCIs; the outcomes may be applicable to people with various levels of neurological damage caused by SCI.

2. Methods

The study was approved by the Human Subjects Ethics Sub-Committee of The Hong Kong Polytechnic University (Reference no: HSEARS20190201002-01).

2.1. Study Participants

Five individuals with chronic SCI with impaired trunk and sitting function were recruited. Each participant had sustained a traumatic complete cervical SCI (Table 1). Participant consent was obtained prior to the start of the experiment.

Table 1. The demographic and clinical characteristics of the participants.

Participants	Age	Gender	Time since Injury (Years)	Type of Injury	NLI	AIS Category
P1	57	F	1.5	Traumatic	C6	A
P2	55	F	19	Traumatic	C7	A
P3	26	F	12	Traumatic	C5	A
P4	40	M	12	Traumatic	C5	A
P5	32	M	2	Traumatic	C4	A

NLI, neurological level of injury; AIS, American Spinal Injury Association Impairment Scale.

As shown in Table 1, the mean age of the participants was 42.0 years (SD = 13.7 years, range = 26–57 years), and the mean time since injury was 9.3 years (SD = 7.4 years, range = 1.5–19 years). Each participant had a traumatic cervical SCI based on the American Spinal Injury Association Impairment Scale (AIS), and the injury level varied from C4–C7. The inclusion criteria were people with SCI aged between 18 and 60 years; a cervical SCI resulting in complete tetraplegia; injury duration of at least 1 year; an injury level between C4–C8; absence of independent trunk control; being unable to sit independently; stable respiratory functions; and no prior neuromodulation treatment. People were excluded if they had received injections (Botox or Dysport) within the previous 6 months, had spasticity (more than grade 1⁺ based on the Modified Ashworth Scale), contracture, pressure injuries, infections, or internal fixations at the site of the injury, transplants (cardiac pacemakers and defibrillators), or had comorbidities (e.g., asthma, hypertension).

2.2. Study Design

This was a case series study design where people with chronic tetraplegia underwent two phases of treatment (first TSCS+TSR followed by TSR alone). The participants were recruited through a convenience sampling method. The treatment frequency was divided into three groups depending on the availability and schedules of the study participants: (a) three sessions per week; (b) two sessions per week; and (c) one session per week. In addition, participants P1, P2, and P4 attended three sessions per week, whereas P3 and P5 attended one session and two sessions per week, respectively. The outline of the study is described in Figure 1.

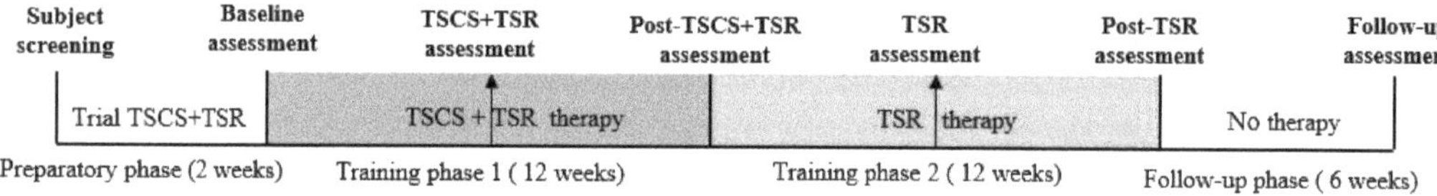

Figure 1. The outline of the study, divided into four phases: (i) The preparatory phase, in which participants were enrolled and screened to ensure their eligibility for the study. During this period, their response to TSCS was tested and optimal stimulation settings were determined. The visual analog scale was used to record their response to the stimulation. After completing a baseline assessment over a period of 2 weeks, the participants received two phases of therapy, each lasting for 12 weeks, with two half-way assessments completed every 6 weeks. (ii) Training phase 1, transcutaneous electrical spinal cord stimulation (TSCS) with task-specific rehabilitation (TSR); (iii) training phase 2, TSR alone; and (iv) the follow-up phase, during which no therapy was performed, and assessments were repeated 6 weeks following the completion of the TSR training period.

2.3. Experimental Protocol

The 12th free rib was palpated and followed to identify the T11 and T12 vertebrae. The iliac crest was palpated and followed to the posterior superior iliac spine, and then to the L1 and L2 levels. Two stimulation electrodes were then attached between T11–T12 and L1–L2 spinous processes in the middle, targeting the spinal cord (hereinafter called T11 and L1 electrodes), while another two referencing electrodes were placed above the iliac crests bilaterally. For the active electrodes, we utilized a pair of self-adhesive electrodes with a size of 3.2 cm (ValuTrode, Axelgaard Manufacturing Co., Ltd., Fallbrook, CA, USA) and another pair of internally linked 6.0×9.0 cm self-adhesive rectangle-shaped electrodes (Guangzhou Jetta Electronic Medical Device Manufacturing Co., Ltd., Guangdong, China) as ground electrodes. The T11 and L1 regions of the study participants were stimulated using two specifically designed constant current stimulators (DS8R, Digitimer, UK). In order to activate the stimulators, a function generator (AFG1022, manufactured by Tektronix, Inc., Beaverton, OR, USA) was used to produce a burst of 10 kHz, which was transmitted at a frequency of 20–30 Hz [41]. The burst configuration was raised to 10 biphasic pulses (1.06 ms burst duration, henceforth referred to as 1 ms), and the pulse lengths of each cycle were maintained constant at 50 µs for both devices. The previous study of [41] showed that a shorter pulse of 0.5 ms was more effective in improving functional movements. In addition, it has been reported that short pulse durations (0.05–0.4 ms) preferentially activate motor axons, while the use of longer pulse durations (0.5–1 ms) activates sensory axons [43]. Therefore, the short pulse duration of 0.05 msec. was chosen in the current study.

To determine the optimal stimulation intensity, the investigators used the Visual Analog Scale to measure the pain intensity. The participants were asked to experience involuntary extension of the trunk (straightening of the spine), and asked to feel when the trunk became more stable in the presence of the stimulation. Based on the individual participant's response, he/she experienced the best trunk stability and tolerable stimulation intensity, which ranged between 95–115 milliamperes (mA), depending on each individual's response. The other stimulation settings (such as frequency, carrier frequency, and burst configuration) were kept constant for all the participants throughout the experiment. The stimulation intensity varied between individual participants, which was set based on their response as described above. During every treatment session, the participant was asked to experience trunk extension with increased trunk stability in sitting as the intensity was gradually increased. A previous study demonstrated that stimulation of the rostral portion of the lumbar enlargement (corresponding to the T11–T12 vertebral level) at a frequency of 30 Hz could specifically facilitate voluntary movements, whereas stimulation delivered over the caudal area of the lumbar enlargement (corresponding to the L1–L2 vertebral level) at a frequency of 15 Hz would facilitate tonic extensor activity specific for postural control [12]. The 10 kHz carrier frequency was found to be suitable for stimulating the spinal circuits of both injured and uninjured individuals [44]. Therefore, the above-mentioned parameters were adopted for the stimulation settings. The TSR included spinal mobility exercises such as flexion, extension, lateral flexion, and rotation, as well as static and dynamic seated balancing exercises. The participants were trained in a variety of experimental settings and positions, ranging from sitting in a wheelchair to lying on a bed to lying on a floor mat, with the help and supervision of a physiotherapist.

For combinational treatment, after the placement of stimulating electrodes, the stimulation parameters were set and the intensity gradually increased until the individual's maximum tolerance was reached. Then the participant was asked to perform various task-specific exercises. Each session lasted for 45–60 min, divided into three sub-sessions of 15–20 min each. A break was given after each sub-session, and the stimulation was turned off during the rest period. The participants were stimulated throughout each session, except during the break. Because all participants in the present study had complete tetraplegia, TSCS+TSR was delivered in the first 12 weeks to determine whether the combinational treatment could improve trunk and sitting functions. Another 12 weeks of TSR alone was followed to determine whether the functional gains could be maintained.

2.4. Outcome Measures

The functional assessment was conducted using the following outcome measures: (1) Modified Functional Reach Test (mFRT) to determine the functional reach distance; (2) Trunk Control Test (TCT) to assess static and dynamic balance (3) Function in Sitting Test (FIST) to measure functional sitting balance; and (4) International Standards for Neurological Classification of Spinal Cord Injury (ISNCSCI) to assess sensory and motor levels on both the right and left sides, the neurological degree of damage, and the completeness of the injury. The mFRT [45], TCT [46], and FIST [47] have proven reliability and validity for assessing people with SCI. The TCT comprises 0–24 points and FIST 0–56 points, where a higher score indicates improvement in function. Additionally, the trunk kinematics (i.e., range of motion, ROM) were assessed using a motion capture system (Vicon Nexus 2.5.1, Vicon Nexus TM, Vicon Motion Systems Ltd., Yarnton, UK) (ROM), while surface electromyography (EMG) (Model DE-2.1; Delsys USA, Inc., Boston, MA, USA) was used to the measure muscle activity of four pairs of trunk muscles (rectus abdominis, external oblique, erector spinae, and latissimus dorsi).

2.5. Vicon Marker and EMG Electrode Placement

The placement of the Vicon reflective markers and EMG electrodes was performed by an experienced physiotherapist. Specific bony landmarks were identified, and Vicon markers were placed as follows (Figure 2A): anteriorly, at the acromia, sternum, right and left anterior superior iliac spine (ASIS), and right and left patella; posteriorly, at the cervical spinous process (C7), thoracic spinous process (T3, T8, T12), lumbar spinous process (L2, L4), sacrum spinous process (S1), and right and left posterior superior iliac spine (PSIS) [12,48]. The EMG electrodes over the rectus abdominis muscle were positioned 5 cm below the xiphoid process, while for the external oblique they were placed 5 cm superior to the anterior superior iliac spine and 10 cm lateral to the umbilicus. Likewise, electrodes for the latissimus dorsi were placed 2 cm inferior and lateral to the scapula's inferior angle, while those for the external oblique were placed 3 cm lateral to the L3 spinous process [49]. The tested movements included trunk flexion, trunk extension, bilateral trunk lateral flexion, and bilateral trunk rotation. Each movement was performed thrice, and the best one was selected for data analysis. All assessments were conducted without using TSCS. The placement of Vicon markers and a flexion movement performed in the assessment are shown in Figure 2.

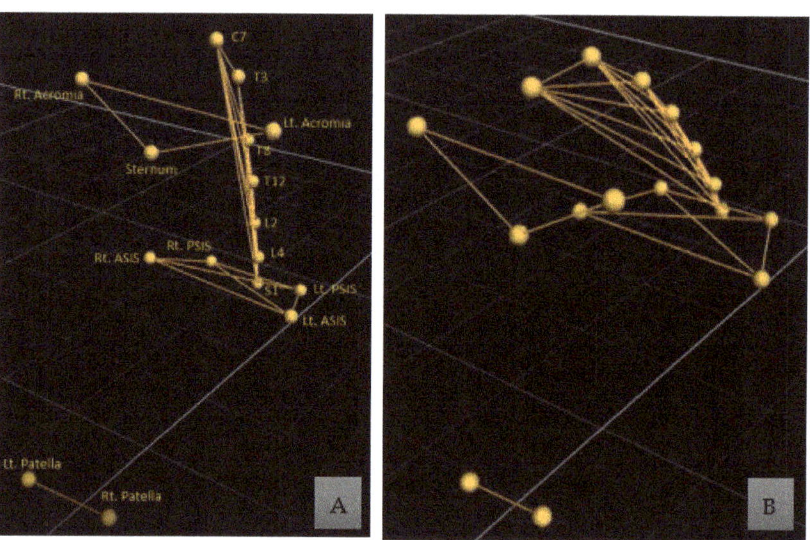

Figure 2. (**A**) Vicon markers positioned over the anatomical bony landmarks of a participant in a normal straight sitting position; (**B**) a flexion movement performed during the assessment.

2.6. Data Processing and Analysis

The functional outcome scores were graphically and statistically analyzed through GraphPad Prism version 9.0. Similarly, the kinesiologic and electrophysiologic data acquired from Vicon and EMG were extracted, processed, and analyzed using MATLAB (version 2016a, The MathWorks Inc., Natick, MA, USA). The sampling frequency was 2000 Hz and the filters applied were 10 Hz to 40 Hz during data collection. Different cutoff frequencies from 10 Hz to 50 Hz had been trialed, and 40 Hz was found to be the best based on visual inspection. The descriptive analysis (mean $\pm$ SD) was calculated for each phase. The trunk ROM angles were calculated by connecting selected markers at specific landmarks (Figure 3). A Friedman one-way repeated measures ANOVA with multiple comparisons was used to compare temporal differences in functional assessment (mFRT, TCT, and FIST) and ROM among the baseline, post-TSCS+TSR, post-TSR, and the final follow-up. The root mean square (RMS) value of EMG of each muscle was calculated in microvolts (μV), and the above-mentioned repeated measures of ANOVA was used to show significant improvements. In addition, Spearman correlation analysis was performed to determine the correlation between these assessment parameters (mFRT, TCT, FIST, EMG, and ROM). A p–value of 0.05 was set as the statistical level of significance.

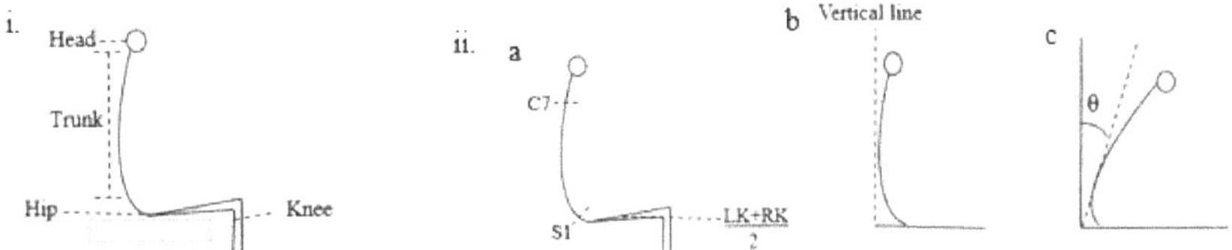

Figure 3. The principle of data analysis process of Vicon data for trunk flexion movement: (**i**) The erect sitting posture with knees in 90 degrees flexion and feet flat on the floor. For measuring flexion angle (**ii-a**) segments C7, S1, and the midline connecting the left knee (LK) and right knee (RK), (**ii-b**) an axis connecting the C7, S1 segments with a perpendicular line through the middle of the knees, (**ii-c**) flexion movement was performed, and angle measured in degrees (θ). The participant was allowed to take the required time to successfully and independently complete the movement. Similarly, measurements for other movements are presented in the Supplementary Materials (Supplementary Figure S1).

3. Results

3.1. Combinational Treatment Initiated Functional Improvements in Reaching, Trunk Control, and Sitting Balance

As shown in Figure 4, the overall (mean $\pm$ SD) forward reach distance (FRD) was 2.0 $\pm$ 1.6 cm at the baseline, which increased by 10.3 $\pm$ 4.5 cm after TSCS+TSR (to 12.3 $\pm$ 6.1 cm), and further slightly raised by 1.4 $\pm$ 0.7 cm during TSR (to 13.7 $\pm$ 6.8 cm), which was maintained throughout the follow-up period in the absence of any intervention (to 13.4 $\pm$ 6.9 cm). Statistical analysis showed significant improvements between the baseline and TSCS+TSR ($p = 0.025$), TSR ($p = 0.024$), and follow-up ($p = 0.026$), respectively. For right lateral reach distance, the baseline mean $\pm$ SD was 0.9 $\pm$ 0.7 cm, which increased by 3.7 $\pm$ 1.8 cm after the introduction of TSCS+TSR (to 4.6 $\pm$ 2.6 cm), followed by a further increment of 1.2 $\pm$ 0.5 cm during the TSR training period (to 5.8 $\pm$ 3.0 cm), and then reduced to 5.7 $\pm$ 3.0 cm at the final follow-up. Significant improvements were found between the baseline and TSCS+TSR ($p = 0.037$), TSR ($p = 0.029$), and follow-up ($p = 0.030$), respectively. At the baseline, the left lateral reach distance was 1.0 $\pm$ 0.8 cm, which increased to 4.0 $\pm$ 1.7 cm after TSCS+TSR, and further increased to 4.5 $\pm$ 1.96 cm throughout TSR, and then remained constant during the follow-up period (4.5 $\pm$ 2.0 cm). Significant improvements were observed between the results of the baseline and TSCS+TSR ($p = 0.014$), TSR ($p = 0.016$), and follow-up ($p = 0.017$), respectively.

As demonstrated in Figure 5A, the overall mean TCT score was 3.0 $\pm$ 0.7 at the baseline, which significantly increased to 11.6 $\pm$ 3.4 after TSCS+TSR administration, which increased further by 1.4 $\pm$ 1.1 during TSR to 13.0 $\pm$ 4.5, followed by a slight reduction to 12.8 $\pm$ 4.1 at the follow-up period. All these values of TCT were significantly greater than the baseline values ($p < 0.01$). Likewise, the overall mean FIST score (Figure 5B) was 12.6 $\pm$ 4.5 at the baseline, which underwent the greatest increase, by 29.0 $\pm$ 8.8, after TSCS+TSR, which further rose to 31.0 ± 9.7 during TSR, with a minor decrease of 0.4 ± 1.3 at the follow-up period (to 30.6 ± 8.4). All the FIST scores at the follow-ups were significantly larger than the baseline values ($p < 0.01$). However, there was no significant difference in the TCT or FIST scores between TSCS+TSR and TSR, between TSCS+TSR and follow-up, or between TSR and follow-up.

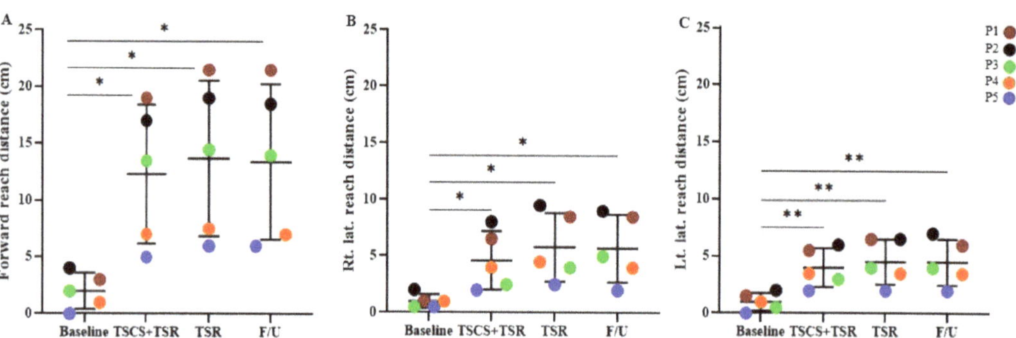

Figure 4. The increased functional scores of mFRT measured from the baseline. Statistical analysis was conducted between the baseline with TSCS+TSR, TSR, and follow-up. (**A**) Forward reach distance, (**B**) right lateral reach distance, and (**C**) left lateral reach distance, respectively, measured for each participant during the study. mFRT, Modified Functional Reach Test; TSCS, transcutaneous electrical spinal cord stimulation; TSR, task-specific rehabilitation; F/U, follow-up; * $p < 0.05$, ** $p < 0.01$, all in comparison to the baseline value.

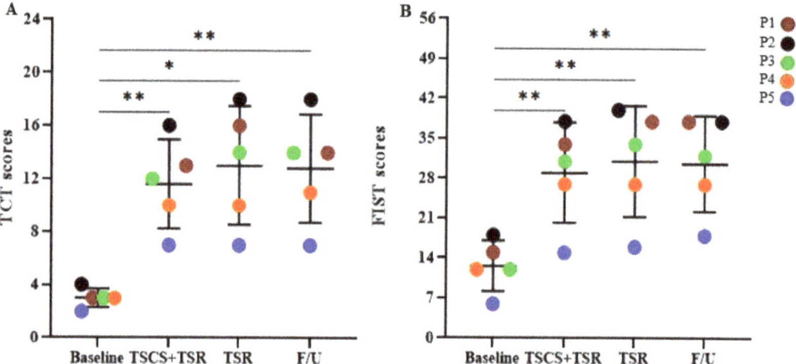

Figure 5. The increased functional scores of (**A**) TCT (0–24 points) and (**B**) FIST (0–56 points) measured from the baseline for each participant during the study. Statistical analysis was conducted between the baseline with TSCS+TSR, TSR, and follow-up. TCT, Trunk Control Test; FIST, Function in Sitting Test; TSCS, transcutaneous electrical spinal cord stimulation; TSR, task-specific rehabilitation; F/U, follow-up; * $p < 0.05$, ** $p < 0.01$, all in comparison to the baseline value.

3.2. Increased Trunk Range of Motion after TSCS+TSR Treatment

Figure 6 shows that trunk flexion, extension, left lateral flexion, and bilateral rotation ROM at TSCS+TSR, TSR, and the final follow-up were significantly higher than the respective baseline values. However, the right trunk lateral flexion underwent significant improvement in ROM only after TSCS+TSR. Specifically, the overall trunk flexion ROM was 12.2° ± 4.7° at the baseline, which significantly increased to 23.1° ± 9.0° after TSCS+TSR, followed by a 3.7° ± 0.4° reduction in motion after TSR (to 19.4° ± 9.37°), which further decreased by 0.3° ± 1.7° to 19.1° ± 7.7° at the final follow-up. The trunk extension ROM at the baseline was 5.7° ± 2.0°, which was significantly increased to 12.4° ± 4.5° after TSCS+TSR, then back to $10.0° \pm 4.0°$, and then increased to $10.4° \pm 3.9°$ at the follow-up. Likewise, the trunk right lateral flexion was $5.8° \pm 5.6°$ at the baseline, which significantly increased to $9.1° \pm 5.4°$ after TSCS+TSR, followed by a decrease to $8.4° \pm 5.3°$ after TSR, and then to $8.7° \pm 5.1°$ at the follow-up. At the baseline, the left lateral flexion ROM increased from $6.0° \pm 2.8°$ to $9.8° \pm 2.9°$ after TSCS+TSR, and then decreased to $8.8° \pm 2.3°$ after TSR, and further dropped to $8.2° \pm 2.6°$ at the follow-up. The right rotation increased from $1.7° \pm 2.3°$ at the baseline to $4.5° \pm 2.7°$ after the TSCS+TSR, then dropped to $3.0° \pm 2.2°$ after TSR, followed by an increase to $4.1° \pm 1.9°$ at the follow-up. The mean $\pm$ SD for left rotation was $18.4° \pm 10.1°$ at the baseline, which increased by $21.2° \pm 3.3°$ to $39.6° \pm 13.4°$ after TSCS+TSR, followed by a decrease of $4.3° \pm 0.2°$ to $35.3° \pm 13.2°$ after TSR, and further reduction to $31.8° \pm 8.2°$ at the follow-up. However, no significant difference was observed in any trunk ROMs between TSCS+TSR and TSR, between TSCS+TSR and follow-up, or between TSR and follow-up.

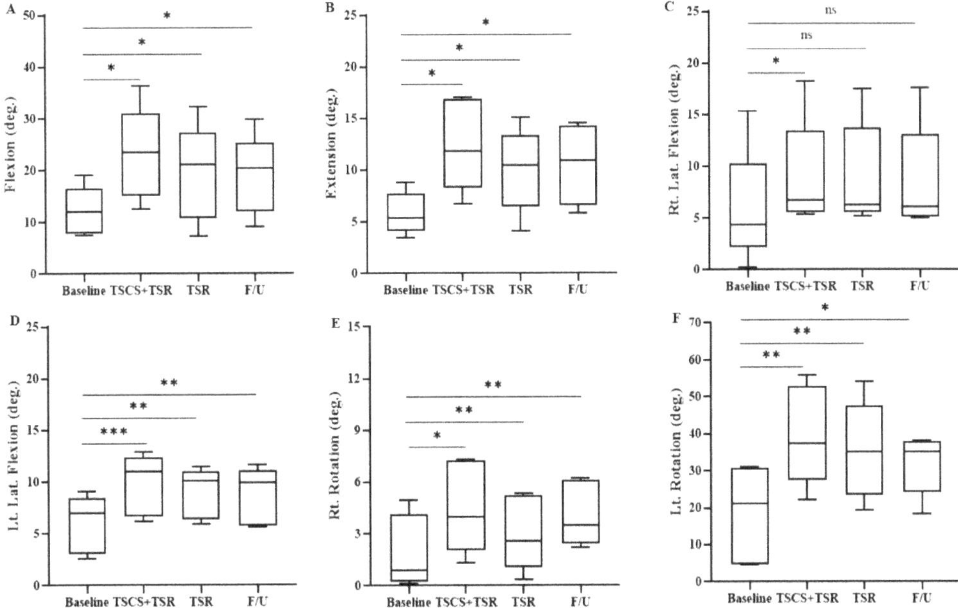

Figure 6. The trunk range of motion measured with Vicon for each participant during the study. Statistical analysis was conducted between the baseline with TSCS+TSR, TSR and follow-up. (**A**) Flexion, (**B**) extension, (**C**) right lateral flexion, (**D**) left lateral flexion, (**E**) right rotation, and (**F**) left rotation, respectively. TSCS, transcutaneous electrical spinal cord stimulation; TSR, task-specific rehabilitation; F/U, follow-up; $*$ $p < 0.05$, $**$ $p < 0.01$, $***$ $p < 0.001$, non-significant (ns) $p > 0.05$, all in comparison to the baseline value.

3.3. Elevated Electromyographic Response of the Trunk Muscles

The right latissimus dorsi (Rt LD) and left latissimus dorsi (Lt LD), as well as the right erector spinae (Rt ES) and left erector spinae (Lt ES) during extension, and the right external oblique (Rt EO) and left external oblique (Lt EO) during left rotation, exhibited the significantly highest EMG amplitude at TSCS+TSR and TSR than the respective baseline values in comparison to other trunk ROMs, i.e., flexion, bilateral lateral flexion, and right rotation. However, there was no significant difference observed in the EMG response at follow-up for Rt LD or Lt LD during extension, as well as for Rt EO and Lt EO for right rotation and left rotation, and Rt ES during left lateral flexion, respectively, compared with the baseline values (Table 2). The EMG responses of other trunk ROMs (excluding extension and left rotation) are presented in the Supplementary Materials (Supplementary Table S1). Specifically, the

EMG response for extension was 2.20 ± 1.60 µV for Rt LD and 2.57 ± 1.81 µV for Lt LD at the baseline, which was significantly increased to 8.86 ± 6.04 µV (Rt LD) and 9.94 ± 6.70 µV (Lt LD), respectively, after TSCS+TSR, which further reduced to 5.01 ± 3.71 µV (Rt LD) and 5.70 ± 4.86 µV (Lt LD) after TSR. In addition, the EMG values remained almost unchanged after the follow-up for Rt LD (5.02 ± 3.03 µV) and Lt LD (6.07 ± 4.45 µV) in comparison with the baseline value, similar for the subsequent descriptions. Likewise, the EMG values during extension were 1.62 ± 0.95 µV (Rt ES) and 1.79 ± 1.25 µV (Lt ES) at the baseline, and increased to 6.93 ± 6.32 µV (Rt ES) and 7.53 ± 5.47 µV (Lt ES), respectively, after TSCS+TSR, followed by a further decrement to 4.11 ± 2.68 µV (Rt ES) and 4.57 ± 3.02 µV (Lt ES) after TSR. Additionally, the EMG amplitude remained almost unchanged after the follow-up for Rt ES (4.19 ± 2.58 µV) and Lt ES (4.69 ± 2.97 µV). For left rotation, the EMG response was 1.55 ± 0.93 µV for Rt EO and 2.07 ± 1.17 µV for Lt EO at the baseline, which significantly increased to 6.86 ± 3.94 µV (Rt EO) and 13.47 ± 7.49 µV (Lt EO) after TSCS+TSR, respectively, and further reduced to 6.06 ± 3.38 µV (Rt EO) and 12.06 ± 6.73 µV (Lt EO) after TSR, which remained almost consistent after the follow-up period for Rt EO (5.84 ± 3.40 µV) and Lt EO (11.19 ± 6.23 µV), respectively. Furthermore, no significant difference was observed in EMG amplitudes for any trunk ROMs between TSCS+TSR and TSR, between TSCS+TSR and follow-up, or between TSR and follow-up.

Table 2. The responses recorded from trunk muscles measured through EMG.

(A)

	Extension							
	Rt. ES		Lt. ES		Rt. LD		Lt. LD	
Study Timeline	Mean $\pm$ SD	p-Value	Mean $\pm$ SD	p-Value	Mean $\pm$ SD	p-Value	Mean $\pm$ SD	p-Value
Baseline	1.62 ± 0.95	-	1.79 ± 1.25	-	2.20 ± 1.60	-	2.57 ± 1.81	-
TSCS+TSR	6.93 ± 6.32	*	7.53 ± 5.47	*	8.86 ± 6.04	***	9.94 ± 6.70	*
TSR	4.11 ± 2.68	***	4.57 ± 3.02	***	5.01 ± 3.71	***	5.70 ± 4.86	*
Follow-up	4.19 ± 2.58	***	4.69 ± 2.97	***	5.02 ± 3.03	ns	6.07 ± 4.45	***

(B)

	Left rotation			
	Rt. EO		Lt. EO	
Study timeline	Mean $\pm$ SD	p-Value	Mean $\pm$ SD	p-Value
Baseline	1.55 ± 0.93	-	2.07 ± 1.17	-
TSCS+TSR	6.86 ± 3.94	*	13.47 ± 7.49	*
TSR	6.06 ± 3.38	*	12.06 ± 6.73	*
Follow-up	5.84 ± 3.40	*	11.19 ± 6.23	*

LD, latissimus dorsi; ES, erector spinae; EO, external oblique; Rt, right; Lt, left; Lat, lateral; SD, standard deviation; TSCS, transcutaneous electrical spinal cord stimulation; TSR, task-specific rehabilitation; * $p < 0.05$, *** $p < 0.001$, non-significant (ns) $p > 0.05$, all in comparison with the baseline value.

3.4. Improvements of Trunk Control and Function in Sitting after Combinational Treatment

The functional improvements of trunk control and function in sitting were correlated with each other (Figure 7), demonstrating a strong positive correlation between the TCT and FIST after TSCS+TSR ($R^2 = 0.916$, $p = 0.001$). Similarly, the TCT and mFRT ($R^2 = 0.774$, $p = 0.017$), TCT and EMG ($R^2 = 0.743$, $p = 0.009$), EMG and ROM ($R^2 = 0.626$, $p = 0.004$), and FIST and EMG ($R^2 = 0.746$, $p = 0.0125$), respectively, exhibited mild positive correlations after the TSCS+TSR. Conversely, FIST and mFRT ($R^2 = 0.305$, $p = 0.009$), EMG and mFRT ($R^2 = 0.233$, $p = 0.047$), and TSR and TSCS+TSR ($R^2 = 0.217$, $p = 0.094$) displayed weak correlations following TSCS+TSR.

3.5. Treatment Effect on Sensorimotor Recovery

The ISNCSCI underwent some changes in the sensorimotor scores, but the neurological level of injury and AIS remained unchanged (Supplementary Table S2). Figure 8 shows that for ISNCSCI, a participant (P1) revealed an increase of 8 points and 18 points in response to light touch and pinprick sensation (68/64 to 76/82), respectively, while P3 showed a 4-point elevation in response to pinprick sensation (64/64 to 64/68) after TSCS+TSR. Moreover, throughout TSR and the follow-up period, the increased motor and sensory scores remained unchanged. However, over the entire study period, there was no change in ISNCSCI scores for P2, P4, or P5.

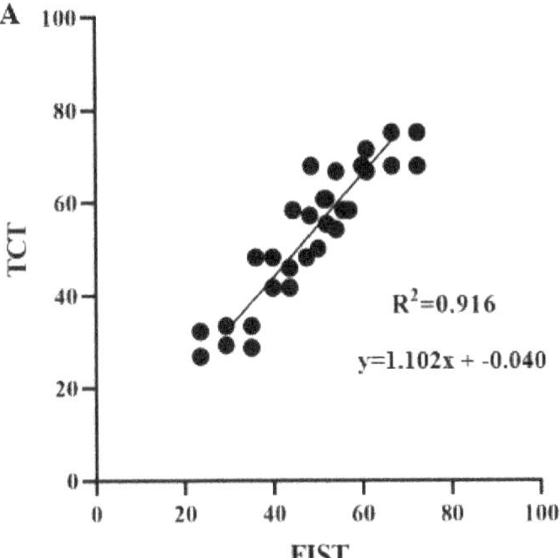

Figure 7. The correlation between the functional improvements regarding the Trunk Control Test (TCT) and Function in Sitting Test (FIST).

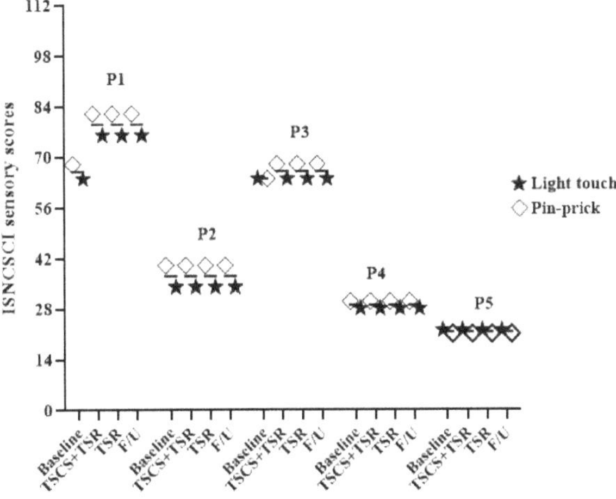

Figure 8. The International Standards for Neurological Classification of Spinal Cord Injury (ISNCSCI) scores with light touch and pinprick sensory sub-scores obtained in this study.

3.6. The Impact of Treatment Frequency on Functional Outcome

The frequency of treatment delivered to the study participants is illustrated in Figure 9, where P1, P2, and P4 received the intervention thrice per week with C6-, C7-, and C5-level cervical cord injuries, respectively. P3, who attended once a week, had a SCI at the C5 level. Furthermore, P5, who participated twice a week, had a C4 SCI. P1 and P2 improved the most in regard to mFRT (forward reach), TCT, and FIST. Additionally, P4, who attended the same treatment sessions as P1 and P2, presented with minimal improvement, relative to P1 and P2, in comparison to the increased functional scores. Interestingly, P3 scored better than P5 in terms of the above functional outcome measures. However, their injury levels differed from each other. Additionally, P3, who received fewer weekly sessions, improved more than P4, although they had the same injury level. Furthermore, P1 and P2 were able to independently perform transfers from their wheelchairs to their beds and vice versa, using a sliding board, under the supervision and assistance of caregivers. They also reported feeling more secure and having less fear of falling.

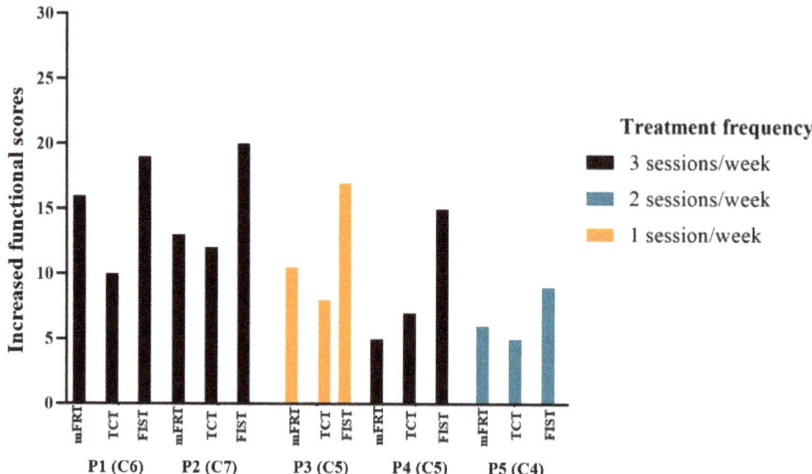

Figure 9. The increased functional scores based on treatment frequency obtained by the participants (cervical cord injury level: C4, C5, C6, C7) during the functional assessment of Modified Functional Reach Test (mFRT), Trunk Control Test (TCT), and Function in Sitting Test (FIST). A higher score indicates an improvement in function.

4. Discussion

The present investigation examined the effects of combining TSCS and TSR treatment on trunk control and sitting function in individuals with complete cervical SCI. The study demonstrates that the combined intervention (TSCS+TSR) could improve trunk stability along with static and dynamic sitting balance in people with complete tetraplegia. The results substantiate that all participants trunk control and sitting function progressively improved during TSCS+TSR, no matter whether they received one, two, or three treatment sessions per week with an injury at C4, C5, C6, or C7 level. In addition, functional improvements were maintained throughout the subsequent TSR and follow-up periods, demonstrating positive long-term treatment effects. Our findings highlight that trunk recovery is possible even in people with chronic complete cervical SCI, and the functional gains can be achieved using TSCS+TSR.

4.1. Functional Reaching Ability in Sitting

TSCS+TSR can enhance the forward reaching ability of people with SCI, which is essential in order for wheelchair users to accomplish everyday tasks [50]. The present study reported an increase in FRD (10.3 ± 4.5 cm) after TSCS+TSR, which enabled the participants to reach objects placed in front of them. Additionally, the lateral reach distances also increased to assist the maintenance of trunk balance. Prior research demonstrated that implanted electrical stimulation in SCI individuals with thoracic-level injuries exhibited improved FRD (5.5 ± 6.6 cm) in sitting [8]. Another study revealed that people with thoracic SCI receiving epidural stimulation displayed increased FRD, while lateral reach lengths remained unchanged [51]. The current results support these findings; additionally, the lateral reach distances also significantly increased. It has been previously revealed that those with higher FRD exhibited greater sitting control and postural stability [50]. Likewise, the current investigation indicates that participants with increased FRD demonstrated improved trunk and sitting function. This may assist clinicians to individualize exercise regimens to improve the wheelchair-based ADL of people with SCI based on their abilities [45]. Prior research showed that "off and on" epidural spinal cord stimulation settings affected the reach lengths, where the individuals reaching abilities returned to normal when the stimulator was switched off, but the FRD increased while the stimulator was on [51]. However, the present study shows that the increased reaching distances following TSCS+TSR were maintained even in the absence of further stimulation during the follow-up period.

4.2. Trunk Control and Seated Postural Stability

While falls have been reported in up to 75% of people with SCI, lack of trunk control is the predominant reason for falls in these individuals [52]. As trunk stability is a significant contributor to falls, rehabilitation programs should aim to increase sitting balance [52]. Interestingly, the present results indicate that TSCS+TSR could improve static trunk control in all participants, with the ability to sit independently with or without support of the upper limbs, while dynamic trunk control was restored in two participants. Therefore, improved trunk control could reduce the risk of falling from a wheelchair. Although a chest strap is beneficial for stabilizing the trunk of individuals with paraplegia in a wheelchair, it is only a temporary external method [53]. Given that TSCS+TSR could improve independent sitting and trunk function with increased FRD in people with tetraplegia, this method may complement or even replace traditional techniques such as chest straps, seating adjustments, and other manual support. The promising results suggest that this intervention may have the potential to help individuals with tetraplegia to turn in bed, which is critical in preventing the development of pressure injuries.

Sitting stability is essential for the functioning of those who cannot stand. The inability to complete transfer tasks may limit an individual's independence in a wheelchair, thereby interfering with ADL [54]. It was thought that a decrease or increase in trunk strength might predict sitting balance in people with SCI, although this premise remains inconclusive [10]. Our outcomes show that individuals with a lower cervical SCI displayed greater recovery in seated function, trunk strength, and sitting stability than those with a higher cervical SCI. A recent study using TSCS also showed an immediate effect on the upright sitting ability of SCI individuals that improved trunk stability and maintained sitting balance [12]. However, participants in that study had less severe injuries, and the level of injury was lower than that of our participants. That said, three of our participants reported improvements in their ability to propel their wheelchairs. They switched from motorized wheelchairs to manual wheelchairs.

4.3. Trunk Muscle EMG Response and Range of Motion

TSCS can increase trunk muscle activity and active ROM. Individuals with SCI above the L1 level experience trunk instability due to impaired trunk musculature [55]. A recent study showed that TSCS significantly increased the EMG response of the ES, RA, and EO muscles, which promoted trunk stability and sitting balance in people with chronic SCI [12]. These findings concur with those of the current study. Although invasive functional electrical stimulation (FES) together with therapeutic exercise increased trunk muscle tone and improved dynamic sitting stability, the effects reverted without FES [56]. Conversely, our participants demonstrated increased EMG responses in ES, EO, LD, and trunk ROM (maximum for flexion and rotation) following 12 weeks of TSCS+TSR. These effects persisted even without further stimulation. Previous research showed that stimulating trunk muscles with TSCS for thoracic-level SCI resulted in an increased mean trunk extension by $9.2° \pm 9.5°$ in sitting [8], whereas the current study yielded an even greater increased mean extension of $12.4° \pm 4.5°$. Collectively, these EMG changes may indicate the initial stage of neuromuscular recovery [36], which may result in further functional improvements with prolonged treatment. Interestingly, the right rotation ($4.5° \pm 2.7°$) was significantly lower than the left rotation ($39.6° \pm 13.4°$). A prior study reported that an individual who had right hand dominance before the SCI showed greater gains in relation to right hand function than the left side after bimanual task-oriented training, likely due to hand dominance [57]. As such, it was possible that hand dominance might contribute to the current findings. However, a recent study on sensorimotor and functional recovery between dominant and non-dominant upper extremities showed no significant differences in outcomes [58]. Therefore, further research is required to determine the veracity of this statement.

4.4. TSCS and TSR for Sensorimotor Improvement

In recent years, TSCS has become increasingly popular as a treatment option for people living with SCI. During locomotion training, TSCS has demonstrated exceptional gains in motor performance that were previously believed to be unattainable for individuals with chronic SCI [59]. Researchers have hypothesized that TSCS may have long-term benefits even in cases of complete SCI [12,42]. The TSCS treatment has shown continual improvements in locomotion [60], standing [44], upper extremity function [61], and sitting posture [12,42] in individuals with SCI. A few studies have also reported the immediate effects of TSCS on maintaining postural stability in people with SCI [12,42]. The present study adds to the findings of previous studies [12,42] by demonstrating that TSCS+TSR induced sustainable improvement in trunk function in people with complete tetraplegia. Our findings may assist clinicians and rehabilitation experts in the planning of trunk rehabilitation programs for individuals with complete tetraplegia to enhance patients' trunk stability and sitting balance.

Furthermore, the effects of TSCS on the sensory functions of the study participants revealed that pinprick and light touch sensations were improved, particularly in P1, with no prognosis at the neurological level. This may be attributed to the fact that individuals with SCI often reach a plateau in their spontaneous physical recovery approximately 1.5 years after SCI [62]. While most of our participants were 2–19 years post-SCI, for P1, post-SCI injury duration was 1.5 years. This could be one of the potential causes of the sensory improvement in P1 only. Further studies are needed to confirm this hypothesis. A prior study reported similar outcomes, where no change in ISNCSCI scores was observed in individuals with complete SCI (AIS A) [44]. Some studies showed improvements in neurological levels. However, the recovery was demonstrated in people with incomplete SCI [61,63]. Simultaneously, motor function improvements have been observed in individuals with complete SCI, but evidence on clinical prognosis is lacking [61].

The present study also shows that people with complete tetraplegia are unlikely to recover by TSR alone. This statement is supported by a review study evaluating the effectiveness of TSR for improving independent sitting and standing function in people with chronic SCI, which highlighted the minimal effect of TSR on functional recovery [29]. Indeed, little benefit to motor improvement has been reported in people with incomplete SCI [30], and another review paper presented little evidence on the benefit of conventional therapeutic exercise for increasing motor strength after SCI [64]. The current study findings support the above statements, where no significant improvements were observed when the TSR treatment was administered alone.

Research suggests that the combination of neuromodulation with a specific rehabilitation approach may recover adequate function after SCI [65]. The current findings support this notion, where significant improvements were seen after TSCS+TSR. Interestingly, the findings indicate that TSCS with TSR increased trunk stability and improved sitting balance, which assisted the study participants to advance the functional tasks (such as being able to perform rolling, accomplish transfer tasks, achieve progression in spinal mobility functions, and maintain erect sitting posture) independently or with minimal assistance. The impact of treatment frequency on functional outcome shows that improvement varied between individuals. These differences in existing functional outcome may be related to SCI level. Participants with lower tetraplegia had significantly more improvement than those with higher tetraplegia, despite attending similar treatment sessions. Given that a participant (P3) with lower cervical SCI, who attended fewer treatment sessions than another participant (P1) with higher cervical SCI, had better improvements, it is possible that improvement depends on the SCI level of neurological damage rather than the frequency of treatment. Future research should evaluate whether continuous intervention sessions may yield further recovery.

4.5. Relationship between Functional Improvement of mFRT, TCT, and FIST

The TCT and FIST were found to have a stronger correlation than other assessment parameters. Both TCT and FIST assessments have several trunk-based static and dynamic balancing assessments. These activities provide a more stable sitting position [46,66]. Therefore, when the trunk control is increased, the sitting function will also improve. This may explain their strong correlation in the current study. The mFRT did not show a strong correlation with other outcome measures. Daily activities or functional movements require a wide range of motions. Forward or lateral body movements alone may not accurately reflect the functional ability of an individual [16]. A prior study also found no correlation between mFRT and the mobility assessment of spinal cord independence measure III in people with chronic SCI [16]. Similarly, the lack of correlation between ROM and other functional assessment parameters in our study may be because SCI individual's ability to flex or extend their trunks had no effect on sitting stability [10]. Specifically, prolonged sitting in a wheel chair may lead to the development of a C-shaped spinal sagittal profile in people with SCI due to hyperkyphosis, hypolordosis, and a posterior pelvic tilt, which could affect sitting stability [9,67]. In fact, trunk muscle strength and endurance [68], pelvic alignment [69], and forward reaching ability together are the determinants of seated performance [70]. Therefore, improved ROM alone may not result in the corresponding improvement in sitting ability.

4.6. Study Limitations and Indications for Future Research

The research described herein has several limitations. First, the participants had varying levels of tetraplegia, which affected their ability to perform motor tasks to different extents. Although motor improvements were observed in all of the study participants regardless of the number of treatment sessions, greater functional gain was observed in those with lower levels of tetraplegia. Second, the participants were recruited via convenience sampling, which might lead to selection bias. Third, because one participant with C4 SCI (AIS A) had a sudden increase in blood pressure, the training was hampered by the need for frequent pauses to avoid autonomic dysreflexia. Fourth, one participant developed skin rashes after stimulation, which required a week to recover, necessitating the suspension of training for a week. To successfully complete all sessions, the treatment time was prolonged for that participant. Fifth, because the data collection was performed by the same investigator, there was a possibility of some bias. It was difficult to blind the participants because they underwent the treatment. Given the difficulty of double blinding, we adopted single blinding, where the rater was blinded in order to reduce the potential bias.

Despite our promising results, the mechanisms underlying the combined treatment effects remain unknown. One possible mechanism could be that supraspinal adaptations significantly improve balance performance following externally challenged balance training which is facilitated by a feed-forward mechanism [11]. Since the posterior root reflexes were not studied, the investigators did not know which structures were activated by the stimulation and parameters used [43]. Future studies should investigate the mechanisms and posterior root reflexes. Furthermore, future research should validate the current findings in a larger sample from a wider geographic region. Randomized controlled trial studies are warranted to compare the effects of TSCS+TSR and TSR alone under different treatment regimens with adjustment for various confounders (e.g., body mass index, duration of injury, residual upper limb function, and grasp function) in order to inform clinical practice. Future work should also evaluate the effects of TSCS+TSR on the quality of life of people with chronic complete cervical SCI.

5. Conclusions

The current study demonstrates that 12 weeks of TSCS+TSR treatment significantly improved functional reach distance and trunk muscle activation in multiple directions among individuals with chronic complete cervical SCI. These improvements persisted even after TSCS had been stopped for 18 weeks. The resulting increase in sitting balance may reduce the risk of falling from a wheelchair and improve functional independence when carrying out the purposeful tasks in sitting. The findings may be considered preliminary, and future studies should determine the optimal treatment regimen and duration to attain the maximum long-term treatment effects.

Supplementary Materials: The following supporting information can be downloaded at: https://www.mdpi.com/article/10.3390/biomedicines11010034/s1, Figure S1: The detailed analysis process of Vicon involves connecting selected markers at specific anatomical landmarks; Table S1: The responses recorded from trunk muscles measured through EMG; Table S2: The ISNCSCI classification of the participants.

Author Contributions: Conceptualization, N.S.T., M.A. and Y.-P.Z.; investigation, N.S.T., Y.T.L., A.Y.W. and M.A.; writing—original draft preparation, N.S.T.; writing—reviewing and editing, N.S.T., M.A., Y.T.L., A.Y.W. and Y.-P.Z.; project administration, Y.-P.Z.; funding acquisition, M.A. and Y.-P.Z. All authors have read and agreed to the final version of the manuscript.

Funding: The work was supported by The Hong Kong Polytechnic University (UAKB) and the Telefield Charitable Fund (ZH3V).

Institutional Review Board Statement: The study was conducted in accordance with the Declaration of Helsinki, and approved by the Human Subjects Ethics Sub-Committee of The Hong Kong Polytechnic University (Reference no: HSEARS20190201002-01).

Informed Consent Statement: Informed consent was obtained from all subjects involved in the study. Written informed consent has been obtained from the patient(s) to publish this paper if applicable.

Data Availability Statement: The data generated from this work can be obtained from the corresponding author upon reasonable request.

Acknowledgments: We are grateful to the participants and their caregivers for their efforts and commitment throughout the experiments. Our special thanks go to Lyn Wong for her support toward the success of our research. We also thank Mohammad Akhlasur Rahman and Vaheh Nazari for their valuable contributions in this study. This paper is an extended version of our conference paper published in Artificial Organs, Volume 45, Issue 3, Page E42.

Conflicts of Interest: The authors declare no conflict of interest.

References

1. Nulle, A.; Tjurina, U.; Erts, R.; Vetra, A. A profile of traumatic spinal cord injury and medical complications in Latvia. *Spinal Cord Ser. Cases* **2017**, *3*, 17088. [CrossRef] [PubMed]
2. Tharu, N.S.; Alam, M.; Bajracharya, S.; Chaudhary, G.P.; Pandey, J.; Kabir, M.A. Caregivers' Knowledge, Attitude, and Practice towards Pressure Injuries in Spinal Cord Injury at Rehabilitation Center in Bangladesh. *Adv. Orthop.* **2022**, *2022*, 8642900. [CrossRef] [PubMed]
3. Furlan, J.C.; Sakakibara, B.M.; Miller, W.C.; Krassioukov, A.V. Global incidence and prevalence of traumatic spinal cord injury. *Can. J. Neurol. Sci.* **2013**, *40*, 456–464. [CrossRef] [PubMed]
4. Post, M.W.; Dallmeijer, A.J.; Angenot, E.L.; van Asbeck, F.W.; van der Woude, L.H. Duration and functional outcome of spinal cord injury rehabilitation in the Netherlands. *J. Rehabil. Res. Dev.* **2005**, *42*, 75. [CrossRef]
5. Wyndaele, M.; Wyndaele, J.-J. Incidence, prevalence and epidemiology of spinal cord injury: What learns a worldwide literature survey? *Spinal Cord* **2006**, *44*, 523–529. [CrossRef]
6. Alizadeh, A.; Dyck, S.M.; Karimi-Abdolrezaee, S. Traumatic spinal cord injury: An overview of pathophysiology, models and acute injury mechanisms. *Front. Neurol.* **2019**, *10*, 282. [CrossRef]
7. Milosevic, M.; Masani, K.; Kuipers, M.J.; Rahouni, H.; Verrier, M.C.; McConville, K.M.; Popovic, M.R. Trunk control impairment is responsible for postural instability during quiet sitting in individuals with cervical spinal cord injury. *Clin. Biomech.* **2015**, *30*, 507–512. [CrossRef]
8. Triolo, R.J.; Bailey, S.N.; Miller, M.E.; Lombardo, L.M.; Audu, M.L. Effects of stimulating hip and trunk muscles on seated stability, posture, and reach after spinal cord injury. *Arch. Phys. Med. Rehabil.* **2013**, *94*, 1766–1775. [CrossRef]

9.   Minkel, J.L. Seating and mobility considerations for people with spinal cord injury. *Phys. Ther.* **2000**, *80*, 701–709. [CrossRef]
10.   Chen, C.-L.; Yeung, K.-T.; Bih, L.-I.; Wang, C.-H.; Chen, M.-I.; Chien, J.-C. The relationship between sitting stability and functional performance in patients with paraplegia. *Arch. Phys. Med. Rehabil.* **2003**, *84*, 1276–1281. [CrossRef]
11.   Potten, Y.; Seelen, H.; Drukker, J.; Reulen, J.; Drost, M. Postural muscle responses in the spinal cord injured persons during forward reaching. *Ergonomics* **1999**, *42*, 1200–1215. [CrossRef] [PubMed]
12.   Rath, M.; Vette, A.H.; Ramasubramaniam, S.; Li, K.; Burdick, J.; Edgerton, V.R.; Sayenko, D.G. Trunk stability enabled by noninvasive spinal electrical stimulation after spinal cord injury. *J. Neurotrauma* **2018**, *35*, 2540–2553. [CrossRef] [PubMed]
13.   Milosevic, M. *Neuromuscular Control of the Trunk during Sitting Balance*; University of Toronto: Toronto, ON, Canada, 2015.
14.   Frasuńska, J.; Tarnacka, B.; Wojdasiewicz, P. Quality of life in patients with tetraplegia and paraplegia after traumatic spinal cord injury. *Adv. Psychiatry Neurol.* **2020**, *29*, 143–153. [CrossRef]
15.   Abou, L.; de Freitas, G.R.; Palandi, J.; Ilha, J. Clinical instruments for measuring unsupported sitting balance in subjects with spinal cord injury: A systematic review. *Top. Spinal Cord Inj. Rehabil.* **2018**, *24*, 177–193. [CrossRef]
16.   Gao, K.L.; Chan, K.; Purves, S.; Tsang, W.W. Reliability of dynamic sitting balance tests and their correlations with functional mobility for wheelchair users with chronic spinal cord injury. *J. Orthop. Transl.* **2015**, *3*, 44–49. [CrossRef]
17.   Elmgreen, S.B.; Krogh, S.; Løve, U.S.; Forman, A.; Kasch, H. Neuromodulation in spinal cord injury rehabilitation. *Ugeskr. Laeger* **2019**, *181*, V02190104.
18.   McDonald, J.W.; Sadowsky, C. Spinal-cord injury. *Lancet* **2002**, *359*, 417–425. [CrossRef]
19.   Chompoonimit, A.; Nualnetr, N. The impact of task-oriented client-centered training on individuals with spinal cord injury in the community. *Spinal Cord* **2016**, *54*, 849–854. [CrossRef]
20.   Lotter, J.K.; Henderson, C.E.; Plawecki, A.; Holthus, M.E.; Lucas, E.H.; Ardestani, M.M.; Hornby, T.G. Task-specific versus impairment-based training on locomotor performance in individuals with chronic spinal cord injury: A randomized crossover study. *Neurorehabilit. Neural Repair* **2020**, *34*, 627–639. [CrossRef]
21.   Cunningham, P.; Turton, A.J.; van Wijck, F.; van Vliet, P. Task-specific reach-to-grasp training after stroke: Development and description of a home-based intervention. *Clin. Rehabil.* **2016**, *30*, 731–740. [CrossRef]
22.   Tamburella, F.; Scivoletto, G.; Molinari, M. Balance training improves static stability and gait in chronic incomplete spinal cord injury subjects: A pilot study. *Eur. J. Phys. Rehabil. Med.* **2013**, *49*, 353–364. [PubMed]
23.   Hubbard, I.J.; Parsons, M.W.; Neilson, C.; Carey, L.M. Task-specific training: Evidence for and translation to clinical practice. *Occup. Ther. Int.* **2009**, *16*, 175–189. [CrossRef] [PubMed]
24.   Kloosterman, M.; Snoek, G.; Jannink, M. Systematic review of the effects of exercise therapy on the upper extremity of patients with spinal-cord injury. *Spinal Cord* **2009**, *47*, 196–203. [CrossRef]
25.   Edgerton, V.R.; Harkema, S. Epidural stimulation of the spinal cord in spinal cord injury: Current status and future challenges. *Expert Rev. Neurother.* **2011**, *11*, 1351–1353. [CrossRef]
26.   Spooren, A.; Janssen-Potten, Y.; Kerckhofs, E.; Bongers, H.; Seelen, H. Evaluation of a task-oriented client-centered upper extremity skilled performance training module in persons with tetraplegia. *Spinal Cord* **2011**, *49*, 1049–1054. [CrossRef]
27.   Huie, J.R.; Morioka, K.; Haefeli, J.; Ferguson, A.R. What is being trained? How divergent forms of plasticity compete to shape locomotor recovery after spinal cord injury. *J. Neurotrauma* **2017**, *34*, 1831–1840. [CrossRef] [PubMed]
28.   Sayenko, D.G.; Alekhina, M.I.; Masani, K.; Vette, A.H.; Obata, H.; Popovic, M.R.; Nakazawa, K. Positive effect of balance training with visual feedback on standing balance abilities in people with incomplete spinal cord injury. *Spinal Cord* **2010**, *48*, 886–893. [CrossRef]
29.   Tse, C.M.; Chisholm, A.E.; Lam, T.; Eng, J.J. A systematic review of the effectiveness of task-specific rehabilitation interventions for improving independent sitting and standing function in spinal cord injury. *J. Spinal Cord Med.* **2018**, *41*, 254–266. [CrossRef]
30.   Jones, M.L.; Evans, N.; Tefertiller, C.; Backus, D.; Sweatman, M.; Tansey, K.; Morrison, S. Activity-based therapy for recovery of walking in chronic spinal cord injury: Results from a secondary analysis to determine responsiveness to therapy. *Arch. Phys. Med. Rehabil.* **2014**, *95*, 2247–2252. [CrossRef]
31.   Fisher, C.G.; Noonan, V.K.; Smith, D.E.; Wing, P.C.; Dvorak, M.F.; Kwon, B. Motor recovery, functional status, and health-related quality of life in patients with complete spinal cord injuries. *Spine* **2005**, *30*, 2200–2207. [CrossRef]
32.   Chan, S.C.C.; Chan, A.P.S. One-year follow-up of Chinese people with spinal cord injury: A preliminary study. *J. Spinal Cord Med.* **2013**, *36*, 12–23. [CrossRef] [PubMed]
33.   Rath, M. Trunk Control and Postural Stability in Individuals with Spinal Cord Injury: Noninvasive Enabling Strategies and Modeling. Ph.D. Thesis, University of California Los Angeles, (UCLA), Los Angeles, CA, USA, 2018.
34.   Rahman, M.A.; Tharu, N.S.; Gustin, S.M.; Zheng, Y.-P.; Alam, M. Trans-Spinal Electrical Stimulation Therapy for Functional Rehabilitation after Spinal Cord Injury. *J. Clin. Med.* **2022**, *11*, 1550. [CrossRef] [PubMed]
35.   Ahmed, Z. Trans-spinal direct current stimulation modifies spinal cord excitability through synaptic and axonal mechanisms. *Physiol. Rep.* **2014**, *2*, e12157. [CrossRef] [PubMed]
36.   Taylor, C.; McHugh, C.; Mockler, D.; Minogue, C.; Reilly, R.B.; Fleming, N. Transcutaneous spinal cord stimulation and motor responses in individuals with spinal cord injury: A methodological review. *PLoS ONE* **2021**, *16*, e0260166. [CrossRef]
37.   Hofstoetter, U.S.; McKay, W.B.; Tansey, K.E.; Mayr, W.; Kern, H.; Minassian, K. Modification of spasticity by transcutaneous spinal cord stimulation in individuals with incomplete spinal cord injury. *J. Spinal Cord Med.* **2014**, *37*, 202–211. [CrossRef]

38. Calvert, J.S.; Manson, G.A.; Grahn, P.J.; Sayenko, D.G. Preferential activation of spinal sensorimotor networks via lateralized transcutaneous spinal stimulation in neurologically intact humans. *J. Neurophysiol.* **2019**, *122*, 2111–2118. [CrossRef]
39. Gerasimenko, Y.; Gorodnichev, R.; Moshonkina, T.; Sayenko, D.; Gad, P.; Edgerton, V.R. Transcutaneous electrical spinal-cord stimulation in humans. *Ann. Phys. Rehabil. Med.* **2015**, *58*, 225–231. [CrossRef]
40. Minassian, K.; Hofstoetter, U.S.; Danner, S.M.; Mayr, W.; Bruce, J.A.; McKay, W.B.; Tansey, K.E. Spinal rhythm generation by step-induced feedback and transcutaneous posterior root stimulation in complete spinal cord–injured individuals. *Neurorehabilit. Neural Repair* **2016**, *30*, 233–243. [CrossRef]
41. Alam, M.; Ling, Y.T.; Wong, A.Y.; Zhong, H.; Edgerton, V.R.; Zheng, Y.P. Reversing 21 years of chronic paralysis via non-invasive spinal cord neuromodulation: A case study. *Ann. Clin. Transl. Neurol.* **2020**, *7*, 829–838. [CrossRef]
42. Keller, A.; Singh, G.; Sommerfeld, J.H.; King, M.; Parikh, P.; Ugiliweneza, B.; D'Amico, J.; Gerasimenko, Y.; Behrman, A.L. Noninvasive spinal stimulation safely enables upright posture in children with spinal cord injury. *Nat. Commun.* **2021**, *12*, 5850. [CrossRef]
43. Manson, G.; Calvert, J.S.; Ling, J.; Tychhon, B.; Ali, A.; Sayenko, D.G. The relationship between maximum tolerance and motor activation during transcutaneous spinal stimulation is unaffected by the carrier frequency or vibration. *Physiol. Rep.* **2020**, *8*, e14397. [CrossRef] [PubMed]
44. Al'Joboori, Y.; Massey, S.J.; Knight, S.L.; Donaldson, N.N.; Duffell, L.D. The effects of adding transcutaneous spinal cord stimulation (tSCS) to sit-to-stand training in people with spinal cord injury: A pilot study. *J. Clin. Med.* **2020**, *9*, 2765. [CrossRef] [PubMed]
45. Lynch, S.M.; Leahy, P.; Barker, S.P. Reliability of measurements obtained with a modified functional reach test in subjects with spinal cord injury. *Phys. Ther.* **1998**, *78*, 128–133. [CrossRef] [PubMed]
46. Quinzaños, J.; Villa, A.; Flores, A.; Pérez, R. Proposal and validation of a clinical trunk control test in individuals with spinal cord injury. *Spinal Cord* **2014**, *52*, 449–454. [CrossRef] [PubMed]
47. Abou, L.; Sung, J.; Sosnoff, J.J.; Rice, L.A. Reliability and validity of the function in sitting test among non-ambulatory individuals with spinal cord injury. *J. Spinal Cord Med.* **2020**, *43*, 846–853. [CrossRef]
48. Pan, G.-S.; Lu, T.-W.; Lin, K.-H. Comparison of arm-trunk movement between complete paraplegic and abled bodied subjects during circle drawing. *J. Spine* **2014**, *3*, 265. [CrossRef]
49. Bjerkefors, A.; Carpenter, M.G.; Cresswell, A.G.; Thorstensson, A. Trunk muscle activation in a person with clinically complete thoracic spinal cord injury. *J. Rehabil. Med.* **2009**, *41*, 390–392. [CrossRef]
50. Sliwinski, M.M.; Akselrad, G.; Alla, V.; Buan, V.; Kaemmerlen, E. Community exercise programing and its potential influence on quality of life and functional reach for individuals with spinal cord injury. *J. Spinal Cord Med.* **2020**, *43*, 358–363. [CrossRef]
51. Gill, M.; Linde, M.; Fautsch, K.; Hale, R.; Lopez, C.; Veith, D.; Calvert, J.; Beck, L.; Garlanger, G.; Edgerton, R.; et al. Epidural electrical stimulation of the lumbosacral spinal cord improves trunk stability during seated reaching in two humans with severe thoracic spinal cord injury. *Front. Syst. Neurosci.* **2020**, *14*, 79. [CrossRef]
52. Brotherton, S.S.; Krause, J.S.; Nietert, P.J. Falls in individuals with incomplete spinal cord injury. *Spinal Cord* **2007**, *45*, 37–40. [CrossRef]
53. Curtis, K.A.; Kindlin, C.M.; Reich, K.M.; White, D.E. Functional reach in wheelchair users: The effects of trunk and lower extremity stabilization. *Arch. Phys. Med. Rehabil.* **1995**, *76*, 360–367. [CrossRef] [PubMed]
54. Thompson, M.; Medley, A. Forward and lateral sitting functional reach in younger, middle-aged, and older adults. *J. Geriatr. Phys. Ther.* **2007**, *30*, 43–48. [CrossRef] [PubMed]
55. Kuipers, M.J. Functional Electrical Stimulation as a Neuroprosthesis for Sitting Balance: Measuring Respiratory Function and Seated Postural Control in Able-Bodied Individuals and Individuals with Spinal Cord Injury. Master's Thesis, University of Toronto, Toronto, ON, Canada, 2013.
56. Bergmann, M.; Zahharova, A.; Reinvee, M.; Asser, T.; Gapeyeva, H.; Vahtrik, D. The effect of functional electrical stimulation and therapeutic exercises on trunk muscle tone and dynamic sitting balance in persons with chronic spinal cord injury: A crossover trial. *Medicina* **2019**, *55*, 619. [CrossRef] [PubMed]
57. Hoffman, L.R.; Field-Fote, E.C. Cortical reorganization following bimanual training and somatosensory stimulation in cervical spinal cord injury: A case report. *Phys. Ther.* **2007**, *87*, 208–223. [CrossRef] [PubMed]
58. Bondi, M.; Kalsi-Ryan, S.; Delparte, J.J.; Burns, A.S. Differences in sensorimotor and functional recovery between the dominant and non-dominant upper extremity following cervical spinal cord injury. *Spinal Cord* **2022**, *60*, 422–427. [CrossRef]
59. Seáñez, I.; Capogrosso, M. Motor improvements enabled by spinal cord stimulation combined with physical training after spinal cord injury: Review of experimental evidence in animals and humans. *Bioelectron. Med.* **2021**, *7*, 16. [CrossRef] [PubMed]
60. Estes, S.; Zarkou, A.; Hope, J.M.; Suri, C.; Field-Fote, E.C. Combined transcutaneous spinal stimulation and locomotor training to improve walking function and reduce spasticity in subacute spinal cord injury: A randomized study of clinical feasibility and efficacy. *J. Clin. Med.* **2021**, *10*, 1167. [CrossRef]
61. Inanici, F.; Brighton, L.N.; Samejima, S.; Hofstetter, C.P.; Moritz, C.T. Transcutaneous spinal cord stimulation restores hand and arm function after spinal cord injury. *IEEE Trans. Neural Syst. Rehabil. Eng.* **2021**, *29*, 310–319. [CrossRef]
62. Samejima, S.; Henderson, R.; Pradarelli, J.; Mondello, S.E.; Moritz, C.T. Activity-dependent plasticity and spinal cord stimulation for motor recovery following spinal cord injury. *Exp. Neurol.* **2022**, *357*, 114178. [CrossRef]

63. McGeady, C.; vu Ckovi, C.A.; Singh Tharu, N.; Zheng, Y.P.; and Alam, M. Brain-Computer Interface Priming for Cervical Transcutaneous Spinal Cord Stimulation Therapy: An Exploratory Case Study. *Front. Rehabilit. Sci.* **2022**, *3*, 896766. [CrossRef]

64. Aravind, N.; Harvey, L.A.; Glinsky, J.V. Physiotherapy interventions for increasing muscle strength in people with spinal cord injuries: A systematic review. *Spinal Cord* **2019**, *57*, 449–460. [CrossRef] [PubMed]

65. Alam, M.; Rodrigues, W.; Pham, B.N.; Thakor, N.V. Brain-machine interface facilitated neurorehabilitation via spinal stimulation after spinal cord injury: Recent progress and future perspectives. *Brain Res.* **2016**, *1646*, 25–33. [CrossRef] [PubMed]

66. Palermo, A.E.; Cahalin, L.P.; Garcia, K.L.; Nash, M.S. Psychometric testing and clinical utility of a modified version of the function in sitting test for individuals with chronic spinal cord injury. *Arch. Phys. Med. Rehabil.* **2020**, *101*, 1961–1972. [CrossRef] [PubMed]

67. Shirado, O.; Kawase, M.; Minami, A.; Strax, T.E. Quantitative evaluation of long sitting in paraplegic patients with spinal cord injury. *Arch. Phys. Med. Rehabil.* **2004**, *85*, 1251–1256. [CrossRef]

68. Gabison, S.; Verrier, M.C.; Nadeau, S.; Gagnon, D.H.; Roy, A.; Flett, H.M. Trunk strength and function using the multidirectional reach distance in individuals with non-traumatic spinal cord injury. *J. Spinal Cord Med.* **2014**, *37*, 537–547. [CrossRef]

69. Hobson, D.A.; Tooms, R.E. Seated lumbar/pelvic alignment: A comparison between spinal cord-injured and noninjured groups. *Spine* **1992**, *17*, 293–298. [CrossRef]

70. Gagnon, D.H.; Roy, A.; Gabison, S.; Duclos, C.; Verrier, M.C.; Nadeau, S. Effects of seated postural stability and trunk and upper extremity strength on performance during manual wheelchair propulsion tests in individuals with spinal cord injury: An exploratory study. *Rehabil. Res. Pract.* **2016**, *2016*, 6842324. [CrossRef]

 **biomedicines**

Review

Advancements in Spinal Cord Injury Repair: Insights from Dental-Derived Stem Cells

Xueying Wen [1], Wenkai Jiang [2], Xiaolin Li [1], Qian Liu [3], Yuanyuan Kang [1] and Bing Song [3,4,*]

1   School and Hospital of Stomatology, China Medical University, Liaoning Provincial Key Laboratory of Oral Diseases, Shenyang 110002, China; wenx14@cardiff.ac.uk (X.W.); lixiaolin@cmu.edu.cn (X.L.); 20082068@cmu.edu.cn (Y.K.)

2   State Key Laboratory of Oral & Maxillofacial Reconstruction and Regeneration, National Clinical Research Center for Oral Diseases, Shaanxi Key Laboratory of Stomatology, Department of Operative Dentistry & Endodontics, School of Stomatology, Fourth Military Medical University, Xi'an 710032, China; jiangw6@cardiff.ac.uk

3   Shenzhen Institutes of Advanced Technology, Chinese Academy of Sciences, Shenzhen 518055, China; liuqian@siat.ac.cn

4   School of Dentistry, Cardiff University, Heath Park, Cardiff CF14 4XY, UK

*   Correspondence: bing.song@siat.ac.cn

Abstract: Spinal cord injury (SCI), a prevalent and disabling neurological condition, prompts a growing interest in stem cell therapy as a promising avenue for treatment. Dental-derived stem cells, including dental pulp stem cells (DPSCs), stem cells from human exfoliated deciduous teeth (SHED), stem cells from the apical papilla (SCAP), dental follicle stem cells (DFSCs), are of interest due to their accessibility, minimally invasive extraction, and robust differentiating capabilities. Research indicates their potential to differentiate into neural cells and promote SCI repair in animal models at both tissue and functional levels. This review explores the potential applications of dental-derived stem cells in SCI neural repair, covering stem cell transplantation, conditioned culture medium injection, bioengineered delivery systems, exosomes, extracellular vesicle treatments, and combined therapies. Assessing the clinical effectiveness of dental-derived stem cells in the treatment of SCI, further research is necessary. This includes investigating potential biological mechanisms and conducting Large-animal studies and clinical trials. It is also important to undertake more comprehensive comparisons, optimize the selection of dental-derived stem cell types, and implement a functionalized delivery system. These efforts will enhance the therapeutic potential of dental-derived stem cells for repairing SCI.

Keywords: spinal cord injury; stem cell therapy; neural repair; dental-derived stem cells; dental pulp stem cells; stem cells from human exfoliated deciduous teeth; stem cells from the apical papilla; dental follicle stem cells; delivery system; tissue regeneration

Citation: Wen, X.; Jiang, W.; Li, X.; Liu, Q.; Kang, Y.; Song, B. Advancements in Spinal Cord Injury Repair: Insights from Dental-Derived Stem Cells. *Biomedicines* **2024**, *12*, 683. https://doi.org/10.3390/biomedicines12030683

Academic Editor: Nicolas Guerout

Received: 5 December 2023
Revised: 14 February 2024
Accepted: 26 February 2024
Published: 19 March 2024

1. Introduction

Spinal Cord Injury (SCI) is a highly intricate and catastrophic neurological disorder that often results in temporary or permanent changes in its function, leading to a range of motor, sensory, and vegetative dysfunctions, making it one of the most complex and devastating nervous system disorders [1]. The causes of SCI encompass a variety of factors, such as traffic accidents, violent injuries, falls, degenerative diseases of the spine, tumors, infections, ischemia-reperfusion injuries, and blood vessel-related injuries [2]. SCI can be categorized into traumatic and non-traumatic SCI, and from a pathophysiological perspective, acute SCI can further be categorized into primary injury and secondary injury. Additionally, the severity of SCI can be distinguished as complete injury and incomplete injury [3].

Despite advancements in modern medicine that have improved the survival rates of SCI patients, progress in alleviating functional impairments related to neurogenic shock, respiratory difficulties, changes in ion and neurotransmitter levels, and inflammation remains limited, resulting in a poor quality of life for SCI patients [4]. Moreover, SCI imposes a substantial economic burden on patients, their families, and communities. Over the past 30 years, the global incidence of SCI has increased from 236 cases per million people to 1298 cases [5]. It is estimated that the annual global incidence of SCI ranges from 250,000 to 500,000 individuals [6]. According to the 2016 SCI Data Sheet published by the National Spinal Cord Injury Statistical Centre in the United States, healthcare costs and living expenses for SCI patients in the first year can amount to as high as $1,065,980, with average annual indirect costs, such as lost wages, additional benefits, and productivity losses, at $72,047 [7].

Alongside the surgical intervention in treating SCI, the current treatment methods for SCI also involve using anti-inflammatory medications such as ketorolac, minocycline, riluzole, magnesium, decompression surgery (decompression and instrumentation) to stabilize the spinal column and proper supportive management to prevent secondary injury [8]. Based on the current literature, no guidelines strongly endorse surgical intervention or pharmacological treatment as the primary method for treating SCI. Nevertheless, the treatment outcomes and prognosis for SCI patients are far from ideal [9]. Consequently, SCI remains a global challenge in clinical settings, presenting a significant hurdle for neuroscientists and neurosurgeons alike [10].

In the last decade, stem cell (SC) therapy has emerged as a novel and promising treatment for SCI. Stem cells derived from dental sources, such as dental pulp stem cells (DPSCs), stem cells from human exfoliated deciduous teeth (SHED), stem cells from the apical papilla (SCAPs), dental follicle stem cells (DFSCs), have garnered significant attention due to their ease of procurement and robust neurogenic differentiation potential. There is a promising outlook for their application in SCI treatment. Within the scope of this review, we comprehensively examined how dental stem cells promote SCI repair through various methods, including stem cell transplantation, conditioned culture medium injection, bioengineered delivery systems, exosomes, and extracellular vesicle treatments, and combined therapies, and discussed the limitations of existing research, offering new insights for future research directions. Through a meticulous examination of methodologies and a critical discussion of existing research, we not only highlight the innovative potential of dental-derived stem cells in SCI treatment but also provide new perspectives for future research directions and treatment methods.

2. Phases and Pathophysiology of SCI

The degree of loss of neurological function is measured using the American Spinal Injury Association (ASIA) impairment scale: A (complete sensory and motor loss below the lesion, including sacral sensation loss), B (sensory function below the lesion with no motor function), C (partial motor function preservation, with most muscles grading below 3), D (more than half of muscles below the lesion grading at three or higher), and E (normal motor and sensory testing) [11]. The AISA impairment scale classifies spinal cord injuries into complete and incomplete types. Complete injuries are Grade A, while incomplete injuries are graded B through D [12]. According to studies reporting complete SCI (grade A), there is little chance of resuming standing or walking with exercise training alone [13]. In chronic patients with incomplete SCI who are grade C or D and more than 2 years post-injury, the results of reconstructing gait with manually assisted exercise training with weight support are also unsatisfactory [14]. At the moment, incomplete tetraplegia is the most common SCI type (45%), followed by incomplete paraplegia (21.3%), complete paraplegia (20%), and complete tetraplegia (13.3%) [15]. Less than 1% of patients achieve complete recovery after discharge [15]. Spontaneous recovery after SCI is exceedingly rare, possibly due to several inhibitory regulatory factors, such as extracellular matrix proteins, which reduce the spinal cord's potential for endogenous repair by lowering its regenerative and plasticity capacities [2].

The pathological mechanisms of SCI involve two primary processes: primary injury and secondary injury (Figure 1). Primary injury occurs when the spinal cord is exposed to external forces such as contusion, ripping, compression, or transection or when it suffers an ischemic infarction due to vascular damage [16]. The features of the primary pathology of SCI are bone fragments and spinal tissue tearing. In primary injury, the pathophysiology includes neural parenchyma, axonal network glial membrane disruption, and hemorrhage (Figure 1) [2,17]. The severity of the injury is determined by the extent of initial destruction and the duration of spinal cord compression. A cascade of events associated with secondary injury is triggered by biochemical, mechanical, and physiological changes within neural tissues [18].

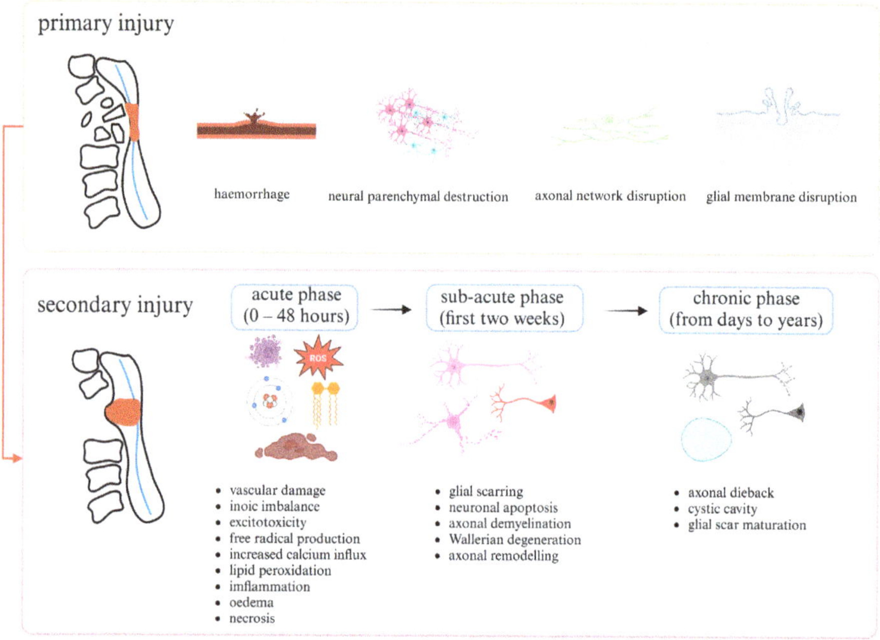

Figure 1. Phases and Pathophysiology of SCI. Created with https://www.biorender.com/ (accessed on 28 February 2024).

Following primary injury, a retrograde process initiates within minutes to hours, and its severity is directly proportional to the extent of the initial injury. Secondary injury is classified into three phases: acute, sub-acute, and chronic (Figure 1). During the acute phase, which occurs within 0–48 h, the spinal cord experiences excitotoxicity, vascular damage, ionic imbalance, increased calcium influx, edema, free radical production, inflammation, lipid peroxidation, and necrosis [2,19]. The acute phase persistence leads to the sub-acute phase, characterized by glial scarring, neuronal apoptosis, axonal demyelination, Wallerian degeneration, and axonal remodeling in the first two weeks [2,20]. During the chronic secondary injury phase of SCI, extending from days to years, cystic cavity forms, axonal dieback occurs, and the glial scar matures [21,22]. These three phases in secondary injury cause damage to underlying nucleic acid, proteins, and phospholipids, resulting in neurological dysfunction [5].

Besides, it is imperative to consider the role of imaging in assessing and prognosticating these injuries. Recent studies have highlighted the utility of magnetic resonance imaging (MRI) in providing valuable insights into the pathophysiology and prognosis of SCI [23]. For instance, MRI intramedullary signal characteristics in the early stages after SCI have been shown to correlate with the severity of injury and predict functional recovery [24].

Models of SCI are classified by the injury mechanism as contusion, compression, distraction, dislocation, transaction, or chemical. Revealing their characteristics helps us understand the unique pathophysiology of various SCI models (Table 1).

Table 1. Characteristics of different SCI models.

Models	Characteristics	Refs.
Contusion	Extensive tissue pathology; White matter apoptosis; Demyelination Incomplete remyelination; Robust macrophage response Early cell apoptosis; Presence of cavities and fibrotic scars Changes extending several millimetres cranial and caudal to the injury epicentre	[25–27]
Compression	Widespread inflammation; Edema; Bleeding Ischemia and demyelination due to venous congestion Changes occur near the injury epicentre	[28,29]
Distraction	No significant vascular damage Membrane damage to neuronal cell bodies and axons extends several vertebral segments rostrally Extracellular space enlargement and white matter structural alterations	[30,31]
Dislocation	Intramedullary bleeding; Early cell apoptosis Membrane damage to neuronal cell bodies and axons extends several vertebral segments rostrally Extensive loss of nerve fibres and accumulation of β-amyloid precursor protein; Greatest loss in ventral and dorsal horn neuron	[27,30,31]
Transaction	Focal tissue pathology with white matter apoptosis; Demyelination Incomplete remyelination; Robust macrophage response at the injury epicentre Absence of apoptosis and demyelination at a distance from the epicentre Damage to the dural membrane, epidural hematoma, and leakage of cerebrospinal fluid due to knife wound, potentially leading to infection	[26,32]
Chemical	Ischemia; Demyelination; Oxidative damage (lipids and proteins) Inflammation; Cellular injury	[33]

Interventions for enhancing SCI recovery aim to minimize the spread of secondary injury (through neuroprotection or inflammation modulation) and to replace lost nerve cells and disrupted neural circuits (through neuroplasticity and regeneration) [34]. These interventions seek to maximize the patient's rehabilitation potential and alleviate the functional impairments caused by SCI.

3. The Neurodegenerative Potential of Dental Stem Cells

Stem cells, with their pluripotent nature, proliferate and self-renew under specific conditions [35]. Mesenchymal stromal cells (MSCs), a diverse group of postnatal stem cells, stand out for their unique characteristics, including self-renewal, multipotent differentiation capabilities, and immune system modulation [36]. Extensive studies in the last two decades have emphasized MSCs' pivotal role in tissue homeostasis, with applications ranging from treating autoimmune diseases to regenerating damaged tissues like SCI [37–40].

MSCs have been successfully isolated from various adult and neonatal tissues, including bone marrow, skin, dental, adipose tissue, umbilical cord, Wharton's jelly, and placenta [38,41–44]. Dental pulp tissue-derived MSCs, notably, offer distinct advantages over other MSC sources, such as bone marrow, adipose tissue, peripheral blood, and umbilical cord blood, due to their ease of accessibility, good proliferation potential, neurogenic differentiation and neurotrophic capabilities, and negligible ethical issues and minimal invasiveness [45]. They are well-suited for tissue engineering and gene therapy due to their high proliferative potential, regenerative capacity, ability to differentiate into multiple cell types and lower inherent immunogenicity [45].

Dental stem cells can be isolated from different tooth regions, including DPSCs from the pulp of third molars, DFSCs from the dental follicle membrane that surrounds developing teeth, SHED from children's shed deciduous teeth, and SCAP from immature teeth (Figure 2) [46–50]. They all possess a certain degree of neurogenic differentiation capability.

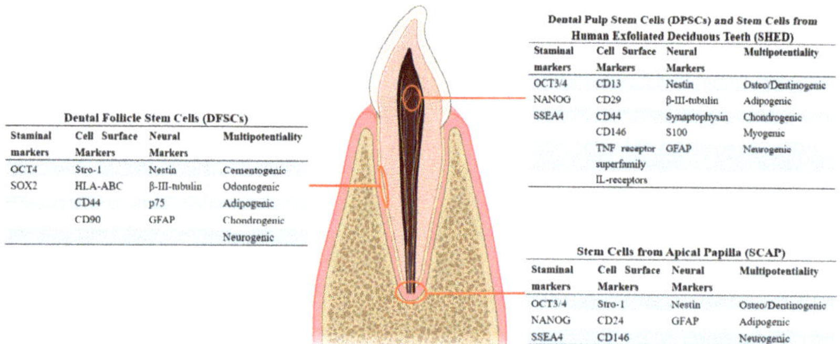

Staminal markers	Cell Surface Markers	Neural Markers	Multipotentiality
OCT4	Stro-1	Nestin	Cementogenic
SOX2	HLA-ABC	β-III-tubulin	Odontogenic
	CD44	p75	Adipogenic
	CD90	GFAP	Chondrogenic
			Neurogenic

Staminal markers	Cell Surface Markers	Neural Markers	Multipotentiality
OCT3/4	CD13	Nestin	Osteo/Dentinogenic
NANOG	CD29	β-III-tubulin	Adipogenic
SSEA4	CD44	Synaptophysin	Chondrogenic
	CD146	S100	Myogenic
	TNF receptor superfamily	GFAP	Neurogenic
	IL-receptors		

Staminal markers	Cell Surface Markers	Neural Markers	Multipotentiality
OCT3/4	Stro-1	Nestin	Osteo/Dentinogenic
NANOG	CD24	GFAP	Adipogenic
SSEA4	CD146		Neurogenic

Figure 2. Anatomical Positions and Characteristics of Dental-Derived Stem Cells. Dental-derived stem cells, including DPSCs, DFSCs, SCAP, and SHED, exhibit diverse anatomical locations. Classification and neurogenic induction properties primarily rely on the expression of neuronal markers on cell surfaces, demonstrating multipotentiality. Created with https://www.biorender.com/ (accessed on 28 February 2024).

3.1. DPSCs

DPSCs, first derived from the dental pulp in 2000 by Gronthos [51], share characteristics with mesenchymal stem cells, including plasticity, adhesiveness, and a fibroblast-like morphology [45]. Notably, DPSCs express neurotrophic and immunomodulatory factors, promoting blood vessel formation and nerve regeneration [52].

Various studies highlight DPSCs' potential for neurodegeneration. Recent research demonstrates that combining vascular endothelial growth factor A (VEGFA)-overexpressing rat dental pulp stem cells (rDPSCs) with a laminin-coated and yarn-encapsulated poly (l-lactide-co-glycolide) (PLGA) nerve guidance conduit (LC-YE-PLGA NGC) enhances myelin sheath quantity, thickness, and axonal diameter, facilitating the repair of facial nerve injuries [53]. Similar experiments show that chitosan tubes inoculated with stem cell factor and DPSCs boost neo-vascularization, providing an effective approach to repairing facial nerve defects [54]. In a rodent model with sciatic nerve deficits, transplantation of neuro-lineage cells (NLC) differentiated from DPSCs revealed substantial improvements in axonal growth, remyelination, electrophysiological activity, and muscle atrophy after 12 weeks [55]. Ben Mead and colleagues' study demonstrates that intravitreal transplants of DPSCs significantly enhance neurotrophin-mediated retinal ganglion cell (RGC) survival and axon regeneration following optic nerve injuries [56].

3.2. SHED

SHED, a unique subtype of pluripotent stem cells, was initially isolated and characterized from the pulp tissue of shed human deciduous teeth by Miura and colleagues [48]. Acknowledged for its highly proliferative capacity and ability to differentiate into various cell types [57]. SHED demonstrates persistence within the mouse brain and expression of neural markers upon in vivo transplantation. Derived from easily accessible tissue sources, SHED provides an abundant cell supply for potential clinical applications [48].

Both SHED and their conditioned media (SHED-CM) effectively address neurodegenerative diseases through mechanisms such as cell replacement, paracrine effects, angiogenesis, synaptogenesis, immunomodulation, and inhibition of apoptosis [58]. SHED's neurogenic differentiation potential and release of bioactive molecules offer promise for neuronal recovery in nerve injury cases, holding potential for disorders like SCI, Alzheimer's disease (AD), and focal cerebral ischemia (FCI) [59]. Research by Sugimura and colleagues highlights SHED-CM's potential in promoting axonal regeneration and functional recovery in a rat model with sciatic nerve deficits. Their findings suggest that SHED-CM facilitates axonal growth, peripheral nerve tissue vascularization, neuronal survival, and Schwann cell migration and proliferation [60].

3.3. SCAP

In 2006, Sonoyama and colleagues cultured cells from apical papilla tissue collected from the roots of young human third molars, identifying them as MSCs and naming them SCAP [61]. SCAP, derived from the apical papilla, is more accessible, separable, and expansible compared to other dental tissues, such as DPSCs [62]. SCAP exhibits superior proliferation rates and expresses classical stem cell markers, outperforming DPSCs in BrdU uptake rate, cell doubling, tissue regeneration capacity, and the number of STRO-1-positive cells [63]. Co-culturing SCAP with trigeminal sensory neurons enhances sustained inward current density induced by ATP, suggesting a positive impact on sensory nerve activity and cold-sensitive ion channels [64]. SCAP also demonstrates low immunogenicity and possesses immunomodulatory characteristics [65]. Studies indicate their role in expediting SCI healing by reducing TNF-α levels and promoting oligodendrocyte progenitor cell differentiation [66]. Consequently, the apical papilla and its resident SCAP offer a unique opportunity for their potential clinical application in neural repair.

3.4. DFSCs

Dental follicle cells (DFCs) were first reported by Wise et al. in 1992, and later, in 2005, a population of cells with colony-forming and plastic-adherent properties was successfully derived from dental follicles, termed dental follicle progenitor/stem cells (DFPCs/DFSCs) [67,68]. As part of the Dental Stem Cell family, DFSCs are obtained during the early stages of development, providing an advantageous cell source for clinical applications due to the larger tissue volume of dental follicles.

DFSCs exhibit a higher proliferation rate, enhanced colony-forming capability, and potent anti-inflammatory characteristics compared to other dental MSCs, making them clinically relevant for treating oral and neurological disorders [69]. Originating from the cranial neural crest, DFSCs express neurogenic membrane markers like NESTIN and TUBULIN IIIβ, and DFCs retain multipotential differentiation, displaying neurogenesis-related behaviors. Obtained easily from third molar extraction or alveolar fossa curettage, human DFCs (hDFCs) are more inclined to express the neurogenic marker MAP2 compared to SHEDs, suggesting their efficiency in neural regeneration [70,71].

In summary, dental-derived stem cells, irrespective of their origin, exhibit rapid proliferation and the ability to differentiate into typical mesenchymal cell lineages, including osteo/dentinogenic, adipogenic, and neurogenic pathways [72]. Derived from the neural crest, dental MSCs uniquely express neural markers, secrete neurotrophic factors, and combine MSC-like characteristics with immunomodulation and neural features [73]. With their neural crest lineage, neuronal markers, and expression of neurotrophic factors, along with the potential for neurogenic differentiation, these cells are actively researched for their applications in treating neuronal diseases and injuries [74,75]. Recent studies highlight their promising role in neural tissue engineering for nerve regeneration, particularly in the treatment of SCI.

4. Treatment Approaches

In the pursuit of effective SCI treatment, diverse approaches leveraging the potential of dental-derived stem cells have surfaced. Current research utilizes various strategies, including stem cell transplantation, conditioned culture medium injection, innovative stem cell delivery systems, exosome-based therapies, and combined interventions with physical methods. Collectively, these approaches constitute a multifaceted framework for SCI treatment (Figure 3).

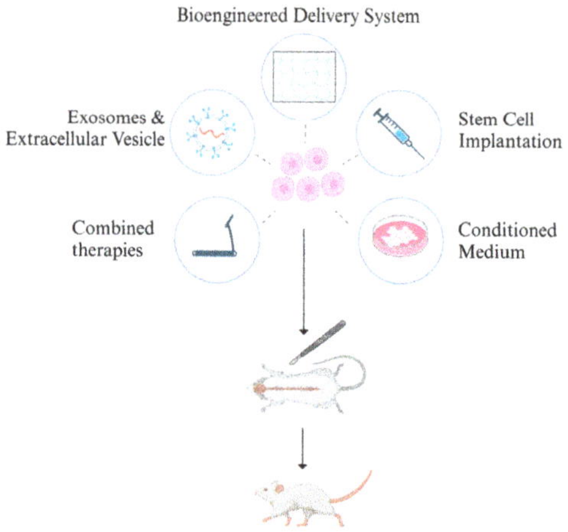

Figure 3. Multifaceted Applications of Oral-Derived Stem Cells in Spinal Cord Injury (SCI) Treatment. This illustration represents therapeutic strategies involving dental-derived stem cells, including stem cell transplantation, conditioned culture medium injection, innovative stem cell delivery systems, exosome-based therapies, and their combination with physical interventions. These strategies synergistically contribute to SCI recovery. Created with https://www.biorender.com/ (accessed on 28 February 2024).

4.1. Stem Cell Implantation

Comparative studies involving mesenchymal stem cells (DP-MSCs, AD-MSCs, and DP-MSCs) reveal that transplanting DP-SMCs into SCI sites effectively promotes neural regeneration and functional recovery [76]. Similarly, DPSCs and SHED transplantation into a completely transected mouse acute SCI model significantly enhanced motor function and promoted spinal cord axon regeneration [77]. DPSCs, through mechanisms like reducing cell apoptosis, promoting axon regeneration, and differentiating into mature oligodendrocytes, contribute to functional recovery after SCI [77].

In addressing the hypoxic environment in SCI, introducing the basic fibroblast growth factor (bFGF) gene into DPSCs via an adeno-associated virus (AAV) vector successfully ameliorates the hypoxic environment, aiding neuron survival and axon regeneration [78].

Additionally, experiments conducted by Fabrício Nicola and their research team have shown that injecting SHED cell suspensions into rodent SCI sites improves motor function, reduces tissue loss, protects neurons, mitigates inflammation, and decreases neuronal apoptosis [79]. SHED transplantation fosters neural precursor cell proliferation, reduces glial scar formation, and slows S100B protein decline in spinal glial cells [80].

Using induced pluripotent stem cells (iPSCs) derived from SHED (iSHED) enhances regenerative potential when transplanted into a rat model of acute SCI [81].

Furthermore, experiments led by Chao Yang and their research team found that various types of dental-derived stem cells, including DFSCs, SCAP, and DPSCs, show their potential to induce neural regeneration, reduce inflammatory responses, promote neural regeneration, and differentiate into mature neurons and oligodendrocytes [82]. DFSCs exhibited particularly significant effects.

In summary, direct transplantation of dental-derived stem cells shows promise in treating acute and chronic SCI, emphasizing their role in promoting neural regeneration and functional recovery. Further research and clinical trials are needed to determine optimal treatment methods and application areas.

4.2. Condition Medium Injection

Cell therapy for SCI poses risks of tumor formation and immune reactions. To overcome these challenges, researchers advocate for conditioned media (CM) from cell sources as an alternative. Dental pulp stem cells' cell source-conditioned medium (DPSC-CM) and human shed deciduous teeth stem cells (SHED-CM) have demonstrated significant potential in restoring cerebellar granule neurons' neurite growth activity, surpassing the effectiveness of conditioned media from fibroblasts or bone marrow mesenchymal stem cells [62].

SHED-CM, in particular, induces an M2-dominant neural repair microenvironment and enhances functional recovery after SCI through factors like MCP-1 and ED-Siglec-9 [83].

To address concerns about the rapid spread of CM through body fluids, a combination method involving biological materials and drug delivery systems has been proposed. A study by Reza Asadi-Golshan's team utilized a collagen hydrogel as a slow-release vehicle for SHED-CM, demonstrating significant advantages in various scores related to spinal cord injury recovery [84]. Another study by the same team found that injecting collagen hydrogel-loaded SHED-CM into the spinal cord after compression SCI in rats prevented tissue loss, including grey and white matter, and protected cells, including neurons and oligodendrocytes [85].

4.3. Bioengineered Delivery System Approaches

While stem cell therapy holds promise for tissue repair and regenerative medicine, challenges persist. Similar to organ transplantation, immunosuppressants are often required for enhanced cell viability and survival. Acute inflammation and immune reactions, along with the absence of extracellular matrix support, contribute to the early demise of transplanted cells [86]. Controlling the fate and integration of transplanted cells within the organism presents further challenges. Researchers are addressing these issues by exploring the immobilization of dental stem cells in biocompatible materials, such as hydrogels, chitosan, PLGA scaffolds, microcapsules, and microspheres, offering promising solutions for SCI treatments.

4.3.1. Hydrogel

After SCI, the local microenvironment becomes detrimental due to the release of excitatory neurotransmitters and toxic substances [87]. To address this challenge, an innovative hydrogel known as TPA@Laponite hydrogel has been developed by Yigo Ying and colleagues. This shear-thinning hydrogel, encapsulating and protecting DPSCs, effectively scavenges harmful oxygen radicals, enhances vascular function, and inhibits lipid peroxidation. DPSCs introduced into this hydrogel adjust the excitatory to inhibitory synapse ratio, reducing muscle spasms and promoting SCI recovery [88].

Moreover, Heng Zhou addressed the loss of zinc ions (Zn^{2+}) post-SCI by introducing ZIF-8 into DPSCs and injecting them into injured rat spinal cords. ZIF-8, a carrier for drug and gene delivery, releases Zn^{2+} in acidic environments. Combined with gelatin methacryloyl (GelMA) hydrogel, this approach promotes motor function recovery, facilitating neural cell repair, inhibiting apoptosis, and enhancing angiogenesis [89].

Pluronic F-127, a synthetic hydrogel, has been explored as an injectable carrier [90]. Lihua Luo's team designed a thermosensitive heparin-poloxamer (HP) hydrogel containing bFGF and DPSCs. This hydrogel, delivered to the spinal cord injury site, sustains DPSC density and bFGF activity, promoting tissue regeneration and improving sensory and motor function recovery [91]. Another study suggests that heparin-based hydrogel containing bFGF and DPSCs (HeP-bFGF-DPSCs) effectively controls inflammation, stabilizes microtubules, regulates the tissue vascular system, and promotes neural regeneration [92].

Calcium alginate hydrogel, a biocompatible material, combined with DPSCs and fibroblast growth factor 21 (FGF21), demonstrated exceptional tissue affinity in Sipin Zhu's study. This hydrogel protects neurons, inhibits apoptosis, promotes autophagy, and enhances recovery post-spinal cord transection surgery by facilitating axon and functional blood vessel regeneration [93].

Additionally, the research attempted to encapsulate the entire human dental pulp in a fibroin hydrogel. Dental pulp implantation in SCI reduced pro-inflammatory markers, inhibited microglia/macrophage activation, and increased immunoreactive 5-HT positive cells [94]. However, a study indicates that, compared to the implantation of the entire dental pulp, the effectiveness of using dental root apical papilla stem cells combined with hydrogel is inferior. Experimental evidence demonstrates that rats receiving dental pulp implantation achieved positive outcomes in terms of recovery, chronic pain, and spinal cord tissue structure [95].

4.3.2. Chitosan

Research into chitosan-bound dental stem cells shows promising prospects. Chitosan, a highly biocompatible and hydrophilic biopolymer extracted from crustaceans' exoskeletons, such as crabs, shrimps, and lobsters, is widely used in tissue engineering and various medical fields. Treatment involving DPSCs combined with a chitosan scaffold has shown enhanced cell viability and neural differentiation. In comparison to the control group without a chitosan scaffold, the DPSCs/chitosan scaffold group exhibited significantly elevated levels of BDNF, GDNF, b-NGF, and NT-3. Transplanting DPSCs with a chitosan scaffold into an SCI rat model resulted in a substantial recovery of hindlimb motor function, with the transplantation group experiencing significantly lower tissue loss, apoptotic cell count, and axon degeneration than other experimental groups. Research also identified the crucial role of the Wnt/β-catenin signaling pathway in neural differentiation when DPSCs were combined with a chitosan scaffold [96].

Additionally, studies have explored the combined use of a chitosan scaffold with bFGF and DPSCs [97]. Results indicate that the treatment group combining DPSCs/chitosan scaffold with bFGF had significantly higher levels of GFAP, S100b, and b-tubulin protein III compared to the control group without bFGF and the group using only DPSC/chitosan scaffold. This suggests that the combined application of a chitosan scaffold with bFGF promotes the neural differentiation of DPSCs. Therefore, the transplantation of DPSCs/chitosan scaffold combined with bFGF presents a potential as a safe and effective treatment for SCI.

4.3.3. PLGA

PLGA (poly-lactic-co-glycolic acid) scaffolds are extensively studied for supporting dental stem cell transplantation in SCI treatment. The limited regeneration of damaged spinal cords is attributed to inadequate vascular supply and neural nutritional support. To address this, researchers led by Shaowei Guo designed a highly vascularized scaffold using biocompatible and biodegradable poly (L-lactic acid) (PLLA)/PLGA scaffolds. Their results demonstrate that this scaffold, containing DPSCs, has the potential to enhance SCI repair through paracrine-mediated angiogenesis and neural regeneration. Implanting these scaffolds into the injured spinal cord of rats with complete spinal cord transection models promotes vascular reconstruction, aiding in axon regeneration, myelin deposition, and sensory recovery. Analysis of the reconstructed spinal cord tissue using 3D micro-computed tomography (micro-CT) imaging and morphometric measurements revealed a substantial presence of regenerating blood vessels, particularly in the sensory tract area, correlating with behavioral recovery after pre-vascularization treatment [98].

For cases involving relatively large spinal cord defects (SCD), which may cause interruptions or blockages in neural pathways requiring reconnection and reconstruction, a study utilized oriented electrospun PCL/PLGA materials (AEM). It was demonstrated that human dental follicle cells (hDFCs) could effectively grow along these oriented fibers. Although observations regarding functional recovery did not significantly differ among the groups, the implanted AEM-hDFCs composite material promoted the migration and growth of neural cells [99].

4.3.4. Microcapsules and Microspheres

To precisely control cell migration, differentiation, and tissue integration, researchers developed microcapsules containing dental stem cells. One study indicated that biocompatible microcapsules could control the fate of dental stem cells in situ to promote the replacement of damaged or lost tissues in SCI. Lorena Hidalgo San Jos and colleagues utilized a microfluidic device to encapsulate DPSCs in alginate-collagen microcapsules, demonstrating cell survival for up to 21 days. The transplanted microcapsules effectively retained DPSCs in an organotypic SCI model, with the cells expressing neural markers after the in situ culture [100].

In another study, researchers encapsulated recombinant BDNF nano-precipitates in PLGA-P188-PLGA microspheres (BDNF-PAM) and implanted SCAP-derived stem cells into PAM. The SCAP BDNF-PAM treatment significantly increased cell retention in the spinal cord, improved motor coordination, reduced inflammation, and promoted axon growth, marking a novel injectable cell delivery system's benefits in the SCI treatment [101].

In summary, combining hydrogels, chitosan scaffolds, PLGA scaffolds, and carriers like microcapsules and microspheres with dental stem cells holds promising prospects for SCI therapy. These methods offer opportunities to enhance cell survival, promote neural regeneration, inhibit inflammation, and facilitate tissue repair, ultimately restoring motor and sensory functions. Their biocompatibility and biodegradability make these carriers ideal for creating a conducive environment in damaged spinal cords, fostering neuron regeneration, and inducing cell differentiation to improve motor function.

4.4. Exosomes and Extracellular Vesicles

Extracellular vesicles derived from stem cells have gained prominence in spinal cord injury treatment. Produced by living cells, these vesicles contain proteins crucial for immune modulation, neuroprotection, and cellular behavior orchestration [102]. Unlike traditional stem cell therapy, stem cell-derived extracellular vesicles offer advantages such as smaller size, reduced tumorigenicity, enhanced membrane transfer efficiency, and the ability to traverse the blood-spinal cord barrier [103].

Chao Liu's work highlights DPSCs-derived exosomes' potential in attenuating M1 macrophage polarization by inhibiting the ROS-MAPK-NFκB P65 signaling pathway. This leads to reduced inflammation and neural damage in SCI, improving the recovery process. DPSCs-derived exosomes emerge as a promising therapeutic modality for spinal cord injuries, offering a potential strategy for mitigating secondary damage by disrupting the ROS and M1 macrophage polarization feedback loop [104].

4.5. Combined Therapies

In the pursuit of effective SCI treatment, dental-derived stem cells have recently gained significant attention as a promising therapeutic tool combined with physical Interventions. This multimodal approach integrates dental-derived stem cells with treatments like underwater treadmill therapy, treadmill training therapy, and electroacupuncture, offering increased hope for SCI patients.

Matheus Levi Tajra Feitosa and colleagues selected three canine breeds with chronic SCI-induced paralysis, meeting specific inclusion criteria like paraplegia, lack of conscious proprioception, exaggerated reflexes, and absence of deep pain sensation. Confirmed through MRI diagnosis of thoracolumbar intervertebral disc disease, the researchers injected immature dental pulp stem cells at three points into the SCI site. Over two months post-surgery, the animals underwent weekly veterinary physical therapy, including hydrotherapy with underwater treadmill sessions. Clinical assessments using the Olby scoring system revealed improvements, indicating the potential positive impact of combining stem cell therapy and physical treatments on chronic SCI recovery. However, precise mechanisms require further investigation [105].

While some studies found that combination therapy did not yield significant effects, César Prado and colleagues evaluated the safety, feasibility, and therapeutic effects of canine exfoliated deciduous tooth stem cell transplantation combined with electroacupuncture for chronic SCI in dogs. Results showed only mild improvements in some animals, with no significant therapeutic effects observed with stem cell treatment, electroacupuncture, or their combination. This could be attributed to the limited number of animals and significant variability in spinal cord injury severity [106]. Additionally, combining SHED transplantation with treadmill training did not yield significant improvements in motor function recovery in traumatic SCI rats compared to using SHED alone, highlighting the need for further research to determine the optimal timing and intensity of exercise training [107].

These studies shed light on the application of dental-derived stem cells in combination with other treatment modalities for SCI therapy, each with its advantages and disadvantages (Table 2). While providing valuable insights, further comprehensive research is essential to fully understand the efficacy and potential of these treatments.

Table 2. Advantages and disadvantages of dental stem cell application in SCI treatment.

Approach	Advantages	Disadvantages	The Role of SCI Repair	Refs.
Stem Cell Transplantation	Easy isolation, Minimal ethical controversy lower immunogenicity risk	Low stem cell survival Risk of immune rejection Cell dedifferentiation Risk of potential tumour formation	Promoting nerve regeneration. Improving motor function. Reducing spinal cord tissue loss. Protecting motor neurons. Mitigating inflammation. Decreasing neuronal apoptosis.	[77–82,108]
Conditioned Medium Injection	Contains growth factors for effective tissue repair. Avoid tumorigenesis and immune issues associated with stem cell transplantation.	The rapid diffusion of culture medium may be uncontrollable.	Establishing a favourable repair microenvironment. Promoting nerve regeneration.	[83–85]
Delivery System Approaches	Provide a suitable environment for stem cell survival, growth, and differentiation. Protects existing cells from apoptosis/necrosis. Allows controlled release of growth factors.	Material instability and potential quick degradation. Risk of cytotoxicity.	Enhancing stem cell viability and neural differentiation. Promoting neuronal regeneration. Inhibiting inflammation. Improving oxygen supply to damaged areas. Reducing gelatinous scarring to aid axonal and vascular regeneration.	[85,88,89,91–101]
Exosomes and Extracellular Vesicle	Exosomes exhibit most of the biological properties of stem cells. Exosomes are small and less likely to block microvessels. Exosomes have a low tumour risk.	Limited exosome production. Sensitivity to microenvironment pH.	Reducing inflammation and nerve damage. Improving spinal cord neuron survival. Enhancing motor function.	[104,109]
Combined Therapies	Electroacupuncture, treadmill training, and physiotherapy have shown some potential for improving functional recovery after SCI.	The timing and intensity of training can affect recovery outcomes.	No significant improvement in combination with treatment.	[105–107]

5. Discussion

5.1. Limitations of Treatment Methods

Current research in dental-derived stem cells for SCI treatment shows promising advances but faces notable constraints, including model choice, the role of stem cells, and delivery system safety.

A Primary limitation is the predominant focus on murine models, with a lack of large-animal studies and clinical trials to validate therapy feasibility and efficacy. For instance, DP-MSC application in porcine SCI models did not yield motor function recovery comparable to murine models [110]. Some studies use small sample sizes, potentially compromising statistical significance. The complexity of SCI models makes them challenging to control, leading to variations and inconsistencies [99].

Different dental stem cell sources, such as DPSCs, SHED, and SCAP, lack distinctly delineated advantages and applicability in SCI treatment. Comparative analysis showed that DFSCs were more effective in SCI repair than DPSCs and SCAP [82]. Disparities between sensory and motor aspects and incongruities between clinical assessments and imaging results pose challenges [98,105]. The nonlinearity of the BBB score complicates discerning improvement degrees, especially in patients with substantial motor function [107]. Despite increased neural markers in some in vitro experiments, more in vivo experiments are needed to confirm neural cell replacement [101]. Further research is needed to elucidate how dental stem cells comprehensively promote neuroregeneration, optimal strategies for combination with other modalities, and the biological safety of co-applying with biomaterials [101] Studies on SCAP and hydrogels show inconclusive results, contrasting with SHED treatment [95].

$CaCl_2$ application in isolation and concentrations in hydrogels need careful assessment [93]. Studies focused on short-term effects necessitate more comprehensive research, including extended follow-ups, to address potential adverse consequences like teratoma and tumorigenesis [78,81,92].

In addition, 11.8% of SCIs and 19.2% of non-traumatic SCIs have been reported to be caused by spinal tumors [111]. As an exceptional case, spinal tumors that cause SCI are particularly challenging to treat, and early diagnosis, multidisciplinary care, and appropriate rehabilitation are essential to improve treatment outcomes and quality of life in all affected patients. More caution and care will be required when applying dental-derived stem cells to the study and treatment of SCI due to spinal tumors in the future. This is because stem cell implantation risks tumor formation in the treatment of SCI [112]. In particular, stem cells, including iPSCs, may be tumorigenic, leading to the formation of teratomas and true tumors [113].

5.2. Future Research Direction

To understand dental stem cells' potential in SCI treatment, future research should focus on key directions.

Large-animal studies and clinical trials are crucial for establishing robust clinical evidence. Comparative studies on different spinal cord injury models and their controllability should be considered [99]. Extensive comparative investigations into distinctions between dental stem cell sources and their efficacy across varying SCI severity levels will optimize stem cell selection. Exploring extracellular vesicles from various sources and optimizing therapy protocols is necessary [104]. Enhancements in the design and characteristics of scaffold materials combined with dental stem cell application should be pursued to improve biocompatibility, controlled and sustained growth factor release, and, ultimately, treatment effectiveness [92,100]. Research into secretomes produced by dental stem cells during SCI repair should be conducted to comprehend their role in tissue regeneration [94].

Furthermore, Biological mechanisms associated with dental stem cells need exploration to understand their role in neural regeneration [77,83,110] Maximizing the therapeutic potential of dental stem cells, effective transplantation methods, timing, and treatment regimens to enhance reparative outcomes, and optimization of differentiation through growth factors require attention [94]. This may require the optimization of differentiation through the use of growth factors and other differentiation-inducing factors [93]. As spinal cord injuries necessitate multiple modalities, research into the combined effects of various approaches, including rehabilitation therapy and pharmacological treatments, is pragmatic [85]. Improved control and adjustment of dental stem cell treatment outcomes for personalized treatment warrant further exploration [78]. Future studies may concentrate on alternative therapies combined with stem cell therapy, such as managing neurite growth inhibitors or using anti-Nogo A antibodies [105].

5.3. Exploration of Critical Questions

For extensive clinical applications, key issues must be addressed, including standardizing dental stem cell collection, expansion, and application procedures for consistent outcomes [106]. Research is needed to determine optimal delivery methods and timeframes for maximizing neural regeneration. An in-depth exploration of the interaction between dental stem cells and the immune system is essential to mitigate rejection risks and enhance treatment efficacy [80,98].

In summary, dental stem cells hold tremendous potential in SCI treatment, but further research is required to address existing limitations and ascertain the optimal strategies for clinical implementation. These efforts will contribute to enhancing the quality of life for SCI patients and realizing broader clinical applications.

6. Conclusions

Dental-derived stem cells represent a promising avenue for SCI therapy that can enhance the quality of life and clinical outcomes for SCI patients. Research has demonstrated their effectiveness in addressing SCI using different treatment strategies and plans. However, the application of dental-derived stem cells for SCI treatment faces technical challenges such as stem cell selection, maintenance of multipotency, cell fate determination, efficacy of delivery systems, and assessment of treatment outcomes. Further research is necessary to optimize treatment protocols and explore differences between various SCI models and biological mechanisms associated with dental-derived stem cells to maximize the therapeutic potential in future clinics.

Author Contributions: Conceptualization, B.S.; resources, B.S. and Y.K.; writing—original draft preparation, X.W.; writing—review and editing, B.S., W.J., Q.L. and X.L.; visualization, X.W.; supervision, B.S. and Y.K.; project administration, B.S.; funding acquisition, B.S. All authors have read and agreed to the published version of the manuscript.

Funding: This research was funded by the National Natural Science Foundation of China, grant number T2350710233.

Conflicts of Interest: The authors declare no conflicts of interest.

References

1. Pajer, K.; Bellák, T.; Nógrádi, A. Stem Cell Secretome for Spinal Cord Repair: Is It More than Just a Random Baseline Set of Factors? *Cells* **2021**, *10*, 3214. [CrossRef] [PubMed]
2. Alizadeh, A.; Dyck, S.M.; Karimi-Abdolrezaee, S. Traumatic Spinal Cord Injury: An Overview of Pathophysiology, Models and Acute Injury Mechanisms. *Front. Neurol.* **2019**, *10*, 282. [CrossRef] [PubMed]
3. Hu, X.; Xu, W.; Ren, Y.; Wang, Z.; He, X.; Huang, R.; Ma, B.; Zhao, J.; Zhu, R.; Cheng, L. Spinal cord injury: Molecular mechanisms and therapeutic interventions. *Signal Transduct. Target. Ther.* **2023**, *8*, 245. [CrossRef] [PubMed]
4. Xu, Y.; Chen, M.; Zhang, T.; Ma, Y.; Chen, X.; Zhou, P.; Zhao, X.; Pang, F.; Liang, W. Spinal cord regeneration using dental stem cell-based therapies. *Acta Neurobiol. Exp.* **2019**.
5. Anjum, A.; Yazid, M.D.; Fauzi Daud, M.; Idris, J.; Ng, A.M.H.; Selvi Naicker, A.; Ismail, O.H.R.; Athi Kumar, R.K.; Lokanathan, Y. Spinal Cord Injury: Pathophysiology, Multimolecular Interactions, and Underlying Recovery Mechanisms. *Int. J. Mol. Sci.* **2020**, *21*, 7533. [CrossRef] [PubMed]
6. Khorasanizadeh, M.; Yousefifard, M.; Eskian, M.; Lu, Y.; Chalangari, M.; Harrop, J.S.; Jazayeri, S.B.; Seyedpour, S.; Khodaei, B.; Hosseini, M.; et al. Neurological recovery following traumatic spinal cord injury: A systematic review and meta-analysis. *J. Neurosurg. Spine* **2019**, *30*, 683–699. [CrossRef]
7. Mackiewicz-Milewska, M.; Newland, P. Spinal Cord Injury (SCI) 2016 Facts and Figures at a Glance. *J. Spinal Cord. Med.* **2016**, *39*, 493–494. [CrossRef]
8. Venkatesh, K.; Ghosh, S.K.; Mullick, M.; Manivasagam, G.; Sen, D. Spinal cord injury: Pathophysiology, treatment strategies, associated challenges, and future implications. *Cell Tissue Res.* **2019**, *377*, 125–151. [CrossRef]
9. Gao, L.; Peng, Y.; Xu, W.; He, P.; Li, T.; Lu, X.; Chen, G. Progress in Stem Cell Therapy for Spinal Cord Injury. *Stem Cells Int.* **2020**, *2020*, 2853650. [CrossRef] [PubMed]
10. Yuan, X.; Wu, Q.; Wang, P.; Jing, Y.; Yao, H.; Tang, Y.; Li, Z.; Zhang, H.; Xiu, R. Exosomes Derived from Pericytes Improve Microcirculation and Protect Blood-Spinal Cord Barrier After Spinal Cord Injury in Mice. *Front. Neurosci.* **2019**, *13*, 319. [CrossRef] [PubMed]
11. Kirshblum, S.C.; Burns, S.P.; Biering-Sorensen, F.; Donovan, W.; Graves, D.E.; Jha, A.; Johansen, M.; Jones, L.; Krassioukov, A.; Mulcahey, M.J.; et al. International standards for neurological classification of spinal cord injury (Revised 2011). *J. Spinal Cord Med.* **2011**, *34*, 535–546. [CrossRef]
12. Roberts, T.T.; Leonard, G.R.; Cepela, D.J. Classifications in Brief: American Spinal Injury Association (ASIA) Impairment Scale. *Clin. Orthop. Relat. Res.* **2017**, *475*, 1499–1504. [CrossRef]

13. Dietz, V.; Colombo, G.; Jensen, L. Locomotor activity in spinal man. *Lancet* **1994**, *344*, 1260–1263. [CrossRef]
14. Piira, A.; Lannem, A.M.; Sørensen, M.; Glott, T.; Knutsen, R.; Jørgensen, L.; Gjesdal, K.; Hjeltnes, N.; Knutsen, S.F. Manually assisted body-weight supported locomotor training does not re-establish walking in non-walking subjects with chronic incomplete spinal cord injury: A randomized clinical trial. *J. Rehabil. Med.* **2019**, *51*, 113–119. [CrossRef] [PubMed]
15. Cao, Y.; Stillman, M.D. SCI Facts and Figures. *J. Spinal Cord. Med.* **2017**, *40*, 126–127. [CrossRef]
16. Lima, R.; Monteiro, A.; Salgado, A.J.; Monteiro, S.; Silva, N.A. Pathophysiology and Therapeutic Approaches for Spinal Cord Injury. *Int. J. Mol. Sci.* **2022**, *23*, 13833. [CrossRef]
17. Katoh, H.; Yokota, K.; Fehlings, M.G. Regeneration of Spinal Cord Connectivity Through Stem Cell Transplantation and Biomaterial Scaffolds. *Front. Cell Neurosci.* **2019**, *13*, 248. [CrossRef] [PubMed]
18. Couillard-Despres, S.; Bieler, L.; Vogl, M. Pathophysiology of Traumatic Spinal Cord Injury. In *Neurological Aspects of Spinal Cord Injury*; Weidner, N., Rupp, R., Tansey, K.E., Eds.; Springer International Publishing: Cham, Switzerland, 2017; pp. 503–528. [CrossRef]
19. Hachem, L.D.; Fehlings, M.G. Pathophysiology of Spinal Cord Injury. *Neurosurg. Clin. N. Am.* **2021**, *32*, 305–313. [CrossRef] [PubMed]
20. Tran, A.P.; Warren, P.M.; Silver, J. New insights into glial scar formation after spinal cord injury. *Cell Tissue Res.* **2022**, *387*, 319–336. [CrossRef] [PubMed]
21. Zhang, Y.; Al Mamun, A.; Yuan, Y.; Lu, Q.; Xiong, J.; Yang, S.; Wu, C.; Wu, Y.; Wang, J. Acute spinal cord injury: Pathophysiology and pharmacological intervention (Review). *Mol. Med. Rep.* **2021**, *23*, 417. [CrossRef]
22. Tran, A.P.; Warren, P.M.; Silver, J. The Biology of Regeneration Failure and Success After Spinal Cord Injury. *Physiol. Rev.* **2018**, *98*, 881–917. [CrossRef]
23. Chay, W.; Kirshblum, S. Predicting Outcomes After Spinal Cord Injury. *Phys. Med. Rehabil. Clin. N. Am.* **2020**, *31*, 331–343. [CrossRef]
24. Bozzo, A.; Marcoux, J.; Radhakrishna, M.; Pelletier, J.; Goulet, B. The Role of Magnetic Resonance Imaging in the Management of Acute Spinal Cord Injury. *J. Neurotrauma* **2011**, *28*, 1401–1411. [CrossRef] [PubMed]
25. Kakulas, B.A. The clinical neuropathology of spinal cord injury a guide to the future. *Spinal Cord.* **1987**, *25*, 212–216. [CrossRef]
26. Siegenthaler, M.M.; Tu, M.K.; Keirstead, H.S. The Extent of Myelin Pathology Differs following Contusion and Transection Spinal Cord Injury. *J. Neurotrauma* **2007**, *24*, 1631–1646. [CrossRef]
27. Choo, A.M.; Liu, J.; Dvorak, M.; Tetzlaff, W.; Oxland, T.R. Secondary pathology following contusion, dislocation, and distraction spinal cord injuries. *Exp. Neurol.* **2008**, *212*, 490–506. [CrossRef]
28. Ropper, A.E.; Ropper, A.H. Acute Spinal Cord Compression. *N. Engl. J. Med.* **2017**, *376*, 1358–1369. [CrossRef] [PubMed]
29. Taylor, J.W.; Schiff, D. Metastatic Epidural Spinal Cord Compression. *Semin. Neurol.* **2010**, *30*, 245–253. [CrossRef]
30. Chen, K.; Liu, J.; Assinck, P.; Bhatnagar, T.; Streijger, F.; Zhu, Q.; Dvorak, M.F.; Kwon, B.K.; Tetzlaff, W.; Oxland, T.R. Differential Histopathological and Behavioral Outcomes Eight Weeks after Rat Spinal Cord Injury by Contusion, Dislocation, and Distraction Mechanisms. *J. Neurotrauma* **2016**, *33*, 1667–1684. [CrossRef]
31. Choo, A.M.-T. Clinically Relevant Mechanisms of Spinal Cord Injury: Contusion, Dislocation, and Distraction. Ph.D. Thesis, University of British Columbia, Vancouver, BC, Canada, 2007. [CrossRef]
32. Wang, F.; Zhang, J.; Tang, H.; Li, X.; Jiang, S.; Lv, Z.; Liu, S.; Chen, S.; Liu, J.; Hong, Y. Characteristics and rehabilitation for patients with spinal cord stab injury. *J. Phys. Ther. Sci.* **2015**, *27*, 3671–3673. [CrossRef]
33. Cheriyan, T.; Ryan, D.J.; Weinreb, J.H.; Cheriyan, J.; Paul, J.C.; Lafage, V.; Kirsch, T.; Errico, T.J. Spinal cord injury models: A review. *Spinal Cord.* **2014**, *52*, 588–595. [CrossRef]
34. Zipser, C.M.; Cragg, J.J.; Guest, J.D.; Fehlings, M.G.; Jutzeler, C.R.; Anderson, A.J.; Curt, A. Cell-based and stem-cell-based treatments for spinal cord injury: Evidence from clinical trials. *Lancet Neurol.* **2022**, *21*, 659–670. [CrossRef]
35. Sybil, D.; Jain, V.; Mohanty, S.; Husain, S.A. Oral stem cells in intraoral bone formation. *J. Oral Biosci.* **2020**, *62*, 36–43. [CrossRef]
36. Bianco, P. "Mesenchymal" Stem Cells. *Annu. Rev. Cell Dev. Biol.* **2014**, *30*, 677–704. [CrossRef]
37. Liu, Q.; Telezhkin, V.; Jiang, W.; Gu, Y.; Wang, Y.; Hong, W.; Tian, W.; Yarova, P.; Zhang, G.; Lee, S.M.; et al. Electric field stimulation boosts neuronal differentiation of neural stem cells for spinal cord injury treatment via PI3K/Akt/GSK-3β/β-catenin activation. *Cell Biosci.* **2023**, *13*, 4. [CrossRef] [PubMed]
38. Levy, O.; Kuai, R.; Siren, E.M.J.; Bhere, D.; Milton, Y.; Nissar, N.; De Biasio, M.; Heinelt, M.; Reeve, B.; Abdi, R.; et al. Shattering barriers toward clinically meaningful MSC therapies. *Sci. Adv.* **2020**, *6*, eaba6884. [CrossRef] [PubMed]
39. Lim, Y.-J.; Jung, G.N.; Park, W.-T.; Seo, M.-S.; Lee, G.W. Therapeutic potential of small extracellular vesicles derived from mesenchymal stem cells for spinal cord and nerve injury. *Front. Cell Dev. Biol.* **2023**, *11*, 1151357. [CrossRef] [PubMed]
40. Song, N.; Scholtemeijer, M.; Shah, K. Mesenchymal Stem Cell Immunomodulation: Mechanisms and Therapeutic Potential. *Trends Pharmacol. Sci.* **2020**, *41*, 653–664. [CrossRef] [PubMed]
41. Chen, P.; Tang, S.; Li, M.; Wang, D.; Chen, C.; Qiu, Y.; Fang, Z.; Zhang, H.; Gao, H.; Weng, H.; et al. Single-Cell and Spatial Transcriptomics Decodes Wharton's Jelly-Derived Mesenchymal Stem Cells Heterogeneity and a Subpopulation with Wound Repair Signatures. *Adv. Sci.* **2023**, *10*, e2204786. [CrossRef] [PubMed]
42. Forte, D.; García-Fernández, M.; Sánchez-Aguilera, A.; Stavropoulou, V.; Fielding, C.; Martín-Pérez, D.; López, J.A.; Costa, A.S.H.; Tronci, L.; Nikitopoulou, E.; et al. Bone Marrow Mesenchymal Stem Cells Support Acute Myeloid Leukemia Bioenergetics and Enhance Antioxidant Defense and Escape from Chemotherapy. *Cell Metab.* **2020**, *32*, 829–843.e9. [CrossRef] [PubMed]

43. Mazini, L.; Rochette, L.; Admou, B.; Amal, S.; Malka, G. Hopes and Limits of Adipose-Derived Stem Cells (ADSCs) and Mesenchymal Stem Cells (MSCs) in Wound Healing. *Int. J. Mol. Sci.* **2020**, *21*, 1306. [CrossRef] [PubMed]

44. Diotallevi, F.; Di Vincenzo, M.; Martina, E.; Radi, G.; Lariccia, V.; Offidani, A.; Orciani, M.; Campanati, A. Mesenchymal Stem Cells and Psoriasis: Systematic Review. *Int. J. Mol. Sci.* **2022**, *23*, 15080. [CrossRef] [PubMed]

45. Jiang, W.; Wang, D.; Alraies, A.; Liu, Q.; Zhu, B.; Sloan, A.J.; Ni, L.; Song, B. Wnt-GSK3 β/β -Catenin Regulates the Differentiation of Dental Pulp Stem Cells into Bladder Smooth Muscle Cells. *Stem Cells Int.* **2019**, *2019*, 1–13. [CrossRef]

46. Stanko, P.; Altanerova, U.; Jakubechova, J.; Repiska, V.; Altaner, C. Dental Mesenchymal Stem/Stromal Cells and Their Exosomes. *Stem Cells Int.* **2018**, *2018*, 8973613. [CrossRef] [PubMed]

47. Mattei, V.; Martellucci, S.; Pulcini, F.; Santilli, F.; Sorice, M.; Delle Monache, S. Regenerative Potential of DPSCs and Revascularization: Direct, Paracrine or Autocrine Effect? *Stem Cell Rev. Rep.* **2021**, *17*, 1635–1646. [CrossRef]

48. Miura, M.; Gronthos, S.; Zhao, M.; Lu, B.; Fisher, L.W.; Robey, P.G.; Shi, S. SHED: Stem cells from human exfoliated deciduous teeth. *Proc. Natl. Acad. Sci. USA* **2003**, *100*, 5807–5812. [CrossRef]

49. Liu, Q.; Gao, Y.; He, J. Stem Cells from the Apical Papilla (SCAPs): Past, Present, Prospects, and Challenges. *Biomedicines* **2023**, *11*, 2047. [CrossRef]

50. Zhang, J.; Ding, H.; Liu, X.; Sheng, Y.; Liu, X.; Jiang, C. Dental Follicle Stem Cells: Tissue Engineering and Immunomodulation. *Stem Cells Dev.* **2019**, *28*, 986–994. [CrossRef]

51. Gronthos, S.; Mankani, M.; Brahim, J.; Robey, P.G.; Shi, S. Postnatal human dental pulp stem cells (DPSCs) in vitro and in vivo. *Proc. Natl. Acad. Sci. USA* **2000**, *97*, 13625–13630. [CrossRef]

52. Fu, J.; Li, X.; Jin, F.; Dong, Y.; Zhou, H.; Alhaskawi, A.; Wang, Z.; Lai, J.; Yao, C.; Ezzi, S.H.A.; et al. The potential roles of dental pulp stem cells in peripheral nerve regeneration. *Front. Neurol.* **2023**, *13*, 1098857. [CrossRef] [PubMed]

53. Xu, W.; Xu, X.; Yao, L.; Xue, B.; Xi, H.; Cao, X.; Piao, G.; Lin, S.; Wang, X. VEGFA-modified DPSCs combined with LC-YE-PLGA NGCs promote facial nerve injury repair in rats. *Heliyon* **2023**, *9*, e14626. [CrossRef]

54. Mu, X.; Liu, H.; Yang, S.; Li, Y.; Xiang, L.; Hu, M.; Wang, X. Chitosan Tubes Inoculated with Dental Pulp Stem Cells and Stem Cell Factor Enhance Facial Nerve-Vascularized Regeneration in Rabbits. *ACS Omega* **2022**, *7*, 18509–18520. [CrossRef] [PubMed]

55. Takaoka, S.; Uchida, F.; Ishikawa, H.; Toyomura, J.; Ohyama, A.; Watanabe, M.; Matsumura, H.; Marushima, A.; Iizumi, S.; Fukuzawa, S.; et al. Transplanted neural lineage cells derived from dental pulp stem cells promote peripheral nerve regeneration. *Hum. Cell* **2022**, *35*, 462–471. [CrossRef] [PubMed]

56. Mead, B.; Logan, A.; Berry, M.; Leadbeater, W.; Scheven, B.A. Intravitreally transplanted dental pulp stem cells promote neuroprotection and axon regeneration of retinal ganglion cells after optic nerve injury. *Investig. Ophthalmol. Vis. Sci.* **2013**, *54*, 7544–7556. [CrossRef] [PubMed]

57. Han, Y.; Zhang, L.; Zhang, C.; Dissanayaka, W.L. Guiding Lineage Specific Differentiation of SHED for Target Tissue/Organ Regeneration. *Curr. Stem Cell Res. Ther.* **2021**, *16*, 518–534. [CrossRef] [PubMed]

58. Anoop, M.; Datta, I. Stem Cells Derived from Human Exfoliated Deciduous Teeth (SHED) in Neuronal Disorders: A Review. *Curr. Stem Cell Res. Ther.* **2021**, *16*, 535–550. [CrossRef]

59. Fuloria, S.; Jain, A.; Singh, S.; Hazarika, I.; Salile, S.; Fuloria, N.K. Regenerative Potential of Stem Cells Derived from Human Exfoliated Deciduous (SHED) Teeth during Engineering of Human Body Tissues. *Curr. Stem Cell Res. Ther.* **2021**, *16*, 507–517. [CrossRef] [PubMed]

60. Sugimura-Wakayama, Y.; Katagiri, W.; Osugi, M.; Kawai, T.; Ogata, K.; Sakaguchi, K.; Hibi, H. Peripheral Nerve Regeneration by Secretomes of Stem Cells from Human Exfoliated Deciduous Teeth. *Stem Cells Dev.* **2015**, *24*, 2687–2699. [CrossRef]

61. Sonoyama, W.; Liu, Y.; Fang, D.; Yamaza, T.; Seo, B.-M.; Zhang, C.; Liu, H.; Gronthos, S.; Wang, C.-Y.; Shi, S.; et al. Mesenchymal Stem Cell-Mediated Functional Tooth Regeneration in Swine. *PLoS ONE* **2006**, *1*, e79. [CrossRef]

62. Nagata, M.; Ono, N.; Ono, W. Unveiling diversity of stem cells in dental pulp and apical papilla using mouse genetic models: A literature review. *Cell Tissue Res.* **2021**, *383*, 603–616. [CrossRef]

63. Calabrese, E.J. Hormesis and dental apical papilla stem cells. *Chem.-Biol. Interact.* **2022**, *357*, 109887. [CrossRef]

64. Eskander, M.A.; Takimoto, K.; Diogenes, A. Evaluation of mesenchymal stem cell modulation of trigeminal neuronal responses to cold. *Neuroscience* **2017**, *360*, 61–67. [CrossRef] [PubMed]

65. Liu, X.M.; Liu, Y.; Yu, S.; Jiang, L.M.; Song, B.; Chen, X. Potential immunomodulatory effects of stem cells from the apical papilla on Treg conversion in tissue regeneration for regenerative endodontic treatment. *Int. Endod. J.* **2019**, *52*, 1758–1767. [CrossRef]

66. Zhou, L.; Liu, W.; Wu, Y.; Sun, W.; Dörfer, C.E.; Fawzy El-Sayed, K.M. Oral Mesenchymal Stem/Progenitor Cells: The Immunomodulatory Masters. *Stem Cells Int.* **2020**, *2020*, 1327405. [CrossRef]

67. Morsczeck, C.; Götz, W.; Schierholz, J.; Zeilhofer, F.; Kühn, U.; Möhl, C.; Sippel, C.; Hoffmann, K.H. Isolation of precursor cells (PCs) from human dental follicle of wisdom teeth. *Matrix Biol.* **2005**, *24*, 155–165. [CrossRef] [PubMed]

68. Wise, G.E.; Lin, F.; Fan, W. Culture and characterization of dental follicle cells from rat molars. *Cell Tissue Res.* **1992**, *267*, 483–492. [CrossRef] [PubMed]

69. Yang, C.; Du, X.-Y.; Luo, W. Clinical application prospects and transformation value of dental follicle stem cells in oral and neurological diseases. *World J. Stem Cells* **2023**, *15*, 136–149. [CrossRef] [PubMed]

70. Morsczeck, C.; Völlner, F.; Saugspier, M.; Brandl, C.; Reichert, T.E.; Driemel, O.; Schmalz, G. Comparison of human dental follicle cells (DFCs) and stem cells from human exfoliated deciduous teeth (SHED) after neural differentiation in vitro. *Clin. Oral Investig.* **2010**, *14*, 433–440. [CrossRef]

71. Bi, R.; Lyu, P.; Song, Y.; Li, P.; Song, D.; Cui, C.; Fan, Y. Function of Dental Follicle Progenitor/Stem Cells and Their Potential in Regenerative Medicine: From Mechanisms to Applications. *Biomolecules* **2021**, *11*, 997. [CrossRef]
72. Mead, B.; Logan, A.; Berry, M.; Leadbeater, W.; Scheven, B.A. Concise Review: Dental Pulp Stem Cells: A Novel Cell Therapy for Retinal and Central Nervous System Repair. *Stem Cells* **2017**, *35*, 61–67. [CrossRef]
73. Bonaventura, G.; Incontro, S.; Iemmolo, R.; La Cognata, V.; Barbagallo, I.; Costanzo, E.; Barcellona, M.L.; Pellitteri, R.; Cavallaro, S. Dental mesenchymal stem cells and neuro-regeneration: A focus on spinal cord injury. *Cell Tissue Res.* **2020**, *379*, 421–428. [CrossRef] [PubMed]
74. Mohebichamkhorami, F.; Niknam, Z.; Zali, H.; Mostafavi, E. Therapeutic Potential of Oral-Derived Mesenchymal Stem Cells in Retinal Repair. *Stem Cell Rev. Rep.* **2023**, *19*, 2709–2723. [CrossRef]
75. Koutsoumparis, A.E.; Patsiarika, A.; Tsingotjidou, A.; Pappas, I.; Tsiftsoglou, A.S. Neural Differentiation of Human Dental Mesenchymal Stem Cells Induced by ATRA and UDP-4: A Comparative Study. *Biomolecules* **2022**, *12*, 218. [CrossRef] [PubMed]
76. Mukhamedshina, Y.; Shulman, I.; Ogurcov, S.; Kostennikov, A.; Zakirova, E.; Akhmetzyanova, E.; Rogozhin, A.; Masgutova, G.; James, V.; Masgutov, R.; et al. Mesenchymal Stem Cell Therapy for Spinal Cord Contusion: A Comparative Study on Small and Large Animal Models. *Biomolecules* **2019**, *9*, 811. [CrossRef] [PubMed]
77. Sakai, K.; Yamamoto, A.; Matsubara, K.; Nakamura, S.; Naruse, M.; Yamagata, M.; Sakamoto, K.; Tauchi, R.; Wakao, N.; Imagama, S.; et al. Human dental pulp-derived stem cells promote locomotor recovery after complete transection of the rat spinal cord by multiple neuro-regenerative mechanisms. *J. Clin. Investig.* **2012**, *122*, 80–90. [CrossRef]
78. Zhu, S.; Ying, Y.; He, Y.; Zhong, X.; Ye, J.; Huang, Z.; Chen, M.; Wu, Q.; Zhang, Y.; Xiang, Z.; et al. Hypoxia response element-directed expression of bFGF in dental pulp stem cells improve the hypoxic environment by targeting pericytes in SCI rats. *Bioact. Mater.* **2021**, *6*, 2452–2466. [CrossRef]
79. do Couto Nicola, F.; Marques, M.R.; Odorcyk, F.; Arcego, D.M.; Petenuzzo, L.; Aristimunha, D.; Vizuete, A.; Sanches, E.F.; Pereira, D.P.; Maurmann, N.; et al. Neuroprotector effect of stem cells from human exfoliated deciduous teeth transplanted after traumatic spinal cord injury involves inhibition of early neuronal apoptosis. *Brain Res.* **2017**, *1663*, 95–105. [CrossRef]
80. Nicola, F.; Marques, M.R.; Odorcyk, F.; Petenuzzo, L.; Aristimunha, D.; Vizuete, A.; Sanches, E.F.; Pereira, D.P.; Maurmann, N.; Gonçalves, C.-A.; et al. Stem Cells from Human Exfoliated Deciduous Teeth Modulate Early Astrocyte Response after Spinal Cord Contusion. *Mol. Neurobiol.* **2019**, *56*, 748–760. [CrossRef]
81. Taghipour, Z.; Karbalaie, K.; Kiani, A.; Niapour, A.; Bahramian, H.; Nasr-Esfahani, M.H.; Baharvand, H. Transplantation of undifferentiated and induced human exfoliated deciduous teeth-derived stem cells promote functional recovery of rat spinal cord contusion injury model. *Stem Cells Dev.* **2012**, *21*, 1794–1802. [CrossRef]
82. Yang, C.; Li, X.; Sun, L.; Guo, W.; Tian, W. Potential of human dental stem cells in repairing the complete transection of rat spinal cord. *J. Neural Eng.* **2017**, *14*, 026005. [CrossRef]
83. Matsubara, K.; Matsushita, Y.; Sakai, K.; Kano, F.; Kondo, M.; Noda, M.; Hashimoto, N.; Imagama, S.; Ishiguro, N.; Suzumura, A.; et al. Secreted Ectodomain of Sialic Acid-Binding Ig-Like Lectin-9 and Monocyte Chemoattractant Protein-1 Promote Recovery after Rat Spinal Cord Injury by Altering Macrophage Polarity. *J. Neurosci.* **2015**, *35*, 2452–2464. [CrossRef]
84. Asadi-Golshan, R.; Razban, V.; Mirzaei, E.; Rahmanian, A.; Khajeh, S.; Mostafavi-Pour, Z.; Dehghani, F. Sensory and Motor Behavior Evidences Supporting the Usefulness of Conditioned Medium from Dental Pulp-Derived Stem Cells in Spinal Cord Injury in Rats. *Asian Spine J.* **2018**, *12*, 785–793. [CrossRef]
85. Asadi-Golshan, R.; Razban, V.; Mirzaei, E.; Rahmanian, A.; Khajeh, S.; Mostafavi-Pour, Z.; Dehghani, F. Efficacy of dental pulp-derived stem cells conditioned medium loaded in collagen hydrogel in spinal cord injury in rats: Stereological evidence. *J. Chem. Neuroanat.* **2021**, *116*, 101978. [CrossRef] [PubMed]
86. Qiao, S.; Liu, Y.; Han, F.; Guo, M.; Hou, X.; Ye, K.; Deng, S.; Shen, Y.; Zhao, Y.; Wei, H.; et al. An Intelligent Neural Stem Cell Delivery System for Neurodegenerative Diseases Treatment. *Adv. Healthc. Mater.* **2018**, *7*, 1800080. [CrossRef] [PubMed]
87. Li, L.; Xiao, B.; Mu, J.; Zhang, Y.; Zhang, C.; Cao, H.; Chen, R.; Patra, H.K.; Yang, B.; Feng, S.; et al. A MnO$_2$ Nanoparticle-Dotted Hydrogel Promotes Spinal Cord Repair via Regulating Reactive Oxygen Species Microenvironment and Synergizing with Mesenchymal Stem Cells. *ACS Nano* **2019**, *13*, 14283–14293. [CrossRef]
88. Ying, Y.; Huang, Z.; Tu, Y.; Wu, Q.; Li, Z.; Zhang, Y.; Yu, H.; Zeng, A.; Huang, H.; Ye, J.; et al. A shear-thinning, ROS-scavenging hydrogel combined with dental pulp stem cells promotes spinal cord repair by inhibiting ferroptosis. *Bioact. Mater.* **2023**, *22*, 274–290. [CrossRef]
89. Zhou, H.; Jing, S.; Xiong, W.; Zhu, Y.; Duan, X.; Li, R.; Peng, Y.; Kumeria, T.; He, Y.; Ye, Q. Metal-organic framework materials promote neural differentiation of dental pulp stem cells in spinal cord injury. *J. Nanobiotechnol.* **2023**, *21*, 316. [CrossRef]
90. Cortiella, J.; Nichols, J.E.; Kojima, K.; Bonassar, L.J.; Dargon, P.; Roy, A.K.; Vacant, M.P.; Niles, J.A.; Vacanti, C.A. Tissue-engineered lung: An in vivo and in vitro comparison of polyglycolic acid and pluronic F-127 hydrogel/somatic lung progenitor cell constructs to support tissue growth. *Tissue Eng.* **2006**, *12*, 1213–1225. [CrossRef]
91. Luo, L.; Albashari, A.A.; Wang, X.; Jin, L.; Zhang, Y.; Zheng, L.; Xia, J.; Xu, H.; Zhao, Y.; Xiao, J.; et al. Effects of Transplanted Heparin-Poloxamer Hydrogel Combining Dental Pulp Stem Cells and bFGF on Spinal Cord Injury Repair. *Stem Cells Int.* **2018**, *2018*, 2398521. [CrossRef] [PubMed]
92. Albashari, A.; He, Y.; Zhang, Y.; Ali, J.; Lin, F.; Zheng, Z.; Zhang, K.; Cao, Y.; Xu, C.; Luo, L.; et al. Thermosensitive bFGF-Modified Hydrogel with Dental Pulp Stem Cells on Neuroinflammation of Spinal Cord Injury. *ACS Omega* **2020**, *5*, 16064–16075. [CrossRef]

93. Zhu, S.; Ying, Y.; Wu, Q.; Ni, Z.; Huang, Z.; Cai, P.; Tu, Y.; Ying, W.; Ye, J.; Zhang, R.; et al. Alginate self-adhesive hydrogel combined with dental pulp stem cells and FGF21 repairs hemisection spinal cord injury via apoptosis and autophagy mechanisms. *Chem. Eng. J.* **2021**, *426*, 130827. [CrossRef]
94. De Berdt, P.; Vanvarenberg, K.; Ucakar, B.; Bouzin, C.; Paquot, A.; Gratpain, V.; Loriot, A.; Payen, V.; Bearzatto, B.; Muccioli, G.G.; et al. The human dental apical papilla promotes spinal cord repair through a paracrine mechanism. *Cell Mol. Life Sci.* **2022**, *79*, 252. [CrossRef] [PubMed]
95. De Berdt, P.; Vanacker, J.; Ucakar, B.; Elens, L.; Diogenes, A.; Leprince, J.G.; Deumens, R.; des Rieux, A. Dental Apical Papilla as Therapy for Spinal Cord Injury. *J. Dent. Res.* **2015**, *94*, 1575–1581. [CrossRef] [PubMed]
96. Zhang, J.; Lu, X.; Feng, G.; Gu, Z.; Sun, Y.; Bao, G.; Xu, G.; Lu, Y.; Chen, J.; Xu, L.; et al. Chitosan scaffolds induce human dental pulp stem cells to neural differentiation: Potential roles for spinal cord injury therapy. *Cell Tissue Res.* **2016**, *366*, 129–142. [CrossRef] [PubMed]
97. Zheng, K.; Feng, G.; Zhang, J.; Xing, J.; Huang, D.; Lian, M.; Zhang, W.; Wu, W.; Hu, Y.; Lu, X.; et al. Basic fibroblast growth factor promotes human dental pulp stem cells cultured in 3D porous chitosan scaffolds to neural differentiation. *Int. J. Neurosci.* **2021**, *131*, 625–633. [CrossRef]
98. Guo, S.; Redenski, I.; Landau, S.; Szklanny, A.; Merdler, U.; Levenberg, S. Prevascularized Scaffolds Bearing Human Dental Pulp Stem Cells for Treating Complete Spinal Cord Injury. *Adv. Healthc. Mater.* **2020**, *9*, 2000974. [CrossRef]
99. Li, X.; Yang, C.; Li, L.; Xiong, J.; Xie, L.; Yang, B.; Yu, M.; Feng, L.; Jiang, Z.; Guo, W.; et al. A therapeutic strategy for spinal cord defect: Human dental follicle cells combined with aligned PCL/PLGA electrospun material. *Biomed. Res. Int.* **2015**, *2015*, 197183. [CrossRef]
100. San Jose, L.H.; Stephens, P.; Song, B.; Barrow, D. Microfluidic Encapsulation Supports Stem Cell Viability, Proliferation, and Neuronal Differentiation. *Tissue Eng. Part. C-Methods* **2018**, *24*, 158–170. [CrossRef]
101. Kandalam, S.; De Berdt, P.; Ucakar, B.; Vanvarenberg, K.; Bouzin, C.; Gratpain, V.; Diogenes, A.; Montero-Menei, C.N.; des Rieux, A. Human dental stem cells of the apical papilla associated to BDNF-loaded pharmacologically active microcarriers (PAMs) enhance locomotor function after spinal cord injury. *Int. J. Pharm.* **2020**, *587*, 119685. [CrossRef]
102. Zhang, Y.; Liu, Y.; Liu, H.; Tang, W.H. Exosomes: Biogenesis, biologic function and clinical potential. *Cell Biosci.* **2019**, *9*, 19. [CrossRef]
103. Fayazi, N.; Sheykhhasan, M.; Soleimani Asl, S.; Najafi, R. Stem Cell-Derived Exosomes: A New Strategy of Neurodegenerative Disease Treatment. *Mol. Neurobiol.* **2021**, *58*, 3494–3514. [CrossRef]
104. Liu, C.; Hu, F.; Jiao, G.; Guo, Y.; Zhou, P.; Zhang, Y.; Yi, J.; You, Y.; Li, Z.; Wang, H.; et al. Dental pulp stem cell-derived exosomes suppress M1 macrophage polarization through the ROS-MAPK-NFκB P65 signaling pathway after spinal cord injury. *J. Nanobiotechnol.* **2022**, *20*, 65. [CrossRef]
105. Feitosa, M.L.T.; Sarmento, C.A.P.; Bocabello, R.Z.; Beltrão-Braga, P.C.B.; Pignatari, G.C.; Giglio, R.F.; Miglino, M.A.; Orlandin, J.R.; Ambrósio, C.E. Transplantation of human immature dental pulp stem cell in dogs with chronic spinal cord injury. *Acta Cir. Bras.* **2017**, *32*, 540–549. [CrossRef]
106. Prado, C.; Fratini, P.; de Sá Schiavo Matias, G.; Bocabello, R.Z.; Monteiro, J.; dos Santos, C.J.; Joaquim, J.G.F.; Giglio, R.F.; Possebon, F.S.; Sakata, S.H.; et al. Combination of stem cells from deciduous teeth and electroacupuncture for therapy in dogs with chronic spinal cord injury: A pilot study. *Res. Vet. Sci.* **2019**, *123*, 247–251. [CrossRef]
107. Nicola, F.C.; Rodrigues, L.P.; Crestani, T.; Quintiliano, K.; Sanches, E.F.; Willborn, S.; Aristimunha, D.; Boisserand, L.; Pranke, P.; Netto, C.A. Human dental pulp stem cells transplantation combined with treadmill training in rats after traumatic spinal cord injury. *Braz. J. Med. Biol. Res.* **2016**, *49*, e5319. [CrossRef] [PubMed]
108. Liu, W.; Rong, Y.; Wang, J.; Zhou, Z.; Ge, X.; Ji, C.; Jiang, D.; Gong, F.; Li, L.; Chen, J.; et al. Exosome-shuttled miR-216a-5p from hypoxic preconditioned mesenchymal stem cells repair traumatic spinal cord injury by shifting microglial M1/M2 polarization. *J. Neuroinflamm.* **2020**, *17*, 47. [CrossRef] [PubMed]
109. Yang, Z.-L.; Rao, J.; Lin, F.-B.; Liang, Z.-Y.; Xu, X.-J.; Lin, Y.-K.; Chen, X.-Y.; Wang, C.-H.; Chen, C.-M. The Role of Exosomes and Exosomal Noncoding RNAs From Different Cell Sources in Spinal Cord Injury. *Front. Cell Neurosci.* **2022**, *16*, 882306. [CrossRef] [PubMed]
110. Kabatas, S.; Demir, C.S.; Civelek, E.; Yilmaz, I.; Kircelli, A.; Yilmaz, C.; Akyuva, Y.; Karaoz, E. Neuronal regeneration in injured rat spinal cord after human dental pulp derived neural crest stem cell transplantation. *Bratisl. Lek. Listy* **2018**, *119*, 143–151. [CrossRef] [PubMed]
111. Ge, L.; Arul, K.; Mesfin, A. Spinal Cord Injury from Spinal Tumors: Prevalence, Management, and Outcomes. *World Neurosurg.* **2019**, *122*, e1551–e1556. [CrossRef] [PubMed]
112. Sahni, V.; Kessler, J.A. Stem cell therapies for spinal cord injury. *Nat. Rev. Neurol.* **2010**, *6*, 363–372. [CrossRef]
113. Deng, J.; Zhang, Y.; Xie, Y.; Zhang, L.; Tang, P. Cell Transplantation for Spinal Cord Injury: Tumorigenicity of Induced Pluripotent Stem Cell-Derived Neural Stem/Progenitor Cells. *Stem Cells Int.* **2018**, *2018*, e5653787. [CrossRef] [PubMed]

Review

Rehabilitation Training after Spinal Cord Injury Affects Brain Structure and Function: From Mechanisms to Methods

Le-Wei He [1], Xiao-Jun Guo [1], Can Zhao [2],* and Jia-Sheng Rao [1],*

[1]   Beijing Key Laboratory for Biomaterials and Neural Regeneration, Beijing Advanced Innovation Center for Biomedical Engineering, School of Biological Science and Medical Engineering, Beihang University, Beijing 100191, China; hlw0118@buaa.edu.cn (L.-W.H.); guoxiaojun0513@163.com (X.-J.G.)
[2]   Institute of Rehabilitation Engineering, China Rehabilitation Science Institute, Beijing 100068, China
*   Correspondence: zhaocan05@163.com (C.Z.); raojschina@126.com (J.-S.R.)

Abstract: Spinal cord injury (SCI) is a serious neurological insult that disrupts the ascending and descending neural pathways between the peripheral nerves and the brain, leading to not only functional deficits in the injured area and below the level of the lesion but also morphological, structural, and functional reorganization of the brain. These changes introduce new challenges and uncertainties into the treatment of SCI. Rehabilitation training, a clinical intervention designed to promote functional recovery after spinal cord and brain injuries, has been reported to promote activation and functional reorganization of the cerebral cortex through multiple physiological mechanisms. In this review, we evaluate the potential mechanisms of exercise that affect the brain structure and function, as well as the rehabilitation training process for the brain after SCI. Additionally, we compare and discuss the principles, effects, and future directions of several rehabilitation training methods that facilitate cerebral cortex activation and recovery after SCI. Understanding the regulatory role of rehabilitation training at the supraspinal center is of great significance for clinicians to develop SCI treatment strategies and optimize rehabilitation plans.

Keywords: spinal cord injury; exercise; brain reorganization; rehabilitation training

Citation: He, L.-W.; Guo, X.-J.; Zhao, C.; Rao, J.-S. Rehabilitation Training after Spinal Cord Injury Affects Brain Structure and Function: From Mechanisms to Methods. *Biomedicines* **2024**, *12*, 41. https://doi.org/10.3390/biomedicines12010041

Academic Editor: Nicolas Guerout

Received: 1 November 2023
Revised: 3 December 2023
Accepted: 12 December 2023
Published: 22 December 2023

1. Introduction

According to the data of the World Health Organization (WHO), 250,000–500,000 people worldwide suffer from spinal cord injury (SCI) every year [1]. SCI typically entails structural damage to the spinal cord, leading to disruption in the transmission of sensory and motor information, thereby causing dysfunction below the affected region. This debilitating condition significantly impacts patients' daily lives and imposes substantial economic burdens. Despite the grave consequences of SCI, a viable clinical treatment approach remains elusive. The repair of the central nervous system (CNS) after SCI has been a focal point of biomedical research.

The nervous system is an essential component of the body. When SCI occurs, it cuts off the crucial information flow between the body and the brain, thereby affecting not only the areas below the level of injury but also the structure and function of the brain [2,3]. The CNS has exceptional plasticity, and this plasticity has the potential to compensate for damage. When the components of the brain are disrupted, the cerebral cortex undergoes neural network reorganization [4]. This process involves changes in neurotrophin secretion, cytokine content, biochemical composition, and nerve cell morphology, all of which contribute to the creation of appropriate new synapses and neural circuits, leading to the reorganization and compensation of the brain [2]. The ability of the brain to reorganize and compensate is crucial to the improvement of brain structure and function after SCI [5].

Rehabilitation training, a type of physical therapy that utilizes physical factors to stimulate adaptive changes in local neural circuits, is used widely in the clinical treatment of SCI to improve the dysfunction caused by SCI [6]. Previous research has shown that rehabilitation training can promote cell activation and functional remodeling in the cerebral cortex [7]. This manuscript will review the effects of rehabilitation training on brain structure and function after SCI, as well as examine how different rehabilitation methods affect brain reorganization.

2. Effect of Rehabilitation Training on Brain Structure and Function

2.1. Physiological Mechanism of Exercise and Brain Structure and Function

Exercise is beneficial for the development of muscle and physical fitness, and it is important for the activation and promotion of neuronal connections [8]. Kobilo et al. [9] found that exercise can increase the volume of the hippocampus and blood flow to this part of the brain through studies of mouse models. In different mouse models, exercise increases synaptic plasticity and induces morphological changes in dendrites [10,11]. When part of the brain structure is damaged, the use of rehabilitation training is currently considered to promote functional recovery. In addition to its effects on the local area, rehabilitation training can also affect neural connections across different brain regions. Batouli et al. [12] found that physical activity can change the brain neural network, which accounts for 82% of the total gray matter volume. Another study involving functional magnetic resonance imaging (fMRI) in patients with cervical SCI showed that the degree of activation of the cerebral cortex related to exercise is closely related to the degree of functional improvement after exercise [13].

Recent studies have shown that skeletal muscle and the brain are connected through a muscle–brain endocrine loop in the human body [14]. Skeletal muscle, acting as a secretory organ, can produce and release hundreds of myokines that communicate with other organs during muscle cell proliferation, differentiation, or muscle contraction [15,16]. Myokines can be transmitted to the liver, pancreas, bones, blood vessels, and other organs and participate in the direct regulation of brain structure and function [17,18]. In addition, skeletal muscle can participate in neural feedback to the brain through the peripheral–central nervous pathway and regulate brain functional activity and network integration through the ascending sensory information flow [14].

2.2. Effect of Rehabilitation Training on the Brain after SCI

2.2.1. Rehabilitation Training Regulates the Secretion of Neurotrophic Factors in the Brain

Brain-derived neurotrophic factor (BDNF) is an essential regulator of neuronal development, synaptic transmission, and cell and synaptic plasticity. It promotes the survival, differentiation, and growth of neurons and glial cells [19]. Rehabilitation training can stimulate the release of BDNF, thereby affecting brain structure and function. When mice engage in voluntary exercise, the level of BDNF in the hippocampus significantly increases compared to mice that do not exercise [20]. This exercise-dependent cell proliferation in the dentate gyrus of the hippocampus plays a role in some forms of memory information and maintenance. Another study showed that rehabilitation training can increase BDNF levels in the brain and spinal cord of mice, alleviate axonal demyelination after SCI, and reduce the size of the injured area [21]. Several clinical studies have also shown that BDNF levels are associated with rehabilitation training; for example, 9 weeks of aerobic and pilates training can increase human serum BDNF [22], while 3 months of aerobic exercise training increased the hippocampal volume of patients with schizophrenia by 16% [23]. A half-year resistance and endurance training can increase the serum BDNF concentration of patients with multiple sclerosis by 13.9% [24].

Aside from directly affecting BDNF concentration, rehabilitation training can regulate the motor–brain pathway related to BDNF [25]. During rehabilitation training, cathepsin B, which are myokines released by skeletal muscle, can affect neurogenesis, learning, and memory. Moon et al. [26] used AMPK agonists to treat myotubes and simulated the effect of exercise in vitro. Results show that AMPK can induce skeletal muscle cells to secrete cathepsin B. Moreover, running can increase the levels of cathepsin B in the gastrocnemius muscle of mice, and the levels of actin and cathepsin B in the plasma of rhesus monkeys and humans. Cathepsin B crosses the blood–brain barrier and enters the CNS to upregulate BDNF concentration [26]. Actin can participate in the cAMP-dependent regulation of BDNF transcription and is the main driving force of dendritic spine remodeling and sustains synaptic plasticity [27,28].

Rehabilitation training can regulate BDNF via the "PGC1α-FNDC5-irisin" pathway. PGC1α has a central role in mediating many of the metabolic effects of exercise locally within the muscle. During rehabilitation training, PGC1α expression in muscle increases and stimulates an increase in the expression of FNDC5 [29]. FNDC5 is a membrane protein that is cleaved and secreted into the circulation as the myokine irisin [30], causing an increase in irisin levels in the blood. And irisin passes through the blood–brain barrier, upregulating BDNF level in the brain [29].

In addition, rehabilitation training can upregulate BDNF expression by stimulating an increase in blood β-hydroxybutyric acid levels [31].

2.2.2. Rehabilitation Training Regulates Cytokines and Biochemical Changes in the Brain

Rehabilitation training affects the level of cytokines in the CNS and tends to create an immunosuppressive environment in the CNS to combat chronic brain inflammation induced by SCI. Rehabilitation training can significantly increase the level of cytokine IL-6 and produce anti-inflammatory effects by inhibiting the increase in circulating TNF levels induced by endotoxin [32]. Additionally, rehabilitation training can reduce the levels of proinflammatory cytokines such as IL-1β and IL-17 and increase the levels of anti-inflammatory cytokines such as IL-10 and transforming growth factor (TGF)-β [33–35]. Nichol et al. [36] found that 3 weeks of voluntary wheel training can effectively reduce the levels of proinflammatory cytokines IL-1β and TNF-α in the hippocampus of rats and make the level of inflammatory cytokines IFN-γ return to normal, which can suppress inflammatory response in the brain [37].

Rehabilitation training can also cause changes in biochemical levels in some parts of the brain. In rats undergoing treadmill rehabilitation training, the nucleolar area of motor neurons increased, and the protein synthesis related to plasticity in the cerebral cortex increased significantly [38]. Rehabilitation training can also promote the ability of motor neurons to transport positive/reverse axonal proteins, such as the highly selective transport of the synaptic protein SNAP25, thus improving the overall adaptive capacity of motor units [39].

Rehabilitation training can also promote the functional regulation of brain mitochondrial machinery. The density and function of mitochondria in the brain of adult mice increase after 4 weeks of voluntary wheel exercise [40]. Another study has shown that endurance training can induce upregulation of PGC-1α and SIRT1, activation of AMPK, decrease in p53 acetylation, and increase in mitochondrial respiratory complex content in the rodent brain [41]. These factors may have an impact on the brain, regulate brain plasticity mechanisms, and induce neuroprotective pathways. PGC-1α can up-regulate BDNF in the brain through the "PGC1α-FNDC5-irisin" pathway [29]. SIRT1 is an NAD+-dependent deacetylase. Evidence indicates that the SIRT1 1 pathway regulates cell survival and rescues proteins in several nervous system diseases [42]. SIRT1 mediates AMPK action on PGC-1α transcriptional activity and the expression of genes involved in mitochondrial activity [43]. Simultaneously, the decrease in p53 acetylation can promote the expression of SIRT1 [44]. The CNS requires a large amount of energy to conduct nerve impulses, and the enhancement of the density and function of mitochondria in the brain is beneficial to neuronal activity [45]. At the same time, the ability of mitochondria to regulate intracellular Ca2+ plays an important role in synaptic maintenance [46]. These mechanisms are beneficial to protect residual nerve tissue and function after SCI and reduce the negative effects of injury.

2.2.3. Effect of Rehabilitation Training on Nerve Cells

Oligodendrocytes, astrocytes, microglia, and other glial cells are widely distributed in the CNS [47]. These cells can connect and support various neural components and play an important role in the structure and function of the brain. SCI can lead to apoptosis or proliferation of nerve cells, thus affecting the original brain structure and function [48]. Therefore, normalizing the type and number of nerve cells in the brain needs to maintain the structure and function of the brain through rehabilitation training.

Oligodendrocyte

Oligodendrocytes are a major type of glial cells in the CNS, which can produce lipid-rich myelin around the axons of the CNS, thereby insulating axonal segments to increase axonal conduction velocity and provide nutritional support for axons by secreting metabolic factors such as lactic acid [49]. Oligodendrocyte progenitor cells (OPCs) originate from neural stem cells in the ventricular region and are evenly distributed in the CNS. In response to local proliferation and differentiation signals at their final location in the CNS, OPCs differentiate into mature oligodendrocytes [50].

Previous studies have shown that, compared with mice without exercise, the number of proliferating OPCs in mice undergoing exercise on a running wheel was significantly higher [51]. Kim et al. [52] found that rehabilitation training promoted the redifferentiation of oligodendrocytes and remyelination of axons by immunolabeling MBP- and CNPase-positive cells in the hippocampus of mice after exercise. Even without CNS injury, exercise is beneficial for oligodendrocyte formation in mice; for instance, 7 days of wheel exercise increased the proliferation of immature oligodendrocytes in the spinal cord [53]. Although exercise can increase the number of oligodendrocytes, the mechanism by which exercise promotes oligodendrocyte regeneration in diseased microenvironments remains to be further clarified.

Astrocytes

Astrocytes are a group of highly heterogeneous glial cells in the CNS. The structural diversity of astrocytes enables astrocytes to perform multiple functions [54]. Astrocytes play a crucial role in synaptic transmission and information processing, as well as the function of neural circuits [54]. Additionally, astrocytes serve important molecular functions in the formation, maintenance, and pruning of synapses, and they can regulate the homeostasis of the CNS by modulating the permeability of the blood–brain barrier and energy homeostasis [55].

After damage to the structure and function of the nervous system, astrocytes can undergo reactive hyperplasia with significant changes in gene expression, cell morphology, and function [56]. This process results in the formation of glial scars, which can promote demyelination and neurodegeneration and facilitate the penetration of peripheral immune cells into the brain through the blood–brain barrier [57].

Although SCI does not directly destroy brain structure, it affects cortical neurons through secondary degeneration; the application of rehabilitation training can help maintain neuronal function and synaptic plasticity by inhibiting astrocyte changes in this process. This is supported by several studies on brain diseases, including those by Zhang et al. [58], who found that long-term running reduced astrocyte proliferation and increased neuronal density in Alzheimer's mice, and Lee et al. [59], who performed treadmill rehabilitation exercises beginning within 24 h after cerebral ischemia injury and found that this process reduced astrocyte formation and promoted the functional recovery of rats.

Exercise can increase shear stress in endothelial cells, regulate the release of soluble factors such as CO, and subsequently regulate Ca2+ in astrocytes. This may be one of the potential mechanisms by which exercise regulates astrocyte injury response [60].

Microglia

Microglia are essential glial cells and the first line of immune defense within the CNS. The microglial population is heterogenous and can be simply classified into pro-inflammatory M1 and anti-inflammatory M2 phenotypes [61]. M1 microglia present antigens to T cells and release proinflammatory cytokines, which destroy oligodendrocytes and myelin, resulting in axonal demyelination. In contrast, M2 microglia participate in reducing inflammation by reducing glutamate excitotoxicity and increasing the release of anti-inflammatory cytokines [61].

Previous studies have shown that rehabilitation training can promote the activation of M2-like microglia in mice, which is beneficial for the protection of the CNS [62]. Subsequent studies have further revealed that rehabilitation training can reduce the number of M1-like microglia expressed in the spinal cord during CNS injury, thereby reducing the degree of inflammatory response in the body [63]. In animal models of demyelination, rehabilitation exercise may increase the proportion of M2 phenotypic microglia in the focal area and alleviate the progress of demyelinating lesions [64].

Therefore, rehabilitation training can reduce inflammation in the brain by activating M2 microglia or inhibiting M1 microglia, thus maintaining the stability of structure and function.

2.2.4. Rehabilitation Training Promotes Cortical Remodeling

The function and structure of the spinal cord and brain are closely related. The dysfunction of conduction after SCI affects the structure and function of the brain, leading to the reorganization of the cerebral cortex [2]. After injury, rehabilitation training can effectively reduce damage to brain nerve cells, promote axonal regeneration, and improve the function of the CNS circuit, thereby promoting structural and functional remodeling of the cerebral cortex and subcortical areas.

Synaptic Plasticity

The synapse is the structural basis of functional connection and information transmission between neurons. Synaptic plasticity refers to sustained changes in synaptic development, morphology, and information transmission function. Synaptic plasticity is realized by enhancing the signal transmission efficiency of synaptic sites between existing neurons, and it is one of the bases for functional recovery after neural damage. Rehabilitation training is beneficial for synaptic plasticity, and it may enhance the stability of new synapses formed in the first few weeks after cerebral infarction in rats [65].

In adults, voluntary exercise can increase the number of dendritic spines [66], as well as the density and length of dendrites in cognitive-related areas, such as the hippocampal horn and dentate gyrus [67]. SCI affects the resting potential of motor neurons in the CNS, and roller rehabilitation exercise in rats with SCI is beneficial for maintaining the resting potential of motor neurons [68]. Many animal models and studies after spinal cord and cerebral nerve injury support the conclusion that rehabilitation training promotes synaptic plasticity and neurogenesis [69].

The effect of rehabilitation training on synaptic plasticity is mainly influenced by various factors at the molecular level of synaptic structure. The mechanism of synaptic plasticity induced by rehabilitation training involves a large number of molecules involved in the maintenance and regulation of brain function, including neurotrophic factors, signal transduction proteins, transcription factors, and synaptic proteins [70,71].

Firstly, rehabilitation training can promote synaptic plasticity by modulating the molecular composition of presynaptic scaffolds. Scaffold proteins play a crucial role in numerous synaptic functions. The scaffold proteins implicated in synaptic plasticity mainly include GAP-43, synaptophysin, and PSD-95. Growth-associated protein (GAP)-43, a neuron-specific protein, is also closely related to neurogenesis, synaptic plasticity, and regeneration [72]. Mizutani et al. discovered that voluntary exercise leads to an upsurge in GAP-43 levels in the ischemic cortex and improves functional recovery [73]. Synaptophysin, a glycoprotein attached to the synaptic vesicle membrane, performs an important function in the biogenesis of synaptic vesicles [74]. Reduced synaptophysin adversely affects synaptic plasticity in the brain [75]. Motor skill training can heighten synaptophysin expression in stroke rats and promote synaptic regrowth in the thalamus [76]. As a postsynaptic marker, PSD-95 is predominantly found at mature glutamate synapses [77]. The decrease in PSD-95 leads to neuronal loss and synaptic disintegration, thereby impacting the motor, sensory, and cognitive function of the brain [77]. After rehabilitation training, the protein levels of PSD-95 in the sensorimotor cortex noticeably increase [78]. Interestingly, in a study of rats after cerebral ischemic injury, the level of PSD-95 in the bilateral hippocampus in the low-intensity exercise group was higher than that in the high-intensity exercise group [79]. Therefore, low-intensity exercise may be more beneficial for synaptic plasticity than high-intensity exercise after cerebral ischemic injury.

Secondly, rehabilitation training can promote synaptic plasticity by regulating synaptic plasticity regulatory proteins. The upregulated expression of neurotrophic factors and certain synaptic-related proteins performs an essential role in the rehabilitation of cerebral ischemic injury [73]. Rehabilitation training can impact the expression of these proteins and regulate synaptic structure and signal transmission. Common neurotrophic factors such as BDNF, GAP-43, and insulin-like growth factor (IGF-1) contribute to the formation of synaptic plasticity. BDNF phosphorylates its TrkB receptor, controls the actin cytoskeleton in dendritic spines [80] and its degeneration [81], and promotes actin polymerization [82]. BDNF may also play a key role in synaptic plasticity by controlling PSD-95 transport [83]. Rehabilitation training can significantly increase BDNF levels in the brain, thereby enhancing synaptic plasticity [84]. Calmodulin-dependent kinase II (CaMKII) is a rich synaptic signal protein molecule. CaMKII is essential for the long-term enhancement of learning and memory and performs a synaptic role in synaptic plasticity [85]. CaMKII can bind to MAP2 and actin to mediate synaptic transmission [86]. The expression of CaMKII in skeletal muscle increases with endurance training [87] and aerobic training [88]. After 10 weeks of aerobic exercise, the number of synapses and the expression of the synaptic plasticity-related protein CaMKIIα in the hippocampus of mice are noticeably increased, and aerobic exercise can reduce neurodegeneration in the hippocampus by regulating the CaMKII signal transduction cascade [89].

In addition, rehabilitation training can contribute to the formation of synaptic plasticity by modulating the molecular organization of postsynaptic membrane receptors. The AMPA and NMDA receptors are the key mediators of excitatory synaptic transmission in the brain. They are glutamate-gated cation channels that convert chemical signals (glutamate released from presynaptic terminals) into electrical signals (changes in membrane voltage) [90]. AMPA receptors can effectively regulate excitatory neurotransmission and activity-dependent plasticity [91]. The expression of AMPA glutamate receptor subunits (GluR1 and GluR2/3) increases with prolonged exercise time, indicating that the enhanced plasticity of GluR is involved in the process of exercise-induced synaptic plasticity [92]. Exercise also increases the expression of GluA2 subunits in postsynaptic proteins in the cerebral basal ganglia and induces changes in the expression of GluA1 and GluA2 subunits of AMPAR [93]. In the subacute stage of stroke, exercise intervention can significantly increase the mRNA levels of GluA1 and GluA4 and improve synaptic transmission and brain plasticity [91]. NMDA receptor is an ionic glutamate-gated receptor, which is composed of NMDA receptor subunits GluN1, GluN2, and GluN3. The composition of NMDA receptors is strictly regulated by activity-dependent synaptic plasticity during development. Treadmill exercise can improve the expression of NMDA receptors in rats [94]. Treadmill exercise inhibits neuronal apoptosis by increasing the expression of NMDA receptors [95].

It should be noted that more synapses do not always mean better brain function. A published study of brain tissue suggests that children with autism have a surplus of syn-apses [96]. Nagata et al. [97] also found that in mouse models of schizophrenia, extra-large synapses are over-represented, which evoked supra-linear dendritic and somatic integration, resulting in increased neuronal firing. The probability of extra-large spines correlated negatively with working memory [97].

Effect of Rehabilitation Training on the Structural Remodeling of the Cerebral Cortex

Following SCI, increased sensitivity of peripheral receptors and neuroinflammation lead to the apoptosis of cortical neurons, and a decrease in the number of cells results in a decrease in the volume of the cerebral cortex [98]. Using fMRI and voxel-based morphometry, Jurkiewicz et al. [99] and Wrigley et al. [100] found that the gray matter volume of the somatosensory cortex and motor cortex in patients with SCI was significantly lower than that in healthy subjects. Freund et al. [101] also confirmed the presence of structural atrophy of the sensory and motor cortex in patients with chronic cervical SCI, and this condition is closely related to the degree of dysfunction and recovery. The deterioration of cortical structural integrity reduces the processing speed and functional integration of the brain, leading to the deterioration of walking function and neuropathic pain after SCI [6,102].

Cerebral cortical plasticity refers to the ability of the CNS to undergo changes in its existing cortical structure and function in response to learning, training, or CNS injury [103]. Synaptic plasticity is one of the main components of cerebral cortex remodeling. Exercise strengthens synaptic connections and can promote synaptic formation or maintain degenerative synapses after injury [4]. The study of synaptic and axon regeneration after peripheral nerve injury also supports the view that the structure of the cerebral cortex is malleable [104]. Rehabilitation training is an important means to promote the structural reorganization of the cerebral cortex after SCI. The improvement of function after exercise therapy is related to the degree of activation of the motor cortex [13]. Studies on the structural remodeling of the cerebral cortex have shown that treadmill training can promote axonal growth [68], sprouting of diseased proximal lateral branches, and synaptic establishment [105]. The activation of dormant axons, which protrude axonal buds and grow new lateral connections, may increase the volume of the cerebral cortex and the structural remodeling of the cerebral cortex [7]. In neonatal rats, treadmill training induced a new type of tissue in both the somatosensory cortex and the motor cortex, which is directly related to the number of weight-bearing steps that animals can take when exercising on the treadmill [106,107]. Exercise can also have a neuroprotective effect on the brains of adult rats [108].

Effect of Rehabilitation Training on Functional Remodeling of the Cerebral Cortex

The spinal cord and brain have a close functional relationship. SCI affects the functional network of the motor sensory brain area and the functional connection between the cortexes [109]. Kaushal et al. [110] tested the resting-state functional connections (rs-FC) of the whole brain in patients with complete SCI and found that compared with the control group, their connectivity in the brain functional network decreased, whereas the rs-FC between the cerebellum and bilateral paracenter lobules increased. Hou et al. [111] first analyzed the whole brain ALFF of 25 patients with SCI. The results showed that FC increased in the motor network in the cerebral hemispheres, while FC in the primary motor cortex between the bilateral hemispheres decreased.

Rehabilitation training affects the structure of the cerebral cortex, as well as the function of specific brain regions or functional networks across brain regions. Considering that the motor cortex is one of the main areas that dominate movement, rehabilitation training is usually closely related to the plasticity of the motor cortex [102]. A longitudinal fMRI study of patients with cervical SCI showed that the improvement of function after exercise therapy was significantly correlated with the degree of activation of the motor cortex [13]. Several SCI rehabilitation training studies have shown that the above correlation is widespread in clinical patients [112].

In addition to changes in cortical functional activity in specific areas, functional compensation of neuronal populations in adjacent cortical areas is also a characteristic of cerebral cortical functional remodeling. Rehabilitation training after SCI can expand the activation range of the cortical area and promote the compensation of the participating function of the adjacent cortex of the denervated brain area, and this finding has been confirmed by electroencephalogram (EEG) [113], fMRI [114], and positron emission tomography studies [115]. Considering the spatial distance, the compensation of participating functions in the adjacent cortex and co-activation of neurons in adjacent areas may result from the establishment of synaptic connections induced by local rehabilitation training [116].

The functional network connections across brain regions are the basis for integrating, processing, and controlling sensorimotor information [117]. Rehabilitation exercise also regulates long-distance functional connections across brain regions, such as interhemispheric connections. Sawada et al. [118] conducted dexterous finger rehabilitation training in rhesus monkeys after SCI and used cortical electroencephalography to reveal the interaction between cortical activation and behavior. The results showed that grasping training promoted the recruitment of related networks in the contralateral hemisphere of rhesus monkeys from PM to M1, as well as the recruitment of related networks between the contralateral PM of SCI and the ipsilateral PM of SCI. The interaction between the cerebral hemispheres provides a compensatory mechanism that facilitates the activation and interaction of neurons in the bilateral motor cortex to promote functional recovery. Chisholm et al. [119] conducted motor resistance training in patients with incomplete SCI. The results showed that the sensory excitability of the body and corticospinal excitability were enhanced, and the FC value of one limb was increased with less functional impairment. This finding shows that functional training causes brain function reorganization in a short time, which is of great significance to its functional recovery. Studies on brain diseases such as stroke have further demonstrated the regulatory effect of rehabilitation exercise on the reorganization of brain functional networks. In patients who experience stroke, the unilateral voluntary handshake task can increase the activation of the M1 cortex on the affected side [120]. At the same time, early high-intensity upper limb training can increase the activation of the ipsilateral premotor area and anterior cingulate cortex and reduce the activation of the contralateral cerebellum [121]. The voluntary unilateral rehabilitation task enhanced the excitatory connection from the affected side to the contralateral M1 cortex [120]. These clinical studies on stroke have confirmed that exercise training can promote functional reorganization between cerebral hemispheres and improve motor ability by increasing the activation of the ipsilateral brain and enhancing the excitatory connection from the ipsilateral to the contralateral M1 cortex.

Maladaptive Plasticity in the Brain

Plasticity refers to the shaping and re-shaping of neural networks at the global level, as well as the remodeling of synaptic contacts at the local level. The plastic reorganization of the brain after CNS lesions plays a crucial role in the recovery and rehabilitation of sensory and motor dysfunction, but it can also be maladaptive [112]. In fact, cortical reorganization is neither beneficial nor harmful: the positive aspect of cortical reorganization can facilitate functional recovery [122,123], while its negative aspect can be maladaptive and lead to phantom sensation and neuropathic pain [112,124,125].

It has been confirmed that brain plasticity responses may result in maladaptive plasticity. Jiang et al. [126] found that reorganization in the primary somatosensory cortex of rats and mice with SCI leads to neuropathic pain. Beeler et al. [127] found that aberrant cortical learning in Parkinson's disease patients impairs cortical compensation mechanisms, exacerbating functional decline. Nudo et al. [128] also found that in the squirrel monkey stroke model, more proximal hand representation areas spread into the former finger representation area but are negatively associated with the recovery of hand function after the lesion. Therefore, a more comprehensive understanding of the effects of plasticity and brain reorganization is necessary.

2.2.5. Summary of This Section

We have briefly evaluated the potential mechanisms of exercise that affect the brain structure and function, as well as the rehabilitation training process for the brain after SCI (Figure 1). Understanding the regulatory role of rehabilitation training at the supraspinal center is of great significance for clinicians to develop SCI treatment strategies and optimize rehabilitation plans.

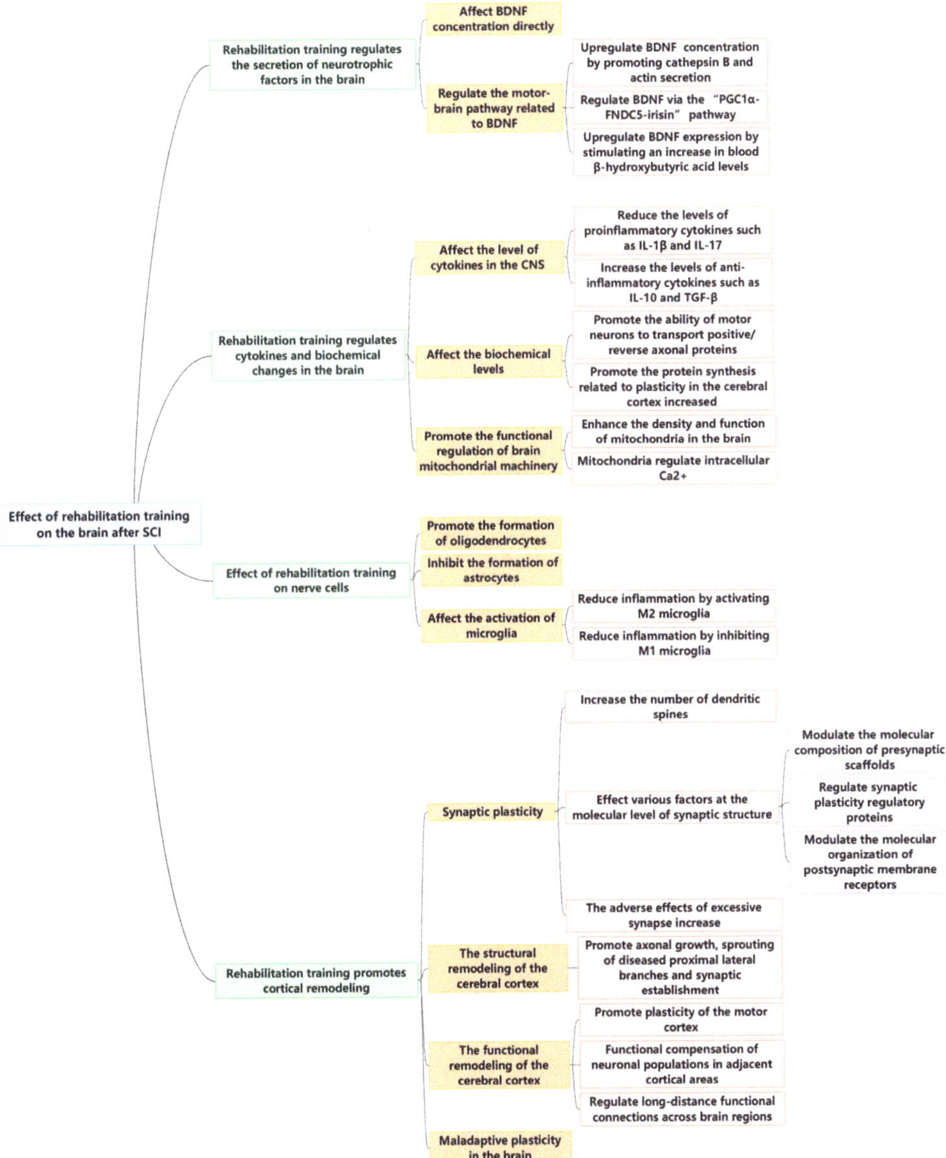

Figure 1. Effect of rehabilitation training on the brain after SCI.

3. Different Rehabilitation Techniques and Their Effects on Brain Reorganization after SCI

3.1. Rehabilitation Exercise Regulates the Brain

Rehabilitation-exercise-induced cortical reorganization mainly depends on the principle of motor learning. Motor learning has benefits for dendritic sprouting, formation of new synapses, modulation of existing synapses, and production of neurochemicals [129]. Sports learning produces remarkable results when carrying out meaningful, repetitive, and intensive rehabilitation exercises [130]. Functional recovery through motor learning can be classified into two types: substantive and compensatory motor recovery. Substantive motor recovery occurs in an uninjured or alternative pathway and transmits nerve impulses to the same muscles used before injury, while compensatory motor recovery uses other muscles different from the original muscles to achieve the same goal [131]. For patients with SCI, substantive motor recovery is the goal of functional recovery; however, it is often difficult to achieve due to damage to conduction pathways and the lack of spontaneous regeneration. Compensatory motor recovery can be achieved by means of residual neural pathways and brain central reorganization, and it is the most important type of functional recovery currently [132].

Many different rehabilitation exercises can be performed after SCI. In animal research, rehabilitation exercises can be divided into voluntary and forced exercises [133]. The most typical voluntary exercise is voluntary wheel running, where animals living alone voluntarily use running wheels for exercise. Forced exercise includes running on a treadmill and swimming in a closed swimming pool. To some extent, these two models are similar to the human rehabilitation exercise model, and they are both beneficial for regulating metabolic protein [134] and BDNF [135] and improving the degree of nervous system injury [133]. Although forced exercise can effectively quantify and reduce isolation stress, some reports suggest that voluntary exercise is beneficial for SCI rats [133] and mice [136].

For human beings, rehabilitation exercises can be classified into anaerobic and aerobic exercise, both of which can promote the change and recovery of brain structure and function after SCI. Chisholm et al. [119] found that anaerobic exercise such as resistance training can improve somatosensory sensation and corticospinal excitability in patients with motor incomplete SCI and regulate brain functional connections at rest. After resistance training, somatosensory excitability and corticospinal excitability were enhanced, and motor-evoked potentials increased [119]. The analysis of fMRI in the resting state of the patients showed that the functional connection of the motor cortex was more obvious in the less affected side after exercise. Aerobic exercise mainly refers to dynamic exercise that relies on aerobic metabolism to promote muscle contraction and the exercise of large muscle groups, including jogging, cycling, and swimming. Aerobic exercise can improve walking and balance function in patients with SCI [137]. It can also improve cerebral blood flow and neurovascular coupling to promote changes in brain structure and function [138]. In addition to SCI, aerobic and anaerobic exercise play an important role in the treatment of brain injury caused by other brain diseases. In individuals who experienced stroke, compared with the impedance exercise group, fixed bicycle aerobic training can improve the speed of information processing [139]. Although both aerobic and anaerobic exercise can independently improve the structure and function of the brain, the effects of the two forms of exercise are distinct. Combining the two modes of exercise is beneficial for improving cognitive function in the brain [139]. During rehabilitation exercises, individuals with severe motor dysfunction generally require one-to-one guidance from rehabilitation therapists to move their paralyzed limbs. This assistance promotes the recovery of muscle strength and the rebuilding of motor nerves [140]. At the same time, various instruments are necessary to support patients during rehabilitation exercises, such as weight-support treadmill training for individuals with SCI. This method reduces the biomechanical and balance constraints of walking by providing weight support through harness systems, allowing for a more normal walking pattern [141].

It is worth noting that exercise may have a negative impact. Chapman et al. [142] discovered that vigorous exercise was connected to a higher risk of amyotrophic lateral sclerosis (ALS). A 2021 study looked at roughly 10% of patients with ALS and found that in these individuals, higher rates of exercise were associated with an earlier onset of the disease [143]. At the same time, some researchers believe that the positive effects of exercise on brain function can be described by an inverted U-shaped curve, suggesting that the appropriate amount of rehabilitation exercise may result in the best possible outcome for the human body [144–146]. However, further evidence is needed to explore the potential mechanism and effect of this theory.

3.2. Epidural Electrical Stimulation Regulates the Brain

For patients with severe SCI and those who are unable to exercise autonomously, additional auxiliary means are often necessary to strengthen limb rehabilitation training, and functional electrical stimulation is proposed. Functional electrical stimulation activates muscle contraction by stimulating currents on human muscles, allowing patients to complete the corresponding movements on their own [147]. Epidural electrical stimulation (EES), which is developed from functional electrical stimulation, is a kind of rehabilitation stimulation after SCI with clinical potential. Its mode of action involves the implantation of an electrode into the epidural space of the spinal canal to make electrical stimulation act on the spinal cord tissue. EES transmits a sequence pulse current through the electrode to induce the depolarization of nerve cells in the stimulated part of the spinal cord, causing the dominant muscle to contract to drive joint movement, thereby improving the motor ability of the limbs of patients with SCI [148,149].

The use of epidural stimulation in combination with rehabilitation exercise after SCI can significantly enhance the functional recovery of patients. The first research on the use of EES technology to assist motor function recovery in patients with SCI began in the early 20th century, in which the mode of open-loop stimulation was applied [150]. When performing open-loop stimulation, the stimulator uses a constant or cyclic stimulation mode regardless of the real-time position of the limb and feedback from the brain or peripheral nervous system. This study proved that the combination of EES with exercise training can improve the motor function of patients with incomplete SCI. Since then, researchers have been exploring the closed-loop stimulation model of EES. In the closed-loop mode, EES stimulation can be triggered based on brain activity [151–153] or movement at different stages of the gait cycle [154,155]. Closed-loop stimulation aims to activate the target muscle while maximizing sensory inputs [156]. Wagner et al. [157] showed that after a period of closed-loop EES stimulation combined with lower limb rehabilitation training, two patients with SCI were able to autonomously control the paralyzed legs and restore certain functional activities without active stimulation. EES stimulation alone can also promote the recovery of motor and autonomic nervous function in patients with SCI [158]. In another clinical trial, seven patients were treated with EES for 5–9 months without intensive exercise, and more than half of them were able to resume autonomous exercise continuously without active stimulation after EES treatment [159]. The above studies show that epidural electrical stimulation can induce limb movement and proprioceptive sensory nerve input, which can promote the recovery of motor sensory function below the injury level and strengthen the autonomic nervous function of patients.

Electrical stimulation at the site of SCI can also regulate the pathway above the injury and activate the motor cortex [160]. Some studies have confirmed that in the mouse SCI model, EES combined with photogenetic stimulation of the motor cortex can immediately restore weight-bearing movement in mice. When the motor cortical stimulation stops, motor function is also blocked, indicating that EES induces cortical activity to control movement [161]. EES can also increase the level of BDNF in patients [162] and inhibit nerve inflammation at the injured site [163]. Although current neuroimaging studies on the effects of EES on brain structure and function in patients with SCI are not perfect, the combination of EES with repeated rehabilitation exercises plays an important role in the recovery of motor sensation after SCI and activates the ascending pathway of the spinal cord and motor cortex.

During the rehabilitation training process, clinicians need to choose the appropriate EES intensity and mode based on the patient's condition. The implanted electrode of EES not only has risks related to surgical damage but also needs to maintain biocompatibility for a long time. Hence, the clinical application of this rehabilitation technique requires further research.

3.3. Exoskeleton Rehabilitation Robot Regulates Brain

An exoskeleton rehabilitation robot is a mechanical device that simulates the structural characteristics of the human body and attaches to the exterior of the patient's body to assist or replace the patient's muscle strength and skeletal joint function, thereby helping patients recover or improve the function of their limbs [164]. Mechanical exoskeletons provide rehabilitation training for patients with CNS injury through their mechanical structure and control system [165], and this technique has the characteristics of high efficiency, quantification, and real-time interaction [166].

Currently, the main training modes of wearable upper limb exoskeleton rehabilitation robots include the passive, cooperative, and active modes. In the passive mode, the patient is in a completely relaxed state, and the exoskeleton drives all parts of the upper limb to the intended target position. Most mechanical exoskeletons are equipped with this mode. The cooperative mode improves upon the traditional passive mode, and the patient can drive the exoskeleton according to the trajectory of movement. When the collaboration mode is used, the exoskeleton can judge whether the exoskeleton is helpful or not based on the motion state of the patient. Patients actively participate to some extent, which plays a better role in promoting the reconstruction of the nervous system and the recovery of motor function [167]. Through rehabilitation-robot-assisted training, patients' walking ability [168], walking speed [169,170], leg muscle strength [171], stride length [172], and gait symmetry [173] have been improved.

Rehabilitation-robot-assisted training is widely used in artificial therapy after SCI, thereby providing early, intensive, task-specific motor sensory training [174], and it can also promote the adaptive plasticity of brain and spinal cord sensorimotor networks [175]. Chintan and Simon [176] compared rats that received only running training with those that received active stepping robot-assisted running training after SCI. The results showed that the motor characteristics of the cortex of rats with robot-assisted rehabilitation training were abundant, and the motor cortex was effectively activated. Effective robot rehabilitation training can induce the reorganization of the motor cortex and partially reverse some plasticity changes. Chintan and Simon [176] used functional near-infrared spectroscopy technology, which utilizes the scattering properties of blood's main components to near-infrared light to obtain the changes in oxygenated hemoglobin and deoxyhemoglobin during brain activity, to explore the role of robot-assisted rehabilitation in the rehabilitation of patients with SCI. The results showed that the activity of the motor cortex increased significantly during treatment. In addition to research on patients with SCI, the effect of rehabilitation robot-assisted training on cerebral cortex reorganization has also been examined. In a clinical study of rehabilitation robot-assisted gait training for 40 patients who experienced stroke, the researchers observed an improvement in the effective connection between the frontal and parietal cortex after eight weeks of Ekso™ rehabilitation robot training compared with regular training [177]. Wagner et al. [178] studied the spectral patterns of active and passive robot-assisted walking through EEG signals. The results indicate that both active and passive robot-assisted training can stimulate the cerebral cortex. Furthermore, during active robot-assisted training, changes in cerebral cortex activation are associated with the gait cycle stage.

The positive role of exoskeleton rehabilitation robots in assisting rehabilitation training has been preliminarily confirmed. In the future, the rehabilitation mechanical exoskeleton will need to further effectively exchange dynamic information with patients to achieve man–machine fusion. This type of rehabilitation training can effectively improve the active participation of patients and significantly improve the rehabilitation effect.

3.4. Motor Imagination Rehabilitation Regulates the Brain

In addition to rehabilitation methods that indirectly stimulate the CNS of the brain by passively acting on the distal limbs, new rehabilitation methods have been developed to directly reshape the cerebral cortex by making full use of the subjective intention of patients in order to improve the efficiency of rehabilitation. Among these methods, motor imagination (also known as psychological imagination) refers to the execution of specific actions or tasks without actual signal output [179]. Motor imagination (MI) can activate areas in the primary motor cortex, cerebellum, and basal ganglion circuits [180]. It can also induce functional redistribution and regulation of neural circuits, and then reshape the brain neural network and improve the relearning ability of motor function [181]. In the early stages of SCI, most of the structures and functions of the brain are preserved, providing a matrix for motor imagination [182]. MI therapy does not rely on the residual function of patients, which can give full play to the subjective initiative of patients with SCI and can run through the whole process of rehabilitation of patients with SCI, making it important for the functional recovery of patients with SCI [183].

Sabbah et al. [184] conducted MI training in patients with complete SCI and healthy individuals. Based on the randomized controlled trial, the activated brain regions of the two groups were generally similar during motor imagination, and the primary motor cortex, supplementary motor area, and premotor cortex were all activated, demonstrating that patients with SCI could carry out motor imagination training. Chen et al. [185] performed the right ankle dorsiflexion metatarsal flexion, motor imagination task, and motor execution task in 17 patients with incomplete SCI with partial motor dysfunction of the right ankle. The results showed that the activated brain areas were generally similar, including the bilateral supplementary motor areas and inferior frontal gyrus. However, compared with the motor execution task, the activation degree of some areas in the motor imagination task was lower. Rienzo et al. [186] found that after adding MI training to the rehabilitation program of patients with SCI, the activation of compensatory brain areas decreased, and the cortical activity was closer to the normal state, suggesting that MI training partially reversed the compensatory neural activation in patients with SCI, which was beneficial to the integration of normal neural networks. The latest study by Wang et al. [187] compared the effect of MI in children with SCI with that in healthy subjects by fMRI. The results showed that the activation degrees of bilateral paracentral lobules, auxiliary motor area, putamen, and cerebellar lobules in patients with SCI were higher than those in healthy ones.

The above studies indicate that motor imagination enhances the motor rehabilitation of patients with SCI. MI can also be combined with brain–computer interface technology, which can restore part of the patient's motor function without relying on the normal nerve conduction pathway of the brain [188]. Motor imagination–brain–computer interface will be an important future direction for the recovery of motor function after SCI. Current research on the effects of MI mainly shows that motor imagination can improve the CNS function of patients with brain injury in the short term, but its long-term effects on brain structure and function reorganization require further investigation.

3.5. Comparison of the Advantages and Disadvantages of Different Rehabilitation Methods

Different rehabilitation training methods have varied characteristics and effects (Table 1), and their effects on brain structure and function also differ. In clinics, the most suitable rehabilitation program is typically chosen based on the patient's status in order to achieve the best possible outcome [189].

Table 1. Characteristics and effects of different rehabilitation training methods.

Types	Characteristics	Advantage	Shortcoming	Effect	Applicability
Rehabilitation exercise	Easy to operate, convenient, and low-cost; the most commonly used clinical method for the rehabilitation of spinal cord injury (SCI).	Practicability, operability, and low cost; the most effective rehabilitation means to promote brain reorganization after SCI at present.	Exercise acts on the distal limbs, indirectly stimulates the CNS of the brain; rehabilitation training takes a long time.	One of the most important quantifiable methods of functional recovery after SCI, which activates the cerebral cortex by acting on the distal limb.	It can be applied to people with various degrees of SCI and at various stages of treatment after SCI, but it is mainly used for incomplete SCI.
Epidural electrical stimulation (EES)	It can induce movement through a sequence of pulsed currents and makes effective use of the residual function of spinal nerves and muscles.	It can enhance the neural plasticity of the motor cortex and activate the excitability of the corticospinal tract.	The use of implantable electrodes may cause postoperative infection. Prolonged application of electrical stimulation can cause pain in patients.	Combined with rehabilitation exercise, patients can improve their exercise ability in a short period of time.	EES is suitable for severe and complete SCI. It can be used for the early rehabilitation of SCI and patients with severe motor dysfunction.
Exoskeleton rehabilitation robot	It can simulate the structure characteristics of the human body and has the characteristics of high strength, repeatability, and interaction.	It can provide repeatable and specific movement therapy and promote the rehabilitation of fine movements after SCI. It can objectively quantify rehabilitation parameters and training output.	The structure is complex and expensive, and the weight and body shape should be considered.	It can be used to provide standard movement training for injured limbs. It has a better effect on brain reorganization than routine exercise training.	For patients with severe and complete SCI, it can provide more effective support. It is used for rehabilitation training of patients with severe motor dysfunction after SCI.
Motor imagination (MI)	Makes full use of the patient's subjective intentions and acts directly on the cerebral cortex.	It has a wide range of applications, motivates patients, and guides patients to achieve appropriate brain activation patterns related to tasks.	The standard of treatment and the best intensity of intervention are not clear. The effect depends on the feedback content and the status of the patient.	It can help patients with SCI activate the activity of the cerebral cortex and realize the self-regulation of the functional brain networks. Combining with other rehabilitation training methods is beneficial to the better effect of MI.	It can be applied to individuals with complete and incomplete SCI and at various stages of treatment after SCI.

4. Summary and Prospects

SCI and rehabilitation training both have the ability to modify the structure and function of the brain. This paper provides an overview of the physiological mechanisms responsible for the effects of rehabilitation training on brain structure and function after SCI and introduces and compares the effects of rehabilitation exercise, epidural electrical stimulation, mechanical exoskeleton, and motor imagination on the activation and reorganization of the cerebral cortex. Although rehabilitation training has been widely used in the clinical treatment of SCI, its impact on promoting brain remodeling to enhance motor function remains to be quantitatively evaluated. Understanding the influence mode and law of rehabilitation training on changes in brain structure and function, as well as the regulation and degree of brain remodeling on motor function after SCI, will help guide the clinical development of personalized rehabilitation strategies. These findings have important implications for the treatment of SCI.

Author Contributions: Conceptualization, L.-W.H. and J.-S.R.; writing—original draft preparation, L.-W.H.; writing—review and editing, L.-W.H., X.-J.G., C.Z. and J.-S.R. All authors have read and agreed to the published version of the manuscript.

Funding: This work was supported by the National Natural Science Foundation of China (Grant nos. 31970970, 31900980), Fundamental Research Funds for Central Public Welfare Research Institutes (Grant nos. 2021CZ-10, 2022CZ-12), and Fundamental Research Funds for the Central Universities (Grant no. YWF-23-YG-QB-010).

Conflicts of Interest: The authors declare no conflict of interest.

References

1. Gross-Hemmi, M.H.; Post, M.W.; Ehrmann, C.; Fekete, C.; Hasnan, N.; Middleton, J.W.; Reinhardt, J.D.; Strom, V.; Stucki, G. Study Protocol of the International Spinal Cord Injury (InSCI) Community Survey. *Am. J. Phys. Med. Rehabil.* **2017**, *96*, S23–S34. [CrossRef] [PubMed]
2. Zhao, C.; Bao, S.S.; Xu, M.; Rao, J.S. Importance of brain alterations in spinal cord injury. *Sci. Prog.* **2021**, *104*, 368504211031117. [CrossRef] [PubMed]
3. Northcutt, R.G. Accommodations of the Nervous System: Body and Brain. A Trophic Theory of Neural Connections. Dale Purves. Harvard University Press, Cambridge, MA, 1988. viii, 231 pp., illus. $35. *Science* **1989**, *244*, 993. [CrossRef]
4. Bilchak, J.N.; Caron, G.; Côté, M.P. Exercise-Induced Plasticity in Signaling Pathways Involved in Motor Recovery after Spinal Cord Injury. *Int. J. Mol. Sci.* **2021**, *22*, 4858. [CrossRef] [PubMed]
5. Henderson, L.A.; Gustin, S.M.; Macey, P.M.; Wrigley, P.J.; Siddall, P.J. Functional reorganization of the brain in humans following spinal cord injury: Evidence for underlying changes in cortical anatomy. *J. Neurosci. Off. J. Soc. Neurosci.* **2011**, *31*, 2630–2637. [CrossRef]
6. Nandakumar, B.; Blumenthal, G.H.; Disse, G.D.; Desmond, P.C.; Ebinu, J.O.; Ricard, J.; Bethea, J.R.; Moxon, K.A. Exercise therapy guides cortical reorganization after midthoracic spinal contusion to enhance control of lower thoracic muscles, supporting functional recovery. *Exp. Neurol.* **2023**, *364*, 114394. [CrossRef]
7. Lourenco, M.V.; Frozza, R.L.; de Freitas, G.B.; Zhang, H.; Kincheski, G.C.; Ribeiro, F.C.; Goncalves, R.A.; Clarke, J.R.; Beckman, D.; Staniszewski, A.; et al. Exercise-linked FNDC5/irisin rescues synaptic plasticity and memory defects in Alzheimer's models. *Nat. Med.* **2019**, *25*, 165–175. [CrossRef]
8. Noakes, T.; Spedding, M. Olympics: Run for your life. *Nature* **2012**, *487*, 295–296. [CrossRef]
9. Kobilo, T.; Liu, Q.R.; Gandhi, K.; Mughal, M.; Shaham, Y.; van Praag, H. Running is the neurogenic and neurotrophic stimulus in environmental enrichment. *Learn. Mem.* **2011**, *18*, 605–609. [CrossRef]
10. Mu, L.; Cai, J.; Gu, B.; Yu, L.; Li, C.; Liu, Q.S.; Zhao, L. Treadmill Exercise Prevents Decline in Spatial Learning and Memory in 3xTg-AD Mice through Enhancement of Structural Synaptic Plasticity of the Hippocampus and Prefrontal Cortex. *Cells* **2022**, *11*, 244. [CrossRef]
11. Shen, W.; Jin, L.; Zhu, A.; Lin, Y.; Pan, G.; Zhou, S.; Cheng, J.; Zhang, J.; Tu, F.; Liu, C.; et al. Treadmill exercise enhances synaptic plasticity in the ischemic penumbra of MCAO mice by inducing the expression of Camk2a via CYFIP1 upregulation. *Life Sci.* **2021**, *270*, 119033. [CrossRef]
12. Batouli, S.A.H.; Saba, V. At least eighty percent of brain grey matter is modifiable by physical activity: A review study. *Behav. Brain Res.* **2017**, *332*, 204–217. [CrossRef]
13. Jurkiewicz, M.T.; Mikulis, D.J.; McIlroy, W.E.; Fehlings, M.G.; Verrier, M.C. Sensorimotor cortical plasticity during recovery following spinal cord injury: A longitudinal fMRI study. *Neurorehabilit. Neural Repair.* **2007**, *21*, 527–538. [CrossRef]
14. Pedersen, B.K. Physical activity and muscle-brain crosstalk. *Nat. Rev. Endocrinol.* **2019**, *15*, 383–392. [CrossRef]
15. Giudice, J.; Taylor, J.M. Muscle as a paracrine and endocrine organ. *Curr. Opin. Pharmacol.* **2017**, *34*, 49–55. [CrossRef]
16. Hoffmann, C.; Weigert, C. Skeletal Muscle as an Endocrine Organ: The Role of Myokines in Exercise Adaptations. *Cold Spring Harb. Perspect. Med.* **2017**, *7*, a029793. [CrossRef] [PubMed]
17. Pesce, M.; La Fratta, I.; Paolucci, T.; Grilli, A.; Patruno, A.; Agostini, F.; Bernetti, A.; Mangone, M.; Paoloni, M.; Invernizzi, M.; et al. From Exercise to Cognitive Performance: Role of Irisin. *Appl. Sci.* **2021**, *11*, 7120. [CrossRef]
18. Severinsen, M.C.K.; Pedersen, B.K. Muscle-Organ Crosstalk: The Emerging Roles of Myokines. *Endocr. Rev.* **2020**, *41*, 594–609. [CrossRef]
19. Kowianski, P.; Lietzau, G.; Czuba, E.; Waskow, M.; Steliga, A.; Morys, J. BDNF: A Key Factor with Multipotent Impact on Brain Signaling and Synaptic Plasticity. *Cell. Mol. Neurobiol.* **2018**, *38*, 579–593. [CrossRef] [PubMed]
20. El Hayek, L.; Khalifeh, M.; Zibara, V.; Assaad, R.A.; Emmanuel, N.; Karnib, N.; El-Ghandour, R.; Nasrallah, P.; Bilen, M.; Ibrahim, P.; et al. Lactate Mediates the Effects of Exercise on Learning and Memory through SIRT1-Dependent Activation of Hippocampal Brain-Derived Neurotrophic Factor (BDNF). *J. Neurosci.* **2019**, *39*, 2369–2382. [CrossRef] [PubMed]

21. Bernardes, D.; Oliveira-Lima, O.C.; da Silva, T.V.; Faraco, C.C.F.; Leite, H.R.; Juliano, M.A.; dos Santos, D.M.; Bethea, J.R.; Brambilla, R.; Orian, J.M.; et al. Differential brain and spinal cord cytokine and BDNF levels in experimental autoimmune encephalomyelitis are modulated by prior and regular exercise. *J. Neuroimmunol.* **2013**, *264*, 24–34. [CrossRef]

22. Ozkul, C.; Guclu-Gunduz, A.; Irkec, C.; Fidan, I.; Aydin, Y.; Ozkan, T.; Yazici, G. Effect of combined exercise training on serum brain-derived neurotrophic factor, suppressors of cytokine signaling 1 and 3 in patients with multiple sclerosis. *J. Neuroimmunol.* **2018**, *316*, 121–129. [CrossRef]

23. Pajonk, F.G.; Wobrock, T.; Gruber, O.; Scherk, H.; Berner, D.; Kaizl, I.; Kierer, A.; Muller, S.; Oest, M.; Meyer, T.; et al. Hippocampal plasticity in response to exercise in schizophrenia. *Arch. Gen. Psychiatry* **2010**, *67*, 133–143. [CrossRef]

24. Wens, I.; Keytsman, C.; Deckx, N.; Cools, N.; Dalgas, U.; Eijnde, B.O. Brain derived neurotrophic factor in multiple sclerosis: Effect of 24 weeks endurance and resistance training. *Eur. J. Neurol.* **2016**, *23*, 1028–1035. [CrossRef] [PubMed]

25. Li, X.; Wu, Q.; Xie, C.; Wang, C.; Wang, Q.; Dong, C.; Fang, L.; Ding, J.; Wang, T. Blocking of BDNF-TrkB signaling inhibits the promotion effect of neurological function recovery after treadmill training in rats with spinal cord injury. *Spinal Cord* **2019**, *57*, 65–74. [CrossRef] [PubMed]

26. Moon, H.Y.; Becke, A.; Berron, D.; Becker, B.; Sah, N.; Benoni, G.; Janke, E.; Lubejko, S.T.; Greig, N.H.; Mattison, J.A.; et al. Running-Induced Systemic Cathepsin B Secretion Is Associated with Memory Function. *Cell Metab.* **2016**, *24*, 332–340. [CrossRef]

27. Chazeau, A.; Giannone, G. Organization and dynamics of the actin cytoskeleton during dendritic spine morphological remodeling. *Cell. Mol. Life Sci.* **2016**, *73*, 3053–3073. [CrossRef] [PubMed]

28. Neasta, J.; Fiorenza, A.; He, D.Y.; Phamluong, K.; Kiely, P.A.; Ron, D. Activation of the cAMP Pathway Induces RACK1-Dependent Binding of β-Actin to BDNF Promoter. *PLoS ONE* **2016**, *11*, e0160948. [CrossRef]

29. Boström, P.; Wu, J.; Jedrychowski, M.P.; Korde, A.; Ye, L.; Lo, J.C.; Rasbach, K.A.; Boström, E.A.; Choi, J.H.; Long, J.Z.; et al. A PGC1-α-dependent myokine that drives brown-fat-like development of white fat and thermogenesis. *Nature* **2012**, *481*, 463–468. [CrossRef]

30. Farshbaf, M.J.; Ghaedi, K.; Megraw, T.L.; Curtiss, J.; Faradonbeh, M.S.; Vaziri, P.; Nasr-Esfahani, M.H. Does PGC1α/FNDC5/BDNF Elicit the Beneficial Effects of Exercise on Neurodegenerative Disorders? *Neuromolecular Med.* **2016**, *18*, 1–15. [CrossRef]

31. Sleiman, S.F.; Henry, J.; Al-Haddad, R.; El Hayek, L.; Abou Haidar, E.; Stringer, T.; Ulja, D.; Karuppagounder, S.S.; Holson, E.B.; Ratan, R.R.; et al. Exercise promotes the expression of brain derived neurotrophic factor (BDNF) through the action of the ketone body beta-hydroxybutyrate. *elife* **2016**, *5*, e15092. [CrossRef]

32. Komine, S.; Akiyama, K.; Warabi, E.; Oh, S.; Kuga, K.; Ishige, K.; Togashi, S.; Yanagawa, T.; Shoda, J. Exercise training enhances in vivo clearance of endotoxin and attenuates inflammatory responses by potentiating Kupffer cell phagocytosis. *Sci. Rep.* **2017**, *7*, 11977. [CrossRef] [PubMed]

33. Bernardes, D.; Brambilla, R.; Bracchi-Ricard, V.; Karmally, S.; Dellarole, A.; Carvalho-Tavares, J.; Bethea, J.R. Prior regular exercise improves clinical outcome and reduces demyelination and axonal injury in experimental autoimmune encephalomyelitis. *J. Neurochem.* **2016**, *136* (Suppl. 1), 63–73. [CrossRef]

34. Einstein, O.; Fainstein, N.; Touloumi, O.; Lagoudaki, R.; Hanya, E.; Grigoriadis, N.; Katz, A.; Ben-Hur, T. Exercise training attenuates experimental autoimmune encephalomyelitis by peripheral immunomodulation rather than direct neuroprotection. *Exp. Neurol.* **2018**, *299*, 56–64. [CrossRef] [PubMed]

35. Pryor, W.M.; Freeman, K.G.; Larson, R.D.; Edwards, G.L.; White, L.J. Chronic exercise confers neuroprotection in experimental autoimmune encephalomyelitis. *J. Neurosci. Res.* **2015**, *93*, 697–706. [CrossRef] [PubMed]

36. Nichol, K.E.; Poon, W.W.; Parachikova, A.I.; Cribbs, D.H.; Glabe, C.G.; Cotman, C.W. Exercise alters the immune profile in Tg2576 Alzheimer mice toward a response coincident with improved cognitive performance and decreased amyloid. *J. Neuroinflammation* **2008**, *5*, 13. [CrossRef]

37. Lee, J.H.; Han, J.H.; Woo, J.H.; Jou, I. 25-Hydroxycholesterol suppress IFN-γ-induced inflammation in microglia by disrupting lipid raft formation and caveolin-mediated signaling endosomes. *Free Radic. Biol. Med.* **2022**, *179*, 252–265. [CrossRef]

38. Cheng, T.; Huang, X.D.; Hu, X.F.; Wang, S.Q.; Chen, K.; Wei, J.A.; Yan, L.; So, W.A.; Yuan, T.F.; Zhang, L. Physical exercise rescues cocaine-evoked synaptic deficits in motor cortex. *Mol. Psychiatry* **2021**, *26*, 6187–6197. [CrossRef]

39. Thomas, B.P.; Tarumi, T.; Sheng, M.; Tseng, B.; Womack, K.B.; Cullum, C.M.; Rypma, B.; Zhang, R.; Lu, H. Brain Perfusion Change in Patients with Mild Cognitive Impairment After 12 Months of Aerobic Exercise Training. *J. Alzheimers Dis.* **2020**, *75*, 617–631. [CrossRef]

40. Dietrich, M.O.; Andrews, Z.B.; Horvath, T.L. Exercise-induced synaptogenesis in the hippocampus is dependent on UCP2-regulated mitochondrial adaptation. *J. Neurosci. Off. J. Soc. Neurosci.* **2008**, *28*, 10766–10771. [CrossRef]

41. Bayod, S.; del Valle, J.; Canudas, A.M.; Lalanza, J.F.; Sanchez-Roige, S.; Camins, A.; Escorihuela, R.M.; Pallàs, M. Long-term treadmill exercise induces neuroprotective molecular changes in rat brain. *J. Appl. Physiol.* **2011**, *111*, 1380–1390. [CrossRef] [PubMed]

42. Hipkiss, A.R. Dietary restriction, glycolysis, hormesis and ageing. *Biogerontology* **2007**, *8*, 221–224. [CrossRef]

43. Cantó, C.; Auwerx, J. PGC-1alpha, SIRT1 and AMPK, an energy sensing network that controls energy expenditure. *Curr. Opin. Lipidol.* **2009**, *20*, 98–105. [CrossRef]

44. Marton, O.; Koltai, E.; Nyakas, C.; Bakonyi, T.; Zenteno-Savin, T.; Kumagai, S.; Goto, S.; Radak, Z. Aging and exercise affect the level of protein acetylation and SIRT1 activity in cerebellum of male rats. *Biogerontology* **2010**, *11*, 679–686. [CrossRef]

45. Ravera, S.; Morelli, A.M.; Panfoli, I. Myelination increases chemical energy support to the axon without modifying the basic physicochemical mechanism of nerve conduction. *Neurochem. Int.* **2020**, *141*, 104883. [CrossRef] [PubMed]
46. Markovinovic, A.; Greig, J.; Martín-Guerrero, S.M.; Salam, S.; Paillusson, S. Endoplasmic reticulum-mitochondria signaling in neurons and neurodegenerative diseases. *J. Cell Sci.* **2022**, *135*, jcs248534. [CrossRef]
47. Orr, M.B.; Gensel, J.C. Spinal Cord Injury Scarring and Inflammation: Therapies Targeting Glial and Inflammatory Responses. *Neurotherapeutics* **2018**, *15*, 541–553. [CrossRef] [PubMed]
48. Wang, H.F.; Liu, X.K.; Li, R.; Zhang, P.; Chu, Z.; Wang, C.L.; Liu, H.R.; Qi, J.; Lv, G.Y.; Wang, G.Y.; et al. Effect of glial cells on remyelination after spinal cord injury. *Neural Regen. Res.* **2017**, *12*, 1724–1732. [CrossRef]
49. Stadelmann, C.; Timmler, S.; Barrantes-Freer, A.; Simons, M. Myelin in the central nervous system: Structure, function, and pathology. *Physiol. Rev.* **2019**, *99*, 1381–1431. [CrossRef]
50. McKenzie, I.A.; Ohayon, D.; Li, H.; de Faria, J.P.; Emery, B.; Tohyama, K.; Richardson, W.D. Motor skill learning requires active central myelination. *Science* **2014**, *346*, 318–322. [CrossRef]
51. Zheng, J.; Sun, X.; Ma, C.; Li, B.-m.; Luo, F. Voluntary wheel running promotes myelination in the motor cortex through Wnt signaling in mice. *Mol. Brain* **2019**, *12*, 85. [CrossRef]
52. Kim, T.W.; Sung, Y.H. Regular exercise promotes memory function and enhances hippocampal neuroplasticity in experimental autoimmune encephalomyelitis mice. *Neuroscience* **2017**, *346*, 173–181. [CrossRef]
53. Krityakiarana, W.; Espinosa-Jeffrey, A.; Ghiani, C.A.; Zhao, P.M.; Topaldjikian, N.; Gomez-Pinilla, F.; Yamaguchi, M.; Kotch-abhakdi, N.; de Vellis, J. Voluntary exercise increases oligodendrogenesis in spinal cord. *Int. J. Neurosci.* **2010**, *120*, 280–290. [CrossRef] [PubMed]
54. Maugeri, G.; D'Agata, V.; Magrì, B.; Roggio, F.; Castorina, A.; Ravalli, S.; Di Rosa, M.; Musumeci, G. Neuroprotective Effects of Physical Activity via the Adaptation of Astrocytes. *Cells* **2021**, *10*, 1542. [CrossRef] [PubMed]
55. Zhou, B.; Zuo, Y.X.; Jiang, R.T. Astrocyte morphology: Diversity, plasticity, and role in neurological diseases. *CNS Neurosci. Ther.* **2019**, *25*, 665–673. [CrossRef] [PubMed]
56. Allen, N.J.; Eroglu, C. Cell Biology of Astrocyte-Synapse Interactions. *Neuron* **2017**, *96*, 697–708. [CrossRef] [PubMed]
57. Sofroniew, M.V.; Vinters, H.V. Astrocytes: Biology and pathology. *Acta Neuropathol.* **2010**, *119*, 7–35. [CrossRef]
58. Zhang, J.; Guo, Y.; Wang, Y.; Song, L.; Zhang, R.; Du, Y. Long-term treadmill exercise attenuates Aβ burdens and astrocyte activation in APP/PS1 mouse model of Alzheimer's disease. *Neurosci. Lett.* **2018**, *666*, 70–77. [CrossRef]
59. Lee, S.U.; Kim, D.Y.; Park, S.H.; Choi, D.H.; Park, H.W.; Han, T.R. Mild to moderate early exercise promotes recovery from cerebral ischemia in rats. *Can. J. Neurol. Sci.* **2009**, *36*, 443–449. [CrossRef]
60. Fiuza-Luces, C.; Santos-Lozano, A.; Joyner, M.; Carrera-Bastos, P.; Picazo, O.; Zugaza, J.L.; Izquierdo, M.; Ruilope, L.M.; Lucia, A. Exercise benefits in cardiovascular disease: Beyond attenuation of traditional risk factors. *Nat. Rev. Cardiol.* **2018**, *15*, 731–743. [CrossRef]
61. Chu, F.; Shi, M.; Zheng, C.; Shen, D.; Zhu, J.; Zheng, X.; Cui, L. The roles of macrophages and microglia in multiple sclerosis and experimental autoimmune encephalomyelitis. *J. Neuroimmunol.* **2018**, *318*, 1–7. [CrossRef]
62. Kohman, R.A.; Bhattacharya, T.K.; Wojcik, E.; Rhodes, J.S. Exercise reduces activation of microglia isolated from hippocampus and brain of aged mice. *J. Neuroinflammation* **2013**, *10*, 114. [CrossRef]
63. Benson, C.; Paylor, J.W.; Tenorio, G.; Winship, I.; Baker, G.; Kerr, B.J. Voluntary wheel running delays disease onset and reduces pain hypersensitivity in early experimental autoimmune encephalomyelitis (EAE). *Exp. Neurol.* **2015**, *271*, 279–290. [CrossRef] [PubMed]
64. Jensen, S.K.; Michaels, N.J.; Ilyntskyy, S.; Keough, M.B.; Kovalchuk, O.; Yong, V.W. Multimodal Enhancement of Remyelination by Exercise with a Pivotal Role for Oligodendroglial PGC1α. *Cell Rep.* **2018**, *24*, 3167–3179. [CrossRef] [PubMed]
65. Clark, T.A.; Sullender, C.; Jacob, D.; Zuo, Y.; Dunn, A.K.; Jones, T.A. Rehabilitative Training Interacts with Ischemia-Instigated Spine Dynamics to Promote a Lasting Population of New Synapses in Peri-Infarct Motor Cortex. *J. Neurosci. Off. J. Soc. Neurosci.* **2019**, *39*, 8471–8483. [CrossRef]
66. Dietrich, A.; McDaniel, W.F. Endocannabinoids and exercise. *Br. J. Sports Med.* **2004**, *38*, 536–541. [CrossRef]
67. Opendak, M.; Gould, E. Adult neurogenesis: A substrate for experience-dependent change. *Trends Cogn. Sci.* **2015**, *19*, 151–161. [CrossRef]
68. Houle, J.D.; Côté, M.P. Axon regeneration and exercise-dependent plasticity after spinal cord injury. *Ann. N. Y. Acad. Sci.* **2013**, *1279*, 154–163. [CrossRef] [PubMed]
69. Guo, L.A.-O.; Lozinski, B.; Yong, V.A.-O. Exercise in multiple sclerosis and its models: Focus on the central nervous system outcomes. *J. Neurosci. Res.* **2020**, *98*, 509–523. [CrossRef]
70. Cassilhas, R.; Lee, K.; Fernandes, J.; Oliveira, M.; Tufik, S.; Meeusen, R.; de Mello, M. Spatial memory is improved by aerobic and resistance exercise through divergent molecular mechanisms. *Neuroscience* **2012**, *202*, 309–317. [CrossRef]
71. Lista, I.; Sorrentino, G. Biological mechanisms of physical activity in preventing cognitive decline. *Cell Mol. Neurobiol.* **2010**, *30*, 493–503. [CrossRef]
72. Oestreicher, A.B.; De Graan, P.N.; Gispen, W.H.; Verhaagen, J.; Schrama, L.H. B-50, the growth associated protein-43: Modulation of cell morphology and communication in the nervous system. *Prog. Neurobiol.* **1997**, *53*, 627–686. [CrossRef]
73. Mizutani, K.; Sonoda, S.; Yamada, K.; Beppu, H.; Shimpo, K. Alteration of protein expression profile following voluntary exercise in the perilesional cortex of rats with focal cerebral infarction. *Brain Res.* **2011**, *1416*, 61–68. [CrossRef] [PubMed]

74. Thiele, C.; Hannah, M.J.; Fahrenholz, F.; Huttner, W.B. Cholesterol binds to synaptophysin and is required for biogenesis of synaptic vesicles. *Nat. Cell Biol.* **2000**, *2*, 42–49. [CrossRef] [PubMed]

75. Rapp, S.; Baader, M.; Hu, M.; Jennen-Steinmetz, C.; Henn, F.A.; Thome, J. Differential regulation of synaptic vesicle proteins by antidepressant drugs. *Pharmacogenomics J.* **2004**, *4*, 110–113. [CrossRef] [PubMed]

76. Li, C.; Ke, C.K.; Su, Y.; Wan, C.X. Exercise Intervention Promotes the Growth of Synapses and Regulates Neuroplasticity in Rats with Ischemic Stroke Through Exosomes. *Front. Neurol.* **2021**, *12*, 752595. [CrossRef] [PubMed]

77. Levy, A.M.; Gomez-Puertas, P.; Tümer, Z. Neurodevelopmental Disorders Associated with PSD-95 and Its Interaction Partners. *Int. J. Mol. Sci.* **2022**, *23*, 4390. [CrossRef] [PubMed]

78. Pagnussat, A.S.; Simao, F.; Anastacio, J.R.; Mestriner, R.G.; Michaelsen, S.M.; Castro, C.C.; Salbego, C.; Netto, C.A. Effects of skilled and unskilled training on functional recovery and brain plasticity after focal ischemia in adult rats. *Brain Res.* **2012**, *1486*, 53–61. [CrossRef] [PubMed]

79. Shih, P.-C.; Yang, Y.-R.; Wang, R.-Y. Effects of exercise intensity on spatial memory performance and hippocampal synaptic plasticity in transient brain ischemic rats. *PLoS ONE* **2013**, *8*, e78163. [CrossRef]

80. Sala, C.; Piech, V.; Wilson, N.R.; Passafaro, M.; Liu, G.; Sheng, M. Regulation of dendritic spine morphology and synaptic function by Shank and Homer. *Neuron* **2001**, *31*, 115–130. [CrossRef]

81. Xu, B.; Zang, K.; Ruff, N.L.; Zhang, Y.A.; McConnell, S.K.; Stryker, M.P.; Reichardt, L.F. Cortical degeneration in the absence of neurotrophin signaling: Dendritic retraction and neuronal loss after removal of the receptor TrkB. *Neuron* **2000**, *26*, 233–245. [CrossRef] [PubMed]

82. Bramham, C.R.; Wells, D.G. Dendritic mRNA: Transport, translation and function. *Nat. Rev. Neurosci.* **2007**, *8*, 776–789. [CrossRef] [PubMed]

83. Andreska, T.; Lüningschrör, P.; Sendtner, M. Regulation of TrkB cell surface expression-a mechanism for modulation of neuronal responsiveness to brain-derived neurotrophic factor. *Cell Tissue Res.* **2020**, *382*, 5–14. [CrossRef] [PubMed]

84. Palasz, E.; Wysocka, A.; Gasiorowska, A.; Chalimoniuk, M.; Niewiadomski, W.; Niewiadomska, G. BDNF as a Promising Therapeutic Agent in Parkinson's Disease. *Int. J. Mol. Sci.* **2020**, *21*, 1170. [CrossRef]

85. Yasuda, R.; Hayashi, Y.; Hell, J.W. CaMKII: A central molecular organizer of synaptic plasticity, learning and memory. *Nat. Rev. Neurosci.* **2022**, *23*, 666–682. [CrossRef] [PubMed]

86. Langnaese, K.; Seidenbecher, C.; Wex, H.; Seidel, B.; Hartung, K.; Appeltauer, U.; Garner, A.; Voss, B.; Mueller, B.; Garner, C.C.; et al. Protein components of a rat brain synaptic junctional protein preparation. *Brain Res. Mol. Brain Res.* **1996**, *42*, 118–122. [CrossRef] [PubMed]

87. Fiorenza, M.; Gunnarsson, T.P.; Hostrup, M.; Iaia, F.M.; Schena, F.; Pilegaard, H.; Bangsbo, J. Metabolic stress-dependent regulation of the mitochondrial biogenic molecular response to high-intensity exercise in human skeletal muscle. *J. Physiol. Lond.* **2018**, *596*, 2823–2840. [CrossRef]

88. Skovgaard, C.; Almquist, N.W.; Bangsbo, J. The effect of repeated periods of speed endurance training on performance, running economy, and muscle adaptations. *Scand. J. Med. Sci. Sports* **2018**, *28*, 381–390. [CrossRef]

89. Liu, W.; Xia, Y.; Kuang, H.; Wang, Z.; Liu, S.; Tang, C.; Yin, D. Proteomic Profile of Carbonylated Proteins Screen the Regulation of Calmodulin-Dependent Protein Kinases-AMPK-Beclin1 in Aerobic Exercise-Induced Autophagy in Middle-Aged Rat Hippocampus. *Gerontology* **2019**, *65*, 620–633. [CrossRef]

90. Xing, Y.; Bai, Y.A.-O. A Review of Exercise-Induced Neuroplasticity in Ischemic Stroke: Pathology and Mechanisms. *Mol. Neurobiol.* **2020**, *57*, 4218–4231. [CrossRef]

91. Schwenk, J.; Boudkkazi, S.; Kocylowski, M.K.; Brechet, A.; Zolles, G.; Bus, T.; Costa, K.; Kollewe, A.; Jordan, J.; Bank, J.; et al. An ER Assembly Line of AMPA-Receptors Controls Excitatory Neurotransmission and Its Plasticity. *Neuron* **2019**, *104*, 680–692.e689. [CrossRef]

92. Real, C.C.; Ferreira, A.F.; Hernandes, M.S.; Britto, L.R.; Pires, R.S. Exercise-induced plasticity of AMPA-type glutamate receptor subunits in the rat brain. *Brain Res.* **2010**, *1363*, 63–71. [CrossRef]

93. Kintz, N.; Petzinger, G.M.; Akopian, G.; Ptasnik, S.; Williams, C.; Jakowec, M.W.; Walsh, J.P. Exercise modifies alpha-amino-3-hydroxy-5-methyl-4-isoxazolepropionic acid receptor expression in striatopallidal neurons in the 1-methyl-4-phenyl-1,2,3,6-tetrahydropyridine-lesioned mouse. *J. Neurosci. Res.* **2013**, *91*, 1492–1507. [CrossRef] [PubMed]

94. Kang, J.; Wang, D.; Duan, Y.C.; Zhai, L.; Shi, L.; Guo, F. Aerobic Exercise Prevents Depression via Alleviating Hippocampus Injury in Chronic Stressed Depression Rats. *Brain Sci.* **2021**, *11*, 9. [CrossRef] [PubMed]

95. Chung, J.W.; Seo, J.H.; Baek, S.B.; Kim, C.J.; Kim, T.W. Treadmill exercise inhibits hippocampal apoptosis through enhancing N-methyl-D-aspartate receptor expression in the MK-801-induced schizophrenic mice. *J. Exerc. Rehabil.* **2014**, *10*, 218–224. [CrossRef]

96. Tang, G.; Gudsnuk, K.; Kuo, S.H.; Cotrina, M.L.; Rosoklija, G.; Sosunov, A.; Sonders, M.S.; Kanter, E.; Castagna, C.; Yamamoto, A.; et al. Loss of mTOR-dependent macroautophagy causes autistic-like synaptic pruning deficits. *Neuron* **2014**, *83*, 1131–1143. [CrossRef] [PubMed]

97. Obi-Nagata, K.; Suzuki, N.; Miyake, R.; MacDonald, M.L.; Fish, K.N.; Ozawa, K.; Nagahama, K.; Okimura, T.; Tanaka, S.; Kano, M.; et al. Distorted neurocomputation by a small number of extra-large spines in psychiatric disorders. *Sci. Adv.* **2023**, *9*, eade5973. [CrossRef] [PubMed]

98. Freund, P.; Weiskopf, N.; Ward, N.S.; Hutton, C.; Gall, A.; Ciccarelli, O.; Craggs, M.; Friston, K.; Thompson, A.J.; Freund, P.; et al. Disability, atrophy and cortical reorganization following spinal cord injury. *Brain J. Neurol.* **2011**, *134*, 1610–1622. [CrossRef] [PubMed]

99. Jurkiewicz, M.T.; Crawley, A.P.; Verrier, M.C.; Fehlings, M.G.; Mikulis, D.J. Somatosensory cortical atrophy after spinal cord injury: A voxel-based morphometry study. *Neurology* **2006**, *66*, 762–764. [CrossRef]

100. Wrigley, P.; Gustin, S.; Macey, P.; Nash, P.; Gandevia, S.; Macefield, V.; Siddall, P.; Henderson, L. Anatomical changes in human motor cortex and motor pathways following complete thoracic spinal cord injury. *Cereb. Cortex* **2009**, *19*, 224–232. [CrossRef]

101. Freund, P.; Weiskopf, N.; Ashburner, J.; Wolf, K.; Sutter, R.; Altmann, D.R.; Friston, K.; Thompson, A.; Curt, A. MRI investigation of the sensorimotor cortex and the corticospinal tract after acute spinal cord injury: A prospective longitudinal study. *Lancet Neurol.* **2013**, *12*, 873–881. [CrossRef] [PubMed]

102. Bao, S.S.; Zhao, C.; Chen, H.W.; Feng, T.; Guo, X.J.; Xu, M.; Rao, J.S. NT3 treatment alters spinal cord injury-induced changes in the gray matter volume of rhesus monkey cortex. *Sci. Rep.* **2022**, *12*, 5919. [CrossRef] [PubMed]

103. Sydnor, V.J.; Larsen, B.; Bassett, D.S.; Alexander-Bloch, A.; Fair, D.A.; Liston, C.; Mackey, A.P.; Milham, M.P.; Pines, A.; Roalf, D.R.; et al. Neurodevelopment of the association cortices: Patterns, mechanisms, and implications for psychopathology. *Neuron* **2021**, *109*, 2820–2846. [CrossRef] [PubMed]

104. Lane, R.D.; Stojic, R.S.; Killackey, H.P.; Rhoades, R.W. Source of inappropriate receptive fields in cortical somatotopic maps from rats that sustained neonatal forelimb removal. *J. Neurophysiol.* **1999**, *81*, 625–633. [CrossRef] [PubMed]

105. Gardiner, P.; Dai, Y.; Heckman, C.J. Effects of exercise training on alpha-motoneurons. *J. Appl. Physiol.* **2006**, *101*, 1228–1236. [CrossRef] [PubMed]

106. Giszter, S.; Davies, M.R.; Ramakrishnan, A.; Udoekwere, U.I.; Kargo, W.J. Trunk sensorimotor cortex is essential for autonomous weight-supported locomotion in adult rats spinalized as P1/P2 neonates. *J. Neurophysiol.* **2008**, *100*, 839–851. [CrossRef]

107. Kao, T.; Shumsky, J.S.; Knudsen, E.B.; Murray, M.; Moxon, K.A. Functional role of exercise-induced cortical organization of sensorimotor cortex after spinal transection. *J. Neurophysiol.* **2011**, *106*, 2662–2674. [CrossRef]

108. Gomez-Pinilla, F.; Ying, Z.; Zhuang, Y. Brain and spinal cord interaction: Protective effects of exercise prior to spinal cord injury. *PLoS ONE* **2012**, *7*, e32298. [CrossRef]

109. Matsubayashi, K.; Nagoshi, N.; Komaki, Y.; Kojima, K.; Shinozaki, M.; Tsuji, O.; Iwanami, A.; Ishihara, R.; Takata, N.; Matsumoto, M.; et al. Assessing cortical plasticity after spinal cord injury by using resting-state functional magnetic resonance imaging in awake adult mice. *Sci. Rep.* **2018**, *8*, 14406. [CrossRef]

110. Kaushal, M.; Oni-Orisan, A.; Chen, G.; Li, W.; Leschke, J.; Ward, B.D.; Kalinosky, B.; Budde, M.D.; Schmit, B.D.; Li, S.J.; et al. Evaluation of Whole-Brain Resting-State Functional Connectivity in Spinal Cord Injury: A Large-Scale Network Analysis Using Network-Based Statistic. *J. Neurotrauma* **2017**, *34*, 1278–1282. [CrossRef]

111. Hou, J.M.; Sun, T.S.; Xiang, Z.M.; Zhang, J.Z.; Zhang, Z.C.; Zhao, M.; Zhong, J.F.; Liu, J.; Zhang, H.; Liu, H.L.; et al. Alterations of resting-state regional and network-level neural function after acute spinal cord injury. *Neuroscience* **2014**, *277*, 446–454. [CrossRef] [PubMed]

112. Moxon, K.A.; Oliviero, A.; Aguilar, J.; Foffani, G. Cortical reorganization after spinal cord injury: Always for good? *Neuroscience* **2014**, *283*, 78–94. [CrossRef]

113. Sato, G.; Osumi, M.; Morioka, S. Effects of wheelchair propulsion on neuropathic pain and resting electroencephalography after spinal cord injury. *J. Rehabil. Med.* **2017**, *49*, 136–143. [CrossRef]

114. Curt, A.; Alkadhi, H.; Crelier, G.R.; Boendermaker, S.H.; Hepp-Reymond, M.C.; Kollias, S.S. Changes of non-affected upper limb cortical representation in paraplegic patients as assessed by fMRI. *Brain J. Neurol.* **2002**, *125*, 2567–2578. [CrossRef] [PubMed]

115. Bruehlmeier, M.; Dietz, V.; Leenders, K.L.; Roelcke, U.; Missimer, J.; Curt, A. How does the human brain deal with a spinal cord injury? *Eur. J. Neurosci.* **1998**, *10*, 3918–3922. [CrossRef]

116. Kerr, A.L.; Cheng, S.Y.; Jones, T.A. Experience-dependent neural plasticity in the adult damaged brain. *J. Commun. Disord.* **2011**, *44*, 538–548. [CrossRef] [PubMed]

117. Feng, T.; Zhao, C.; Rao, J.S.; Guo, X.J.; Bao, S.S.; He, L.W.; Zhao, W.; Liu, Z.; Yang, Z.Y.; Li, X.G. Different macaque brain network remodeling after spinal cord injury and NT3 treatment. *iScience* **2023**, *26*, 106784. [CrossRef]

118. Sawada, M.; Kato, K.; Kunieda, T.; Mikuni, N.; Miyamoto, S.; Onoe, H.; Isa, T.; Nishimura, Y. Function of the nucleus accumbens in motor control during recovery after spinal cord injury. *Science* **2015**, *350*, 98–101. [CrossRef]

119. Chisholm, A.E.; Peters, S.; Borich, M.R.; Boyd, L.A.; Lam, T. Short-term cortical plasticity associated with feedback-error learning after locomotor training in a patient with incomplete spinal cord injury. *Phys. Ther.* **2015**, *95*, 257–266. [CrossRef]

120. Peng, Y.; Liu, J.; Hua, M.; Liang, M.; Yu, C. Enhanced Effective Connectivity from Ipsilesional to Contralesional M1 in Well-Recovered Subcortical Stroke Patients. *Front. Neurol.* **2019**, *10*, 909. [CrossRef]

121. Hubbard, I.J.; Carey, L.M.; Budd, T.W.; Levi, C.; McElduff, P.; Hudson, S.; Bateman, G.; Parsons, M.W. A Randomized Controlled Trial of the Effect of Early Upper-Limb Training on Stroke Recovery and Brain Activation. *Neurorehabilit. Neural Repair* **2015**, *29*, 703–713. [CrossRef]

122. Hoffman, L.R.; Field-Fote, E.C. Functional and corticomotor changes in individuals with tetraplegia following unimanual or bimanual massed practice training with somatosensory stimulation: A pilot study. *J. Neurol. Phys. Ther. JNPT* **2010**, *34*, 193–201. [CrossRef]

123. Lotze, M.; Ladda, A.M.; Stephan, K.M. Cerebral plasticity as the basis for upper limb recovery following brain damage. *Neurosci. Biobehav. Rev.* **2019**, *99*, 49–58. [CrossRef]
124. Gustin, S.M.; Peck, C.C.; Cheney, L.B.; Macey, P.M.; Murray, G.M.; Henderson, L.A. Pain and plasticity: Is chronic pain always associated with somatosensory cortex activity and reorganization? *J. Neurosci. Off. J. Soc. Neurosci.* **2012**, *32*, 14874–14884. [CrossRef] [PubMed]
125. Makin, T.R.; Scholz, J.; Filippini, N.; Henderson Slater, D.; Tracey, I.; Johansen-Berg, H. Phantom pain is associated with preserved structure and function in the former hand area. *Nat. Commun.* **2013**, *4*, 1570. [CrossRef] [PubMed]
126. Jiang, L.; Voulalas, P.; Ji, Y.; Masri, R. Post-translational modification of cortical GluA receptors in rodents following spinal cord lesion. *Neuroscience* **2016**, *316*, 122–129. [CrossRef] [PubMed]
127. Beeler, J.A.; Petzinger, G.; Jakowec, M.W. The Enemy within: Propagation of Aberrant Corticostriatal Learning to Cortical Function in Parkinson's Disease. *Front. Neurol.* **2013**, *4*, 134. [CrossRef] [PubMed]
128. Nudo, R.J.; Milliken, G.W.; Jenkins, W.M.; Merzenich, M.M. Use-dependent alterations of movement representations in primary motor cortex of adult squirrel monkeys. *J. Neurosci. Off. J. Soc. Neurosci.* **1996**, *16*, 785–807. [CrossRef] [PubMed]
129. Chen, K.; Zheng, Y.; Wei, J.A.; Ouyang, H.; Huang, X.; Zhang, F.; Lai, C.S.W.; Ren, C.; So, K.F.; Zhang, L. Exercise training improves motor skill learning via selective activation of mTOR. *Sci. Adv.* **2019**, *5*, eaaw1888. [CrossRef]
130. Charalambous, C.C.; Helm, E.E.; Lau, K.A.; Morton, S.M.; Reisman, D.S. The feasibility of an acute high-intensity exercise bout to promote locomotor learning after stroke. *Top. Stroke Rehabil.* **2018**, *25*, 83–89. [CrossRef]
131. Krakauer, J.W. Motor learning: Its relevance to stroke recovery and neurorehabilitation. *Curr. Opin. Neurol.* **2006**, *19*, 84–90. [CrossRef] [PubMed]
132. Olafson, E.R.; Jamison, K.W.; Sweeney, E.M.; Liu, H.; Wang, D.; Bruss, J.E.; Boes, A.D.; Kuceyeski, A. Functional connectome reorganization relates to post-stroke motor recovery and structural and functional disconnection. *Neuroimage* **2021**, *245*, 118642. [CrossRef]
133. Belviranli, M.; Okudan, N. Differential effects of voluntary and forced exercise trainings on spatial learning ability and hippocampal biomarkers in aged female rats. *Neurosci. Lett.* **2022**, *773*, 136499. [CrossRef]
134. Gumus, H.; Ilgin, R.; Koc, B.; Yuksel, O.; Kizildag, S.; Guvendi, G.; Karakilic, A.; Kandis, S.; Hosgorler, F.; Ates, M.; et al. A combination of ketogenic diet and voluntary exercise ameliorates anxiety and depression-like behaviors in Balb/c mice. *Neurosci. Lett.* **2022**, *770*, 136443. [CrossRef]
135. Hayashi, N.; Himi, N.; Nakamura-Maruyama, E.; Okabe, N.; Sakamoto, I.; Hasegawa, T.; Miyamoto, O. Improvement of motor function induced by skeletal muscle contraction in spinal cord-injured rats. *Spine J.* **2019**, *19*, 1094–1105. [CrossRef]
136. Sánchez-Ventura, J.; Giménez-Llort, L.; Penas, C.; Udina, E. Voluntary wheel running preserves lumbar perineuronal nets, enhances motor functions and prevents hyperreflexia after spinal cord injury. *Exp. Neurol.* **2021**, *336*, 113533. [CrossRef]
137. Evans, N.H.; Suri, C.; Field-Fote, E.C. Walking and Balance Outcomes Are Improved Following Brief Intensive Locomotor Skill Training but Are Not Augmented by Transcranial Direct Current Stimulation in Persons with Chronic Spinal Cord Injury. *Front. Hum. Neurosci.* **2022**, *16*, 849297. [CrossRef]
138. Ozturk, E.D.; Lapointe, M.S.; Kim, D.I.; Hamner, J.W.; Tan, C.O. Effect of 6-Month Exercise Training on Neurovascular Function in Spinal Cord Injury. *Med. Sci. Sports Exerc.* **2021**, *53*, 38–46. [CrossRef]
139. Quaney, B.M.; Boyd, L.A.; McDowd, J.M.; Zahner, L.H.; He, J.; Mayo, M.S.; Macko, R.F. Aerobic exercise improves cognition and motor function poststroke. *Neurorehabilit. Neural Repair.* **2009**, *23*, 879–885. [CrossRef]
140. Rademeyer, H.J.; Gauthier, C.; Zariffa, J.; Walden, K.; Jeji, T.; McCullum, S.; Musselman, K.E. Using activity-based therapy for individuals with spinal cord injury or disease: Interviews with physical and occupational therapists in rehabilitation hospitals. *J. Spinal Cord Med.* **2023**, *46*, 298–308. [CrossRef]
141. Angeli, C.A.; Boakye, M.; Morton, R.A.; Vogt, J.; Benton, K.; Chen, Y.S.; Ferreira, C.K.; Harkema, S.J. Recovery of Over-Ground Walking after Chronic Motor Complete Spinal Cord Injury. *N. Engl. J. Med.* **2018**, *379*, 1244–1250. [CrossRef]
142. Chapman, L.; Cooper-Knock, J.; Shaw, P.J. Physical activity as an exogenous risk factor for amyotrophic lateral sclerosis: A review of the evidence. *Brain J. Neurol.* **2023**, *146*, 1745–1757. [CrossRef] [PubMed]
143. Julian, T.H.; Glascow, N.; Barry, A.D.F.; Moll, T.; Harvey, C.; Klimentidis, Y.C.; Newell, M.; Zhang, S.; Snyder, M.P.; Cooper-Knock, J.; et al. Physical exercise is a risk factor for amyotrophic lateral sclerosis: Convergent evidence from Mendelian randomisation, transcriptomics and risk genotypes. *EBioMedicine* **2021**, *68*, 103397. [CrossRef] [PubMed]
144. Chang, Y.K.; Labban, J.D.; Gapin, J.I.; Etnier, J.L. The effects of acute exercise on cognitive performance: A meta-analysis. *Brain Res.* **2012**, *1453*, 87–101. [CrossRef] [PubMed]
145. Lambourne, K.; Tomporowski, P. The effect of exercise-induced arousal on cognitive task performance: A meta-regression analysis. *Brain Res.* **2010**, *1341*, 12–24. [CrossRef]
146. McMorris, T. The acute exercise-cognition interaction: From the catecholamines hypothesis to an interoception model. *Int. J. Psychophysiol. Off. J. Int. Organ. Psychophysiol.* **2021**, *170*, 75–88. [CrossRef]
147. Marquez-Chin, C.; Popovic, M.R. Functional electrical stimulation therapy for restoration of motor function after spinal cord injury and stroke: A review. *Biomed. Eng. Online* **2020**, *19*, 34. [CrossRef] [PubMed]
148. Choi, E.H.; Gattas, S.; Brown, N.J.; Hong, J.D.; Limbo, J.N.; Chan, A.Y.; Oh, M.Y. Epidural electrical stimulation for spinal cord injury. *Neural Regen. Res.* **2021**, *16*, 2367–2375. [CrossRef]

149. Eisdorfer, J.T.; Smit, R.D.; Keefe, K.M.; Lemay, M.A.; Smith, G.M.; Spence, A.J. Epidural Electrical Stimulation: A Review of Plasticity Mechanisms That Are Hypothesized to Underlie Enhanced Recovery from Spinal Cord Injury with Stimulation. *Front. Mol. Neurosci.* **2020**, *13*, 163. [CrossRef]
150. Carhart, M.R.; He, J.P.; Herman, R.; D'Luzansky, S.; Willis, W.T. Epidural spinal-cord stimulation facilitates recovery of functional walking following incomplete spinal-cord injury. *IEEE Trans. Neural Syst. Rehabilitation Eng.* **2004**, *12*, 32–42. [CrossRef]
151. Capogrosso, M.; Wenger, N.; Raspopovic, S.; Musienko, P.; Beauparlant, J.; Luciani, L.B.; Courtine, G.; Micera, S. A computational model for epidural electrical stimulation of spinal sensorimotor circuits. *J. Neurosci. Off. J. Soc. Neurosci.* **2013**, *33*, 19326–19340. [CrossRef]
152. Nishimura, Y.; Perlmutter, S.I.; Fetz, E.E. Restoration of upper limb movement via artificial corticospinal and musculospinal connections in a monkey with spinal cord injury. *Front. Neural Circuits* **2013**, *7*, 57. [CrossRef]
153. Zimmermann, J.B.; Jackson, A. Closed-loop control of spinal cord stimulation to restore hand function after paralysis. *Front. Neurosci.* **2014**, *8*, 87. [CrossRef] [PubMed]
154. Holinski, B.J.; A Mazurek, K.; Everaert, D.G.; Toossi, A.; Lucas-Osma, A.M.; Troyk, P.; Etienne-Cummings, R.; Stein, R.B.; Mushahwar, V.K. Intraspinal microstimulation produces over-ground walking in anesthetized cats. *J. Neural Eng.* **2016**, *13*, 056016. [CrossRef] [PubMed]
155. Wenger, N.; Moraud, E.M.; Raspopovic, S.; Bonizzato, M.; DiGiovanna, J.; Musienko, P.; Morari, M.; Micera, S.; Courtine, G. Closed-loop neuromodulation of spinal sensorimotor circuits controls refined locomotion after complete spinal cord injury. *Sci. Transl. Med.* **2014**, *6*, 255ra133. [CrossRef] [PubMed]
156. Formento, E.; Minassian, K.; Wagner, F.; Mignardot, J.B.; Le Goff-Mignardot, C.G.; Rowald, A.; Bloch, J.; Micera, S.; Capogrosso, M.; Courtine, G. Electrical spinal cord stimulation must preserve proprioception to enable locomotion in humans with spinal cord injury. *Nat. Neurosci.* **2018**, *21*, 1728–1741. [CrossRef] [PubMed]
157. Wagner, F.B.; Mignardot, J.B.; Le Goff-Mignardot, C.G.; Demesmaeker, R.; Komi, S.; Capogrosso, M.; Rowald, A.; Seáñez, I.; Caban, M.; Pirondini, E.; et al. Targeted neurotechnology restores walking in humans with spinal cord injury. *Nature* **2018**, *563*, 65–71. [CrossRef]
158. Darrow, D.; Balser, D.; Netoff, T.I.; Krassioukov, A.; Phillips, A.; Parr, A.; Samadani, U. Epidural Spinal Cord Stimulation Facilitates Immediate Restoration of Dormant Motor and Autonomic Supraspinal Pathways after Chronic Neurologically Complete Spinal Cord Injury. *J. Neurotrauma* **2019**, *36*, 2325–2336. [CrossRef]
159. Pena Pino, I.; Hoover, C.; Venkatesh, S.; Ahmadi, A.; Sturtevant, D.; Patrick, N.; Freeman, D.; Parr, A.; Samadani, U.; Balser, D.; et al. Long-Term Spinal Cord Stimulation After Chronic Complete Spinal Cord Injury Enables Volitional Movement in the Absence of Stimulation. *Front. Syst. Neurosci.* **2020**, *14*, 35. [CrossRef] [PubMed]
160. Siu, R.; Brown, E.H.; Mesbah, S.; Gonnelli, F.; Pisolkar, T.; Edgerton, V.R.; Ovechkin, A.V.; Gerasimenko, Y.P. Novel Noninvasive Spinal Neuromodulation Strategy Facilitates Recovery of Stepping after Motor Complete Paraplegia. *J. Clin. Med.* **2022**, *11*, 3670. [CrossRef]
161. Asboth, L.; Friedli, L.; Beauparlant, J.; Martinez-Gonzalez, C.; Anil, S.; Rey, E.; Baud, L.; Pidpruzhnykova, G.; Anderson, M.A.; Shkorbatova, P.; et al. Cortico-reticulo-spinal circuit reorganization enables functional recovery after severe spinal cord contusion. *Nat. Neurosci.* **2018**, *21*, 576–588. [CrossRef] [PubMed]
162. Ghorbani, M.; Shahabi, P.; Karimi, P.; Soltani-Zangbar, H.; Morshedi, M.; Bani, S.; Jafarzadehgharehziaaddin, M.; Sadeghzadeh-Oskouei, B.; Ahmadalipour, A. Impacts of epidural electrical stimulation on Wnt signaling, FAAH, and BDNF following thoracic spinal cord injury in rat. *J. Cell. Physiol.* **2020**, *235*, 9795–9805. [CrossRef] [PubMed]
163. Sivanesan, E.; Stephens, K.E.; Huang, Q.; Chen, Z.; Ford, N.C.; Duan, W.; He, S.Q.; Gao, X.; Linderoth, B.; Raja, S.N.; et al. Spinal cord stimulation prevents paclitaxel-induced mechanical and cold hypersensitivity and modulates spinal gene expression in rats. *Pain Rep.* **2019**, *4*, e785. [CrossRef] [PubMed]
164. Stampacchia, G.; Gazzotti, V.; Olivieri, M.; Andrenelli, E.; Bonaiuti, D.; Calabro, R.S.; Carmignano, S.M.; Cassio, A.; Fundaro, C.; Companini, I.; et al. Gait robot-assisted rehabilitation in persons with spinal cord injury: A scoping review. *NeuroRehabilitation* **2022**, *51*, 609–647. [CrossRef] [PubMed]
165. Morone, G.; Cocchi, I.; Paolucci, S.; Iosa, M. Robot-assisted therapy for arm recovery for stroke patients: State of the art and clinical implication. *Expert. Rev. Med. Devices* **2020**, *17*, 223–233. [CrossRef] [PubMed]
166. Banala, S.K.; Kim, S.H.; Agrawal, S.K.; Scholz, J.P. Robot assisted gait training with active leg exoskeleton (ALEX). *IEEE Trans. Neural Syst. Rehabil. Eng.* **2009**, *17*, 2–8. [CrossRef]
167. Meng, Q.; Zeng, Q.; Xie, Q.; Fei, C.; Kong, B.; Lu, X.; Wang, H.; Yu, H. Flexible lower limb exoskeleton systems: A review. *NeuroRehabilitation* **2022**, *50*, 367–390. [CrossRef]
168. Fang, C.Y.; Tsai, J.L.; Li, G.S.; Lien, A.S.Y.; Chang, Y.J. Effects of Robot-Assisted Gait Training in Individuals with Spinal Cord Injury: A Meta-analysis. *BioMed Res. Int.* **2020**, *2020*, 2102785. [CrossRef]
169. Benito-Penalva, J.; Edwards, D.J.; Opisso, E.; Cortes, M.; Lopez-Blazquez, R.; Murillo, N.; Costa, U.; Tormos, J.M.; Vidal-Samsó, J.; Valls-Solé, J.; et al. Gait training in human spinal cord injury using electromechanical systems: Effect of device type and patient characteristics. *Arch. Phys. Med. Rehabil.* **2012**, *93*, 404–412. [CrossRef]
170. Yoshikawa, K.; Mutsuzaki, H.; Koseki, K.; Endo, Y.; Hashizume, Y.; Nakazawa, R.; Aoyama, T.; Yozu, A.; Kohno, Y. Gait Training Using a Wearable Robotic Device for Non-Traumatic Spinal Cord Injury: A Case Report. *Geriatr. Orthop. Surg. Rehabil.* **2020**, *11*, 2151459320956960. [CrossRef]

171. Bersch, I.; Alberty, M.; Fridén, J. Robot-assisted training with functional electrical stimulation enhances lower extremity function after spinal cord injury. *Artif. Organs* **2022**, *46*, 2009–2014. [CrossRef]
172. Alwardat, M.; Etoom, M.; Al Dajah, S.; Schirinzi, T.; Di Lazzaro, G.; Salimei, P.S.; Mercuri, N.B.; Pisani, A. Effectiveness of robot-assisted gait training on motor impairments in people with Parkinson's disease: A systematic review and meta-analysis. *Int. J. Rehabil. Res.* **2018**, *41*, 287–296. [CrossRef]
173. Seo, J.S.; Yang, H.S.; Jung, S.; Kang, C.S.; Jang, S.; Kim, D.H. Effect of reducing assistance during robot-assisted gait training on step length asymmetry in patients with hemiplegic stroke: A randomized controlled pilot trial. *Medicine* **2018**, *97*, e11792. [CrossRef]
174. Calabrò, R.S.; Cacciola, A.; Bertè, F.; Manuli, A.; Leo, A.; Bramanti, A.; Naro, A.; Milardi, D.; Bramanti, P. Robotic gait rehabilitation and substitution devices in neurological disorders: Where are we now? *Neurol. Sci.* **2016**, *37*, 503–514. [CrossRef] [PubMed]
175. Turner, D.L.; Ramos-Murguialday, A.; Birbaumer, N.; Hoffmann, U.; Luft, A. Neurophysiology of robot-mediated training and therapy: A perspective for future use in clinical populations. *Front. Neurol.* **2013**, *4*, 184. [CrossRef] [PubMed]
176. Oza, C.S.; Giszter, S.F. Trunk robot rehabilitation training with active stepping reorganizes and enriches trunk motor cortex representations in spinal transected rats. *J. Neurosci. Off. J. Soc. Neurosci.* **2015**, *35*, 7174–7189. [CrossRef] [PubMed]
177. Calabrò, R.S.; Naro, A.; Russo, M.; Bramanti, P.; Carioti, L.; Balletta, T.; Buda, A.; Manuli, A.; Filoni, S.; Bramanti, A. Shaping neuroplasticity by using powered exoskeletons in patients with stroke: A randomized clinical trial. *J. Neuroeng. Rehabil.* **2018**, *15*, 35. [CrossRef] [PubMed]
178. Wagner, J.; Solis-Escalante, T.; Grieshofer, P.; Neuper, C.; Müller-Putz, G.; Scherer, R. Level of participation in robotic-assisted treadmill walking modulates midline sensorimotor EEG rhythms in able-bodied subjects. *Neuroimage* **2012**, *63*, 1203–1211. [CrossRef]
179. Savaki, H.E.; Raos, V. Action perception and motor imagery: Mental practice of action. *Prog. Neurobiol.* **2019**, *175*, 107–125. [CrossRef]
180. Tong, Y.; Pendy, J.T., Jr.; Li, W.A.; Du, H.; Zhang, T.; Geng, X.; Ding, Y. Motor Imagery-Based Rehabilitation: Potential Neural Correlates and Clinical Application for Functional Recovery of Motor Deficits after Stroke. *Aging Dis.* **2017**, *8*, 364–371. [CrossRef] [PubMed]
181. Gowda, A.S.; Memon, A.N.; Bidika, E.; Salib, M.; Rallabhandi, B.; Fayyaz, H. Investigating the Viability of Motor Imagery as a Physical Rehabilitation Treatment for Patients with Stroke-Induced Motor Cortical Damage. *Cureus J. Med. Sci.* **2021**, *13*, e14001. [CrossRef]
182. Carvalho, R.; Azevedo, E.; Marques, P.; Dias, N.; Cerqueira, J.J. Physiotherapy based on problem-solving in upper limb function and neuroplasticity in chronic stroke patients: A case series. *J. Eval. Clin. Pract.* **2018**, *24*, 552–560. [CrossRef]
183. Grangeon, M.; Charvier, K.; Guillot, A.; Rode, G.; Collet, C. Using sympathetic skin responses in individuals with spinal cord injury as a quantitative evaluation of motor imagery abilities. *Phys. Ther.* **2012**, *92*, 831–840. [CrossRef]
184. Sabbah, P.; de Schonen, S.; Leveque, C.; Gay, S.; Pfefer, F.; Nioche, C.; Sarrazin, J.-L.; Barouti, H.; Tadie, M.; Cordoliani, Y.-S.; et al. Sensorimotor cortical activity in patients with complete spinal cord injury: A functional magnetic resonance imaging study. *J. Neurotrauma* **2002**, *19*, 53–60. [CrossRef] [PubMed]
185. Chen, X.; Wan, L.; Qin, W.; Zheng, W.; Qi, Z.; Chen, N.; Li, K. Functional Preservation and Reorganization of Brain during Motor Imagery in Patients with Incomplete Spinal Cord Injury: A Pilot fMRI Study. *Front. Hum. Neurosci.* **2016**, *10*, 46. [CrossRef] [PubMed]
186. Di Rienzo, F.; Guillot, A.; Mateo, S.; Daligault, S.; Delpuech, C.; Rode, G.; Collet, C. Neuroplasticity of prehensile neural networks after quadriplegia. *Neuroscience* **2014**, *274*, 82–92. [CrossRef] [PubMed]
187. Wang, L.; Zheng, W.; Liang, T.; Yang, Y.; Yang, B.; Chen, X.; Chen, Q.; Li, X.; Lu, J.; Li, B.; et al. Brain Activation Evoked by Motor Imagery in Pediatric Patients with Complete Spinal Cord Injury. *AJNR Am. J. Neuroradiol.* **2023**, *44*, 611–617. [CrossRef] [PubMed]
188. Mokienko, O.A.; Chernikova, L.A.; Frolov, A.A.; Bobrov, P.D. Motor Imagery and its Practical Application. *Zhurnal Vysshei Nervnoi Deyatelnosti Imeni IP Pavlova* **2013**, *63*, 195–204. [CrossRef]
189. Sims, C.; Waldron, R.; Marcellin-Little, D.J. Rehabilitation and Physical Therapy for the Neurologic Veterinary Patient. *Vet. Clin. N. Am. Small Anim. Pract.* **2015**, *45*, 123–143. [CrossRef]

Review

Differentiating Lumbar Spinal Etiology from Peripheral Plexopathies

Marco Foreman [1], Krisna Maddy [2], Aashay Patel [1], Akshay Reddy [1], Meredith Costello [2] and Brandon Lucke-Wold [1,*]

[1] Department of Neurosurgery, University of Florida, Gainesville, FL 32610, USA
[2] Department of Neurosurgery, University of Miami, Miami, FL 33136, USA
* Correspondence: brandon.lucke-wold@neurosurgery.ufl.edu

Abstract: Clinicians have managed and treated lower back pain since the earliest days of practice. Historically, lower back pain and its accompanying symptoms of radiating leg pain and muscle weakness have been recognized to be due to any of the various lumbar spine pathologies that lead to the compression of the lumbar nerves at the root, the most common of which is the radiculopathy known as sciatica. More recently, however, with the increased rise in chronic diseases, the importance of differentially diagnosing a similarly presenting pathology, known as lumbosacral plexopathy, cannot be understated. Given the similar clinical presentation of lumbar spine pathologies and lumbosacral plexopathies, it can be difficult to differentiate these two diagnoses in the clinical setting. Resultingly, the inappropriate diagnosis of either pathology can result in ineffective clinical management. Thus, this review aims to aid in the clinical differentiation between lumbar spine pathology and lumbosacral plexopathy. Specifically, this paper delves into spine and plexus anatomy, delineates the clinical assessment of both pathologies, and highlights powerful diagnostic tools in the hopes of bolstering appropriate diagnosis and treatment. Lastly, this review will describe emerging treatment options for both pathologies in the preclinical and clinical realms, with a special emphasis on regenerative nerve therapies.

Keywords: lumbar spine pathology; sciatica; peripheral plexopathy; lumbosacral plexopathy; regenerative nerve therapy

Citation: Foreman, M.; Maddy, K.; Patel, A.; Reddy, A.; Costello, M.; Lucke-Wold, B. Differentiating Lumbar Spinal Etiology from Peripheral Plexopathies. *Biomedicines* **2023**, *11*, 756. https://doi.org/10.3390/biomedicines11030756

Academic Editor: Nicolas Guerout

Received: 4 February 2023
Revised: 20 February 2023
Accepted: 25 February 2023
Published: 2 March 2023

1. Introduction

In the most recent epidemiological overview on lower back pain (LBP), it was found that up to 80% of individuals worldwide will present with an episode of LBP throughout their lifetime [1]. More specifically, LBP can present as pain, muscle tension, or stiffness localized below the costal margin and above the inferior gluteal sulcus, with or without associated leg or foot pain [2]. Regarding the latter major symptom of LBP, known as sciatica, this is the most common sequelae of lumbar spinal pathologies and is reported to affect as much as 40% of the adult population at some time [3]. Clinically, sciatica is diagnosed due to its characteristic presentation of radiating pain and is the principal radiculopathy affecting the lumbar spine [4].

A separate yet related set of more rare disorders are known as peripheral plexopathies. In this paper, we focus on disorders involving the lumbosacral plexus and its tributaries such as the femoral and sciatic nerves. In contrast to radiculopathies such as sciatica, which commonly only affect a single nerve root, plexopathies are clinically defined as the involvement of at least two different root levels and from at least two additional peripheral nerves [5]. As such, the clinical signs and symptoms of plexopathy—weakness, loss of tendon reflexes, sensory deficits, and often pain—tend to follow a dermatomal distribution, localized to the involved neural structures providing peripheral nervous system innervation to the afflicted muscle groups [6,7].

However, differentiating lumbar spinal pathologies from peripheral plexopathies involving the lumbosacral plexus has historically proven to be clinically challenging. Particularly, this fact holds true because neurologic findings are not always well defined as described above and because there is considerable overlap in the sensory and motor territories supplied by the individual lumbar nerve roots, with the lumbosacral plexus or the nerves derived from the lumbosacral plexus [6,8]. Thus, as two separate pathologic entities that affect the population with opposing levels of burden, it is crucial to elucidate their differentiation to increase diagnostic efficacy and appropriate treatment.

2. Anatomy of the Lumbar Spine and Lumbosacral Plexus

A comprehensive understanding of the functional anatomy of the lumbar spine and the various plexuses that supply the lower limbs and pelvic girdle is warranted when discussing their respective etiology and differentiation in a clinical setting.

The lumbar spine consists of five bones in the lower back, known as the L1–L5 vertebrae, which are composed of two parts, the vertebral body and the vertebral (neural) arch [9]. Further, each of these vertebrae have numerous osseous structures, as seen in Figure 1, and articulate to form distinct anatomical structures such as the intervertebral foramen, which transmit the lumbar spinal nerves and the associated radicular arteries that supply the spinal cord [10]. The lumbar spine is segmented into its five component vertebrae by structures known as intervertebral discs, with the vertebral bodies increasing in size as the column descends [11]. These discs are composed of a centrally located nucleus pulposus concentrically encircled by the annulus fibrosis and a cartilaginous endplate that forms the interface between the intervertebral disc and the adjacent vertebrae [12]. Together, these fibro-cartilaginous structures impart the lumbar spine with multiples ranges of motion—including flexion, extension, lateral bending and rotation—as well as the ability to transfer and distribute spinal loads [13].

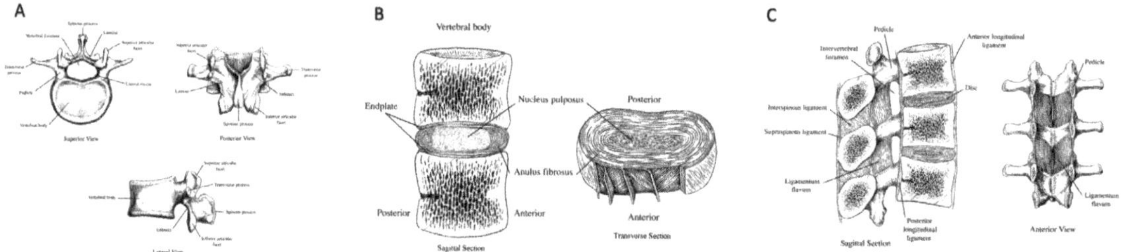

Figure 1. Anatomy of the Lumbar Spine. (**A**) Superior, posterior and lateral aspect of the lumbar vertebrae and the associated osseous structures. (**B**) The sagittal and transverse cross-sections of the lumbar intervertebral disc and its three component parts—the nucleus pulposus, annulus fibrosus, and endplate. (**C**) The sagittal and anterior aspect of the articulating lumbar vertebrae—note the intervertebral foramen formed between the pedicles of the neighboring vertebrae [9]. Adapted from Ebraheim et al. [9].

On the whole, the lumbar spine is responsible for several functions including upper body support and weight distribution, protection of the spinal cord/cauda equina, and—by extension of the nerves that branch off of the lower spinal cord and cauda equina—leg sensation and movement [14]. Regarding the latter functions, it is important to note that there are five pairs of corresponding lumbar spinal nerves (L1–L5) that innervate the lower limbs. Moreover, as seen in Figure 2, these innervations generally follow skin- and muscle-specific distributions supplied by the dorsal and ventral root fibers of a given spinal nerve, known as dermatomes and myotomes, respectively [14]. Specifically, the L1 spinal nerve provides sensation to the groin and genital region as well as motor innervation to the hip; the L2 spinal nerve similarly provides motor innervation to muscles of the hip for movement such as abduction; together, spinal nerves L2–L4 give sensation to the anterior aspect of the thigh and the medial aspect of the lower leg, with spinal nerves L3 and L4 providing motor innervation for movements such as knee extension; finally, spinal nerve L5 provides sensation to the lateral aspect of the lower leg, the dorsum of the foot and the space between the first and second toe, as well as corresponding motor innervation for movements including knee flexion and great toe extension [14–17].

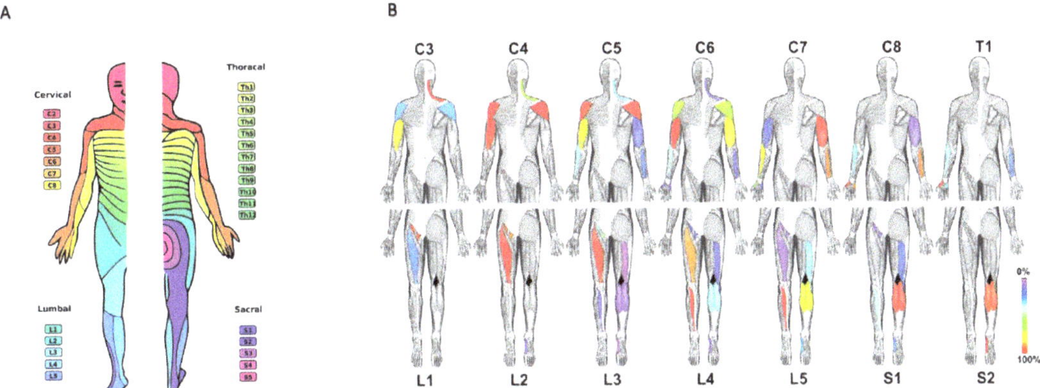

Figure 2. Overview of the Dermatomal and Myotomal Innervations of the Spinal Nerves. (**A**) Diagram of the respective dermatomal distributions for the 31 pairs of spinal nerves as they travel along their distinct paths from posterior to anterior [15]. (**B**) Myotomal distributions of the lumbosacral spinal nerves. The muscle groups are color-mapped to reflect the rate of response upon individual intraoperative nerve root stimulation [16]. Adapted from Whitman et al. and Schirmer et al. [15,16].

Additionally, it is pertinent to underscore that once the lumbar spinal nerves exit the intervertebral foramen, they become a related yet clinically distinct set of anatomical structures known as the lumbar and sacral plexuses—which are together known as the lumbosacral plexus (LSP). The LSP is a complex peripheral nervous system structure that similarly supplies sensory and motor innervation to the lower limbs and is derived from the anterior rami of the L1-S4 nerve roots, and a small contribution from T12 [5]. As illustrated in Figure 3, the LPS has many component nerves and penetrates the psoas major muscle, later emerging on the lateral aspect of the pelvis [18].

Summarized in Table 1, the LSP has an inherently complex set of dermatomal and myotomal innervations and serves as the extension to the spinal nerves with which they are derived from. As a result, at the clinical level, insult to any given component(s) of the LSP can have overlapping sensory and/or motor deficits as those seen in lumbar spinal nerve etiology. In this paper, we will identify common pathologies of the lumbar spine and the lumbosacral plexus, as well as compare their presentation, evaluation, treatment, and management.

Table 1. Nerves of the lumbosacral plexus and their respective muscular and sensory innervations [20].

	Nerve	Muscles	Sensory Distribution
Lumbar Plexus	Iliohypogastric (L1-2)	-	Inferior abdominal wall
	Ilioinguinal (L1-2)	-	Medial Groin
	Genitofemoral (L1-2)	-	-
	Lateral femoral cutaneous (L3-4)	-	Anterolateral thigh

Table 1. *Cont*

	Nerve	Muscles	Sensory Distribution
	Obturator (L2,3,4)	Adductor long Adductor magnus Gracilis	-
	Femoral (L2,3,4)	Quadriceps	-
	Saphenous (L2,3,4)	-	Medial leg and foot
Sacral Plexus	Sup. Gluteal (L4-5)	Gluteus medius Tensor fascia lata	-
	Inf. Gluteal (L4-S1)	Gluteus maximus	-
	Sciatic (L4-S2)	Anterior tibialis Peroneus longus Gastrocnemius Soleus Foot muscles	Foot Lateral leg
	Pudendal (S2,3,4)	External anal sphincter	Perineal

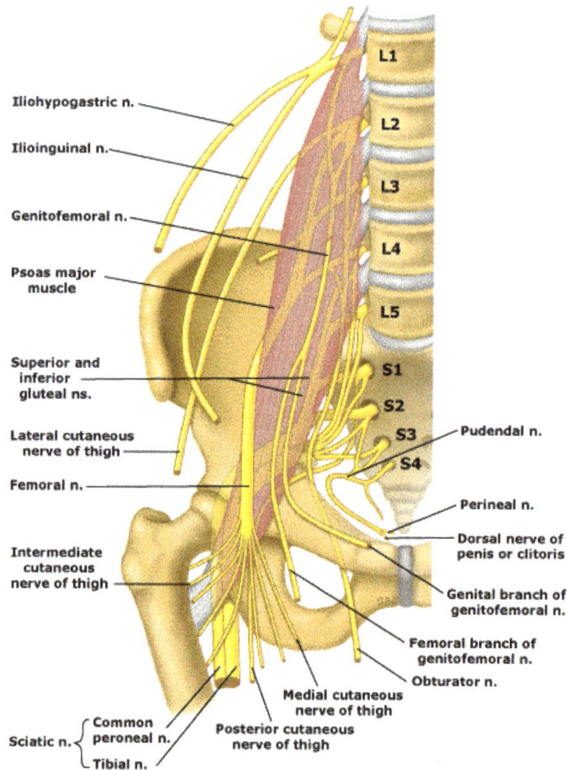

Figure 3. Anatomy of the Lumbosacral Plexus. Coronal view of the lumbosacral plexus and its branches [19].

3. Etiology

Upon extensive review of the literature, the most common etiology of sciatica due to lumbar spinal pathology is caused by lumbar spinal stenosis. As seen in Figure 4 below, however, the pathogenesis that underlie said stenosis cannot be simply traced back to one causative agent, but a multiplicity of pathologies. Thus, the most common mechanisms that lead to the development of lumbar spinal stenosis in the general population will be detailed in this section. Similarly, LS plexopathies also have a complex set of etiologies and thus the most prevalent and clinically significant causes will be discussed.

3.1. Lumbar Spine Pathology

3.1.1. Herniated Nucleus Pulposus

Lumbar disc herniation occurs when the annulus fibrosus ruptures leading to the nucleus pulposus protruding from the intervertebral disc space, also known as herniated nucleus pulposus (HNP) [21]. HNP is one of the most common lumbar spinal pathologies, with 95% of cases occurring at the L4-L5 or L5-S1 level and is one of the main contributors to lumbar spinal stenosis [22,23]. Peak incidence is between ages 30–50, with the prevalence doubled in males [24]. Because of the thicker anterior longitudinal ligament, herniation is most likely to occur in the posterior direction—more specifically the posterolateral direction—leading to compression of the nerve root. With each nerve root having a specific motor and sensory area of innervation, pain and/or paresthesia corresponding with the effected nerve root follow the same distribution [25]. Coughing, sneezing, or straining may worsen symptoms by increasing pressure in the surrounding areas.

3.1.2. Osteoarthritis

A second major contributor to lumbar spine stenosis can be found in the degenerative process known as osteoarthritis. Although arthritis can affect any joint, weight-bearing joints are most susceptible to arthritic changes including the spinal column [26]. The spinal column is composed of three joints: one intervertebral disc and two facet joints, which are all susceptible to degenerative changes [27–29]. This degeneration leads to disc space narrowing, the formation of osteophytes, facet joint osteoarthritis, and eventually foraminal narrowing [29,30]. Further, though osteoarthritis was traditionally thought to be caused from "wear and tear", recent advances in technology now point to a multifactorial pathogenesis including genetic predisposition, epigenetics, and sex, as well as lifestyle factors such as diet, physical activity and work-related habits [31]. Particularly, diet and obesity are significant players in the development of osteoarthritis risk factors because of their role in the pathogenesis of metabolic syndrome, which includes central obesity, diabetes, high blood pressure, and hyperlipidemia, and have been shown to be independently associated with the onset and progression of osteoarthritis [32].

3.1.3. Ligamentum Flavum Thickening

As previously mentioned, both disc bulging and arthritic changes can contribute to lumbar spinal stenosis; however, ligamentum flavum (LF) hypertrophy is believed to be the main contributor [33,34]. The pathophysiology of LF hypertrophy remains unclear. Prior research has demonstrated that tissue thickening may be due to buckling [35]. However, other studies have concluded that thickening may occur mainly in the extracellular matrix indicating hypertrophy is to blame. Age, disc degeneration, and hemodialysis have been indicated as possible risk factors for LF thickening [35–41].

3.1.4. Other Causes

Other lumbar spinal pathologies include spondylolisthesis, space occupying lesions, and infectious processes. Spondylolisthesis is the anterior displacement of a vertebrae and is associated with a bilateral defect of the pars interarticularis. Severity is based on the degree of displacement, and the need for surgical intervention is guided by symptom severity and the concern for instability [42]. Lesions such as cysts, masses, or abscesses can lead to the development of symptoms secondary to spinal cord compression. If symptomatic, compression of the spinal cord is a surgical emergency requiring immediate intervention to prevent permanent deficits. Infectious processes in the spine such as spondylodiscitis and spinal osteomyelitis also lead to various neurological sequalae. Often caused by the spread of infection from adjacent soft tissues or directly from spinal procedures, osteomyelitis/discitis can be a challenge to diagnose and adequately treat [43].

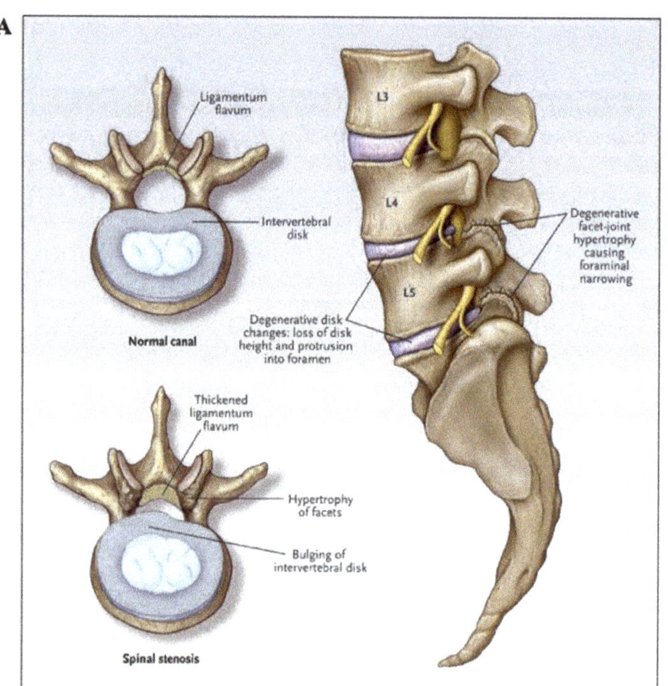

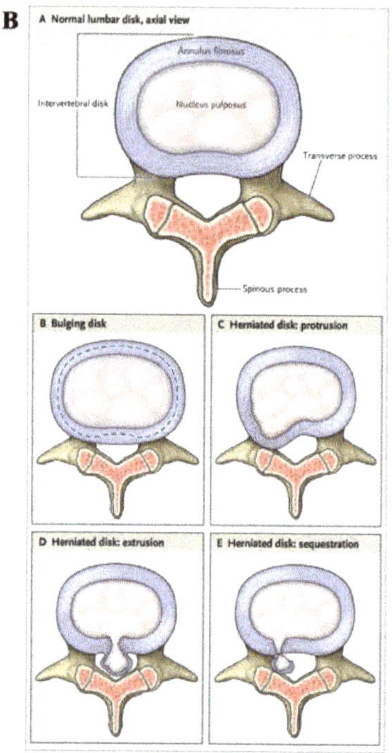

Figure 4. Characteristics of Lumbar Spinal Stenosis and Herniated Disks. (**A**) Axial views of the lumbar spine showing both normal anatomy and spinal stenosis. Depiction demonstrates intervertebral disk bulging, a thickened ligamentum flavum, and hypertrophied fact joints. Additionally, on sagittal view, there is loss of disk height and protrusion, and facet-joint osteoarthritis [44]. (**B**) Panel A demonstrates a non-displaced lumbar intervertebral disk. Panel B shows a bulging disk. Panel C demonstrates disk protrusion. Panel D demonstrates disk extrusion. Panel E demonstrates disk sequestration [45]. Adapted from Deyo et al. and Katz et al. [44,45].

3.2. Lumbosacral Plexopathy

3.2.1. Trauma

Due to the LSP's anatomic location deep within the pelvis, it is relatively shielded from direct injury. Because of this, traumatic plexopathies most likely result from penetration injuries such as a gunshot or puncture wound [46,47]. Typically, traumatic LS plexopathies are commonly associated with pelvic or hip fractures, however, the incidence of LSP complication in patients treated for these types of fractures is low (0.7%) [47]. Contrastingly, in cases involving sacral fractures or sacroiliac dislocations, the frequency is significantly higher (2.03%) due to the proximity of the plexus to both the sacral bone and its articulating sacroiliac joint [47].

3.2.2. Metabolic and Inflammatory Causes

The LSP may be involved alone or along with nearby roots and nerves in immune-mediated or inflammatory disorders [48]. Although lumbosacral radiculoplexus neuropathy (LSPRN) most frequently occurs in patients with diabetes, it can also be found in non-diabetics. In a recent study, incidence of LSRPN was 4.16/100,000 overall. Further, diabetics had an incidence of 2.79/100,000 while non-diabetics had an incidence of

1.27/100,000 [49]. Amyloidosis and sarcoidosis can also cause lumbosacral plexopathies secondary to local inflammatory changes [50].

3.2.3. Neoplastic

According to recent literature, for LS plexopathies caused by neoplasms, the L4-S1 segment is affected in more than 50% of cases, the L1–L4 in 31% of cases, and pan-plexopathy in approximately 10% of cases [51]. Plexopathies were found to occur within the first year after cancer diagnosis in one-third of cases and be the presenting symptom in 15% of cases [52,53]. Pain is the most common initial symptom followed by sensory or motor problems beginning within weeks to months [51].

3.2.4. Other Causes

Other causes of lumbosacral plexopathies include chronic infections such as tuberculosis, fungal infections, lyme, HIV/AIDS, and herpes zoster [5]. Additionally, because of the proximity of the psoas major and minor muscle groups to the plexus, psoas abscesses can lead to various neurologic symptoms. Radiation can also lead to plexopathies and can present without pain and occur bilaterally even years after radiation [54]. Vascular causes include femoral catheterization and ischemic insults from direct compression caused by arterial pseudoaneurysms, aortic dissections, and retroperitoneal hematomas. Postoperative plexopathies can result from direct trauma or prolonged retraction.

4. Neurologic Consultation

With a foundational understanding of the etiologies of lumbar spinal pathology and LS plexopathies, we can better elucidate differences in clinical presentations and some of the current diagnostic modalities that serve to better characterize these pathologies.

4.1. Clinical Presentationand Physical Examination
4.1.1. Distinctions in Symptomatic Presentation

In the initial evaluation of a patient presenting with symptoms of LBP, it is pertinent to evaluate differential etiologies and—for the purpose of this discussion—distinguish between lumbar spinal pathology and LS plexopathy.

Lumbar spinal pathologies typically present with the main complaint of chronic back pain. Important differences in a patient's symptomatic presentation are relevant to note as these allow us to distinguish within specific pathological processes. For example, in lumbar spinal stenosis (LSS), patients complain of neurogenic claudication and pain exacerbated by prolonged ambulation, standing, and lumbar extension—which is relieved by forward flexion and rest [55]. While LSS presents with bilateral pain, sciatica tends to be unilateral radicular pain to the ipsilateral lower extremity [56]. Additionally, the pain is typically "burning" in nature with occasional associated parathesia [56]. Further, lumbar disc herniation can also present with pain following an acute event or spinal movement and, thus, may be indicative of instability or degenerative fracture of the pars at L5 [57].

Given the diverse etiologies of LS plexopathy, this pathology can present variably in the clinical space and can prove difficult to diagnose for said reason. Nonetheless, it is important to note that the symptoms of LS plexopathies tend to reflect the level of anatomical involvement of the plexus and the temporality of the injury. The most common presenting symptoms of LS plexopathy include LBP with unilateral radiation and possible association with positionality [5]. In contrast to LSS, patients typically complain of bilateral leg pain [58]. Furthermore, specific etiologies of LS plexopathies may be associated with more unique symptoms. This exceptionality in presentation is showcased when comparing diabetic LS plexopathy to LS plexopathy secondary to radiotherapy, in which one classically presents with proximal thigh pain and the other is often unremarkable for pain [59]. Lastly, symptoms of muscle weakness and atrophy often occur in more severe cases of disease [60].

4.1.2. Physical Examination

Pertinent physical examination findings in LS plexopathy include a positive straight leg raise test, as well as asymmetric lower limb muscle weakness with asymmetrically absent or reduced deep tendon reflexes [5]. Whereas in some lumbar spinal pathologies such as LSS, as little as 10% of patients demonstrate positive straight leg testing [61]. Similarly, the straight leg raise test has variable specificity for diagnosis of sciatica but may be a more useful indication of lumbar disc herniation [56]. Interestingly, lumbar disc herniations can present with a contralateral positive straight leg test as a supplement to a positive ipsilateral straight leg test, increasing the sensitivity from 84% to 96% [62].

Another important component of the physical examination includes reflex testing as changes to the deep tendon reflexes can be present in both LS plexopathies and lumbar spine pathologies. Specifically, lumbar plexopathy can present with knee jerk reflex abnormalities, while sacral plexopathy can present with ankle jerk reflex abnormalities [5]. Like lumbar plexopathies, an absent or diminished patellar tendon reflex, also known as the Westphal Sign, is commonly found in lumbar spine pathology such as sciatica [63].

Additionally, thorough evaluation of the patient's gait and functionality can provide further insight into pathologic differentiation [57]. The aforementioned holds true because while LS plexopathy can present with more of a dermatomal pattern of symptoms, some lumbar spine pathologies, such as lumbar HNP, must be examined for more mechanical changes affecting function. Additionally, changes will be seen in ankle dorsiflexion or plantarflexion depending on the positionality of an L5/S1 disc herniation laterally or centrally into the neural foramen, further elucidating any functional changes [57].

Further, demonstration of sensory loss and/or weakness upon physical examination also allows us to distinguish between LS plexopathy and lumbar spine pathologies. In LS plexopathy, sacral involvement may present with sensory loss of the medial thigh, anterior thigh, dorsum of the foot, and perineum [5]. More specifically, the sensory and motor loss seen in femoral and sciatic neuropathies—disorders of two major components of the LSP—are crucial to note as they can present similarly to lumbar spine pathologies, albeit there are some notable distinctions upon physical examination. Typically, patients with femoral neuropathy present with sensory loss of the anterior and medial leg [64]. Additionally, there may be weakness involving the quadriceps and iliopsoas muscles on examination [64]. In correlation with these physical examination findings, patients may also complain of difficulty with stairs, frequent falling secondary to knee weakness, and acute pain in the groin, back and lower abdomen [65]. The etiology of this femoral nerve injury is varied and can occur secondary to retroperitoneal hematomas during abdominal surgery, or after total hip arthroplasties, to name a few [65–67]. Moreover, sciatic neuropathies present with more robust foot changes on physical examination, namely foot drop pain, foot inversion, and toe flexion, as well as sensory loss of the upper third of the lateral leg [68]. In contrast, in LSS for example, sensory losses on physical examination in addition to numbness and tingling usually involve the entire lower extremity, rather than following a dermatomal distribution that is characteristic of LS plexopathy [55]. While correlation with symptoms and history is important, the physical examination findings associated with femoral and sciatic neuropathies provide adequate insight to avoid the inaccurate diagnosis of a lumbar spine pathology.

4.2. Diagnostic Evaluation

4.2.1. Radiological Imaging

Imaging for both lumbar spine pathology and LS plexopathy typically involves an MRI with gadolinium contrast, and a computed tomography (CT) scan as a second line option when MRI is contraindicated [69]. In LSS, CT or MRI may demonstrate a diminished intraspinal canal area and anteroposterior spinal canal diameters to confirm diagnosis [44]. Whereas in some lumbar spinal pathologies, such as sciatica, imaging offers little clinical utility, as it is primarily a clinical diagnosis [56]. While CT and MRI offer more insight for some lumbar spine pathologies that may be associated with surgical pathology or structural deformities, LS plexopathy requires more nuanced imaging modalities [70,71]. MR neurography is especially useful for patients presenting with symptoms reflecting lumbosacral plexus or sciatic nerve involvement to better examine the lumbosacral plexus or sciatic nerve [72]. Extraforaminal lesions can better be revealed on MR neurography compared to traditional MR imaging [72]. Findings using this technique can include: fibrous and muscular entrapment, vascular compression, posttraumatic lesions, neuropathy secondary to ischemia, neoplastic and granulomatous infiltration, neural sheath tumors, scar tissue secondary to radiation, and neuropathy that is hypertrophic in nature [72]. Additionally, Gupta et al. noted the presence of a "trident sign" on contrast-enhanced CT and MRI of the lumbosacral region. This "trident sign" is associated with perineural cancer dissemination throughout the lumbosacral plexus, with diffuse thickening of the nerve plexus, and is predictive of adverse survival rates [73].

4.2.2. Electroneurography and Electromyography Studies

Sensory conduction studies are especially useful in distinguishing these two pathologies. A key point in distinguishing LS plexopathy and lumbar spine pathology is the effect on sensory nerve action potentials. If this potential is not reduced in sensory nerve conduction studies, this implies a preganglionic process such as a radiculopathy [74]. Whereas, if the sensory nerve action potential is reduced, it is a postganglionic process, which would imply a plexopathy or mononeuropathy [74]. Motor nerve conduction to diagnose pure LS plexopathy should be focused to the femoral nerve [74]. Peroneal and tibial motor studies may also assess axonal loss in the lumbosacral plexus, however, the usefulness in elderly patients is unlikely given diminished or absent responses in this population [74].

Electromyography (EMG) is also helpful as part of the neurological evaluation to differentiate LS plexopathy and lumbar spinal pathology. Patients with LS plexopathy with referral for EMG studies are typically type 2 diabetes mellitus patients with weight loss, and anterior and posterior thigh and buttock pain [75]. LS plexopathy presents with electrophysiologic abnormalities in the distribution of, at minimum, 2 different peripheral nerves in a minimum of 2 different nerve root distributions [74]. Determining the extent of abnormalities in LS plexopathy requires that needle EMG examination cover L2-S1 innervated muscles and muscles innervated by the same root but not the same peripheral nerves [74]. Pure lumbosacral plexopathy can be further identified when the paraspinals are not affected on needle examination [76]. In contrast to LSS, EMG examination is often normal [77].

In sum, notable differences in clinical presentation and diagnostic evaluation can be used to offer patients accurate diagnosis and subsequently more tailored treatment for their respective disease pathology. Chiefly, these interventions include conservative and surgical approaches, which will be highlighted in the next section.

5. Treatment and Management

The overarching aim of management of pathologies related to the lumbar spine and lumbosacral plexus is to primarily mitigate the progression of the disease, restore functional status of the lower limbs, and preserve the quality of life for the patient [4,60,78,79]. As mentioned above, a shared characteristic between both classes of pathologies is the experience of intense pain, thus highlighting the necessity of alleviatory management care.

5.1. Conservative Approaches

As generally common for spinal pathologies, symptomatic management begins with conservative therapies [4,78,80]. For both lumbar spine pathologies and LS plexopathies, these therapies mainly consist of physical therapy and medications focused on providing pain relief. Physical therapy techniques consist of exercises reinforcing core strength, improving lumbar musculature flexibility, and correcting posture [81]. Limited evidence supports the effectiveness of physical therapy regimens in terms of restoring functionality and alleviating pain [82] and is so far confined to low power studies with small samples sizes [83,84]. In addition to the lack of empirical support for alleviatory physical therapy alone, studies have shown that physical therapy does not consistently provide adequate pain relief and functional strengthening following invasive decompressive procedures [85]. Alongside physical therapy, medications such as analgesics—NSAIDs, opioids, and gabapentin—are prescribed to further alleviate pain symptoms and restore quality of life, although, again, are met with variable effectiveness contingent on the initial intensity of pain, comorbidities, and patient age [86–88]. Lastly, patients with lumbar spine pathologies can undergo selective nerve root injections and translaminar epidural steroid injections to locally deliver long-lasting steroids to irritated spinal nerves. These procedures are typically more favorable in outcome compared to the alternative conservative approaches to pain management, with approximately 60–75% of patients reporting a significant reduction in pain [89].

5.2. Surgical Approaches

Surgical intervention is deemed necessary once it has been sufficiently determined that conservative therapies fail to yield substantial benefits to the patient. As with any form of surgery, it is of the utmost importance to consider potential complications, including repeat post-operative endotracheal intubation, cardiopulmonary resuscitation, pulmonary embolism, and many more. Furthermore, these complications can increase in frequency with various commonly encountered comorbidities such as advanced age, obesity, or diabetes [90].

5.2.1. Laminectomy for Lumbar Spinal Stenosis

Typically, the primary goal of surgical intervention for patients with lumbar spinal stenosis is to improve lower limb functionality, and it is often decided upon following a three-to-six-month period of conservative treatments met with failure to reduce pain and further neurological impairment [91]. The most common surgical procedures indicated for lumbar spinal stenosis are laminectomies [92]. Decompressive lumbar laminectomy procedures entail the excision of the lamina and spinous process at the level at which nerve compression occurs. A successful procedure with resection of the lamina and excessive facet joints results in the respective decompression of the central canal and neural foramina, thus relieving pressure and preventing further nerve impingement [93].

A retrospective analysis performed by Bydon and colleagues demonstrated significant positive outcomes with regards to neurogenic claudication and motor function in patients following lumbar laminectomy, highlighting its relative effectiveness compared to the inconsistent outcomes observed with conservative approaches [94]. In fact, the study emphasizes that the percentage of patients who reported initial lower back pain decreased from 57.40% to 25.40% ($p < 0.001$) following the procedure [94]. Despite its effectiveness for the treatment of lumbar spinal stenosis, the laminectomy procedure is not without complications. A primary concern for laminectomy for lumbar spinal stenosis involves the comorbidity of preoperative spondylolisthesis, as well as iatrogenic spondylolisthesis following the procedure [95]. A systematic review including 2496 patients determined that 1.8% patients that underwent decompressive lumbar laminectomy developed instability that warranted a reoperation [95]. This phenomenon was observed more frequently with patients with pre-existing spondylolisthesis prior to operation [95]. Currently, there is thought that with minimally invasive techniques, the risk of instability and therefore reoperation deceases, although it may achieve less than desirable decompression. Additionally, the field still requires larger prospective studies to verify the effectiveness of these minimally invasive techniques [95]. Otherwise, instability could potentially be avoided through the combination of lumbar laminectomy with spinal fusion [96,97].

5.2.2. Discectomy for Lumbar Herniated Nucleus Pulposus

For patients that fail to experience symptomatic benefits from conservative treatments, lumbar discectomy is typically indicated [98]. These procedures are described by excision of the herniated disk following an opening of the spine and dissection of nearby nerve roots if necessary [99]. The incremental excision process does not typically remove the entire disk, as it is terminated after two to three failed attempts to remove any additional material with the forceps. There are also microendoscopic alternatives to this approach (Figure 5B) [45,99]. However, studies fail to demonstrate a difference in postoperative pain outcomes between the two versions of the procedure [99]. Compared to conservative approaches, lumbar discectomies demonstrate better outcomes regarding pain. Furthermore, the location of the discectomy has a remarkable effect on the favorability of the patient outcomes, as it was demonstrated that upper-level lumbar herniation repairs are slightly better in outcome than lower-level lumbar repairs [100]. Most patients experience an alleviation of back and leg pain following procedure, yet one study determined that 28% of patients still experience pain symptoms [101]. Furthermore, there is risk for recurrent disc herniation following procedure, warranting reoperation (7.3% of patients) [101].

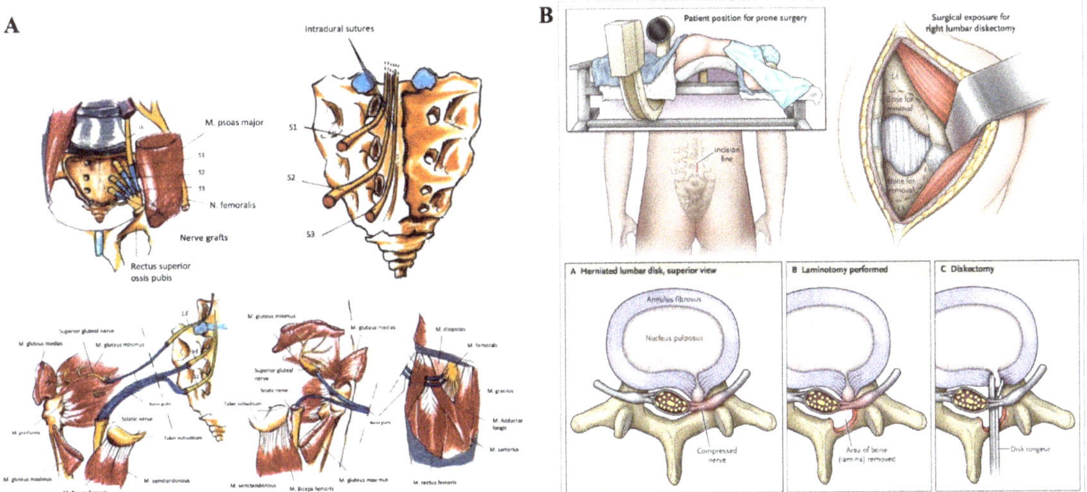

Figure 5. Repair strategies for lumbar spinal pathology and lumbosacral plexopathy. (**A**) Repair strategies used in lumbosacral plexopathy. Upper Left: Intrapelvic repair of the sacral plexus. Upper Right: Intraspinal repair of ruptured ventral roots with sutured nerve grafts. Lower Left: Intraspinal and extrapelvic reconstruction of the sacral plexus by means of nerve grafts to gluteal and sciatic nerves. Lower Right: Nerve transfers by means of nerve grafts from fascicles of femoral nerve to gluteal and sciatic nerves [60]. (**B**) Technique of microdiscectomy is shown. Panel A shows a posterolateral disk herniation. Panel B shows a small incision made with a surgical microscope and a small laminotomy. Panel C shows a discectomy being performed [45]. Adapted from Deyo et al. and Lang et al. [45,60].

5.2.3. Surgical Intervention Strategies for Lumbosacral Plexopathy

In direct contrast to lumbar spine injuries, LS plexopathy lacks clearly established surgical guidelines, let alone decompressive techniques, due to its inherent locational and functional complexity [60]. Further complicating matters, LS plexopathies can secondarily arise from a variety of primary conditions, such as diabetes mellitus, neoplastic developments, or complications in radiotherapy [5]. Surgical intervention for LS plexopathies is typically very involved on the part of the surgeon, requiring extensive navigation such as extraperitoneal lumbotomy, transperitoneal xifopubical incision, or posterior laminectomy, to reach the targeted nervous structure depending on the level location of the structure [102].

Following these procedures to expose the targeted nerve, surgeons typically opt for neurolysis to mitigate pain symptoms related to the pathology of that peripheral nerve, which often is the femoral, obturator, and the sciatic nerve when considering LS plexopathies [103]. As discussed above, pathologies related to these nerves often result in deambulation and the radiation of burning pain [7,58]. As such, the overall goal of these surgeries is to alleviate these symptoms directly related to those nerves. Neurolysis entails the circumferential dissection of the targeted nerve, with the goal of removing any physical obstruction (i.e., scar tissue) that may impact the route and thus transmission of the nerve [104,105]. This peripheral approach to treatment, when targeting the femoral nerve in patients experiencing LS plexopathy, has been found to attain positive results with various patients, including a reduction in severe pain and restored motor function to the gluteal muscles [102,106]. Furthermore, in studies following patients with plexopathy affecting on the obturator nerve, and thus experiencing limited hip adduction, it was determined that neurolysis on the obturator nerve improved hip adduction from Medial Research Council (MRC) grade 2 to MRC grade 5 along with a reduction in pain [103,107].

Further, repair strategies used in LS plexopathies with extensive neuronal damage include procedural nerve grafts to restore function to affected peripheral nerves (Figure 5A) [60,103]. Nerve grafts can either entail an end-to-end nerve suture or suturing an autologous donor nerve to replace the damaged portion of the recipient nerve [108,109]. Various case reports highlight the effectiveness of this strategy, namely restoring the functionality of the femoral, obturator, and sciatic nerves, as well as for multilevel injuries involving nerve roots [102,103,107,110,111]. Across these various procedures, it was reported that related motor and sensory function was successfully restored, including hip adduction, hip flexion, and knee extension [102,103,107,110,111].

6. Emerging Treatment Options

As aforementioned, treatment for lumbar spine pathologies and LS plexopathies include both conservative and surgical approaches. More recently, emerging treatments have taken advantage of new approaches or modified previous approaches, specifically with the aim of nerve regeneration.

6.1. Novel Conservative and Surgical Techniques

6.1.1. Approaches for Common Lumbar Spinal Pathologies

Regarding sciatica, the most common pathology of the lumbar region, a nerve root foramen opening protocol was recently used as conservative treatment [112]. This approach involves placing patients in a side lying position to invoke opening of the foramen. Results have showed further alleviation of pain and reduction of opioid consumption with this approach [112]. Further, this approach is significantly simpler than previous therapies and allows the luxury for patients to perform this manually at home, reducing costs of care. Although further research is needed to prove the effectiveness of this regimen, it is a step in a new direction for sciatica conservative care.

Sciatica can often be caused by lumbar spondylisthesis (LS), which is defined by the "slippage" of a vertebral body [113]. Recently, several clinical trials are in place to find the most optimal and effective treatment for LS. The Spine Patient Outcomes Research Trial (SPORT) has shown the greater effectiveness of surgery in relation to nonoperative care for LS [114]. However, the lack of focus on the superiority of laminectomy plus fusion in relation to just laminectomy in treatment of LS has led to the Spinal Laminectomy versus Instrumented Pedicle Screw (SLIP) trial [96]. Results show that utilization of lumbar spinal fusion in combination with laminectomy showed clinically meaningful improvement in quality of life of patient's post-surgery [96].

Additionally, LSS also contributes to the pathogenesis of lumbar sciatica [115]. Typically, patients with LSS have a variety of conservative options including pain medications and physical therapy. Surgical intervention for LSS is common in cases when patients begin to see neurological deterioration and symptoms [115,116]. The standard of care has been decompression surgery, requiring significant manipulation of spinal muscles [116]. This approach has been seen to increase the risk of postoperative complications [117]. Emerging treatments for lumbar stenosis have adopted a minimally invasive surgical approach using endoscopy, with the aim of decreasing surgical trauma and maximizing preservation of the spinal structure. Recently, unilateral biportal endoscopic spinal surgery (UBESS) has been a promising new approach [118]. The design includes two portals in comparison to traditional spinal endoscopy with one portal [118]. This allows UBESS a wider range of movement and accounts for the shortcomings of traditional endoscopy of a lack of proper visualization. More recent studies have shown the effectiveness of UBESS for LSS, but further research is needed to show further efficacy.

6.1.2. Approaches for Lumbosacral Plexopathies

As previously mentioned, LS plexopathies have historically not had a definitive standard surgical guideline due to their inherent complexity [28]. Surgical options such as extraperitoneal lumbotomy, and posterior laminectomy tend to be the approach, followed by neurolysis [71]. In cases of extensive neuronal damage, opting for procedural nerve grafts is also feasible to restore nerve function [28,72]. More recently, the pararectus approach has been utilized for both anterior exposure and neurolysis of lumbar nerve roots 4/5 [119]. Originally, this technique used in acetabular fracture repairs, but has now has utility in wide arrange of procedures including resection of musculoskeletal tumors [120,121]. This intrapelvic technique involves specific positioning of the patient in a supine position that puts the groin, umbilicus, and anterior superior iliac spine (ASIS) in a triangular arrangement [119]. In comparison to the initial use of the pararectus approach, the neurolysis for LSP involves an incision being roughly 4–5 cm closer towards the cranium [119]. After superficial and deep dissection to visualize the anterior abdominal wall and tranversalis fascia, respectively, surgeons work through the retroperitoneal space bluntly [119]. Following the iliac vessels gently and reaching the L4 and L5 nerve roots along with the obturator nerve allows for proper neurolysis [119]. Although scarce research is available regarding this approach for LS plexopathy, the exposure of the iliac vessels for effective visualization of the LSP combined with and lack of osteotomy demonstrates its procedural advantages. However, further research is needed to smooth some limitations such as the difficulty in treating obese patients and those with previous scarring [119]. Nonetheless, results in cases thus far have shown to be effective with good outcomes [119].

Less invasive approaches for malignant lumbosacral plexopathy have recently been explored as well. Specifically, in the rare malignancy of the LSP, radiation therapy has proven to provide a route for significant pain relief in patients with lumbosacral syndrome and should be investigated further as a viable treatment option [51,122]. More recently, MR neurography (MRN) has also proven valuable in visualization of the peripheral nerves to allow for easier nerve tracking [123]. Due to this radiotherapy potentially causing neurotoxicity at higher doses, accurate contouring becomes an important aspect of treatment [124]. MRN's visualization in combination with CT allows for personalization of contouring by the physician and less need for anatomical knowledge [123]. This is especially made possible through utilizing a 3D MRN sequence such as Lr_NerveVIEW, which has advantageous features that improve LSP nerve contouring [123]. Further, with the likelihood of advanced imaging techniques on the horizon, it only makes this approach more intriguing. However, more research is needed in a larger sample size before this is implemented on a greater scale.

6.2. Nerve Regeneration Strategies

Further, focus has also been on emerging treatment options for nerve regeneration and the coordination of these procedures with the standard surgical options previously mentioned. Although peripheral nerves can regenerate, this is a slow process and patients with neurotmesis or axonotmesis typically experience an impaired quality of life [125]. Conservative approaches to nerve regeneration have primarily included physical rehabilitation using various exercises [125]. This approach has been the standard non-invasive strategy and has been seen to improve peripheral nerve injury (PNI) recovery [126,127]. Recent advances in stem cell therapy have shown the potential of cell transplantation therapy for nerve regeneration. Strategically reconstructing lesion sites with stem cell grafts, which will later differentiate into neurons, has been a valuable approach to nerve regeneration [128]. Utilization of this strategy entails the transplantation of embryonic and pluripotent neuronal stem/progenitor cells (NSCs/NPCs) into lesion sites, where they will eventually develop into self-integrating grafts within the tissue [128,129]. The connection of the implanted neurons with the hosts circuitry provides a route for new sensory connections. Further, axons have been shown to extend into the graft and further synapse with the graft neurons allowing for the formation of a new circuit [130]. Although, this approach is promising and exciting, many axons are at the superficial surface of the graft (1 mm) and will not significantly integrate into the graft. Further research exploring adverse effects and consistency of this strategy must be done prior to having a clinical impact. More recently, mesenchymal stem cells (MSC) have been identified as potentially valuable for nerve regeneration and lack the ethical concern component of their aforementioned counterparts [131]. Particularly, research into MSCs ability to integrate into host lesions has shown superior migration ability due to various factors [132,133]. The ability for MSCs to release neuroprotective factors and control glial scarring makes them particularly attractive, but further research must be done with the aim of translating these findings to clinical practice.

Finally, the most recent development surrounding novel nerve regeneration therapeutics lies in the electrical stimulation of transplanted neural stem cells and the surrounding musculature via "nerve conduits". These conduits, or "nerve guidance channels", are complex biomaterials that offer injured peripheral nerves a scaffolding to promote guided axonal growth through physical, chemical, and electrical means [134]. Particularly, newer conduit materials such as polypyrrole (PPy) have come under extensive investigation for their unique conductive properties under physiologic conditions and their excellent biocompatibility in vivo [135]. These capacities as mentioned above of PPy are of the utmost importance in novel peripheral nerve regeneration because it has been found that the direct electrical stimulation of peripheral nerves leads to an increased release of neurotropic chemicals such as brain-derived neurotrophic factor (BDNF)—an essential small molecule involved in axonal regeneration and re-myelination [136]. Furthermore, BDNF has modulatory effects on tropomyosin receptor kinase (Trk), whose pathway activation has multiple downstream effects involved in stimulating the "growth cone" during nerve repair processes [137]. Cumulatively, by manipulating the physiological microenvironment and providing a robust biocompatible framework using electrical stimulation and nerve conduits, this is a step towards amplifying the therapeutic effect of stem cell-based treatments on neuronal regeneration [138].

7. Conclusions

In this review, we introduced a common ailment—lower back pain and its classically associated symptoms of radiating lower limb pain and muscle weakness—and explored two groups of peripheral neurologic disorders that can lead to said symptomatic manifestations. The first of these disorders, lumbar spine pathology, was found to be comprised of various disease states such as HNP, osteoarthritis, and ligamentum flavum thickening, all with a common etiology of lumbar spine stenosis. The rarer of the two, LS plexopathies, were described as less mechanically involved and were principally caused by metabolic and inflammatory processes. Clinically, since the presentation of lumbar spine pathology and LS plexopathies tend to be indistinguishable upon superficial assessment, we highlighted distinctive symptomatic presentations such as the bilateral leg pain most often reported in patients with a lumbar spine pathology, in contrast to the unilateral radiation found in LS plexopathy patients. Of importance, we also identified key physical examination findings that underscored the differences in sensory and motor deficits, such as the involvement of the entire lower extremity versus following a dermatomal distribution, in lumbar spine pathology and LS plexopathy, respectively. In conjunction with the latter clinical forms of assessment, we described useful diagnostic testing such as contrast MRI and CT to grossly examine nervous structures, as well as detailed the more nuanced techniques of sensory and EMG studies to further elucidate specific involvements. Additionally, we discussed the utility of conservative and surgical approaches for the purpose of alleviatory care in the treatment of both pathologies, although the literature was much scarcer regarding surgical interventions in the treatment of LS plexopathies. Finally, we looked to the future of treatment for these pathologies in the context of novel nerve regeneration therapeutics and their integration into standard-of-care approaches via the employment of NSCs/NPCs, their mesenchymal counterparts, and their combined use with conductive nerve conduits.

Author Contributions: Conceptualization, M.F. and B.L.-W.; methodology, K.M. and A.P.; investigation, A.R. and M.C.; resources, M.F.; writing—original draft preparation, M.F., K.M., A.P., A.R. and M.C.; writing—review and editing, M.F. and B.L.-W.; supervision, B.L.-W. All authors have read and agreed to the published version of the manuscript.

Funding: This research received no external funding.

Informed Consent Statement: Not applicable.

Data Availability Statement: No new data were created or analyzed in this study. Data sharing is not applicable to this article.

Conflicts of Interest: The authors declare no conflict of interest.

References

1. Mattiuzzi, C.; Lippi, G.; Bovo, C. Current Epidemiology of Low Back Pain. *J. Hosp. Manag. Health Policy* **2020**, *4*, 15. [CrossRef]
2. Koes, B.W.; van Tulder, M.W.; Thomas, S. Diagnosis and Treatment of Low Back Pain. *BMJ* **2006**, *332*, 1430–1434. [CrossRef] [PubMed]
3. Frymoyer, J.W. Lumbar Disk Disease: Epidemiology. *Instr. Course Lect.* **1992**, *41*, 217–223.
4. Koes, B.W.; van Tulder, M.W.; Peul, W.C. Diagnosis and Treatment of Sciatica. *BMJ* **2007**, *334*, 1313–1317. [CrossRef]
5. Dydyk, A.M.; Hameed, S. Lumbosacral Plexopathy. In *StatPearls*; StatPearls Publishing: Treasure Island, FL, USA, 2022.
6. Wilbourn, A.J. Plexopathies. *Neurol. Clin.* **2007**, *25*, 139–171. [CrossRef]
7. Van Alfen, N.; Malessy, M.J.A. Chapter 18—Diagnosis of Brachial and Lumbosacral Plexus Lesions. In *Handbook of Clinical Neurology*; Said, G., Krarup, C., Eds.; Peripheral Nerve Disorders; Elsevier: Amsterdam, The Netherlands, 2013; Volume 115, pp. 293–310. [CrossRef]
8. Brejt, N.; Berry, J.; Nisbet, A.; Bloomfield, D.; Burkill, G. Pelvic Radiculopathies, Lumbosacral Plexopathies, and Neuropathies in Oncologic Disease: A Multidisciplinary Approach to a Diagnostic Challenge. *Cancer Imaging* **2013**, *13*, 591–601. [CrossRef]
9. Ebraheim, N.A.; Hassan, A.; Lee, M.; Xu, R. Functional Anatomy of the Lumbar Spine. *Semin. Pain Med.* **2004**, *2*, 131–137. [CrossRef]
10. Intervertebral Foramina—An Overview. ScienceDirect Topics. Available online: https://www.sciencedirect.com/topics/veterinary-science-and-veterinary-medicine/intervertebral-foramina (accessed on 9 January 2023).
11. Waxenbaum, J.A.; Reddy, V.; Williams, C.; Futterman, B. Anatomy, Back, Lumbar Vertebrae. In *StatPearls*; StatPearls Publishing: Treasure Island, FL, USA, 2022.
12. Walter, B.A.; Torre, O.M.; Laudier, D.; Naidich, T.P.; Hecht, A.C.; Iatridis, J.C. Form and Function of the Intervertebral Disc in Health and Disease: A Morphological and Stain Comparison Study. *J. Anat.* **2015**, *227*, 707–716. [CrossRef]
13. Lundon, K.; Bolton, K. Structure and Function of the Lumbar Intervertebral Disk in Health, Aging, and Pathologic Conditions. *J. Orthop. Sports Phys. Ther.* **2001**, *31*, 291–303; discussion 304–306. [CrossRef]
14. Cleveland Clinic. Lumbar Spine: What It Is, Anatomy & Disorders. Available online: https://my.clevelandclinic.org/health/articles/22396-lumbar-spine (accessed on 8 January 2023).
15. Whitman, P.A.; Adigun, O.O. Anatomy, Skin, Dermatomes. In *StatPearls*; StatPearls Publishing: Treasure Island, FL, USA, 2022.
16. Schirmer, C.M.; Shils, J.L.; Arle, J.E.; Cosgrove, G.R.; Dempsey, P.K.; Tarlov, E.; Kim, S.; Martin, C.J.; Feltz, C.; Moul, M.; et al. Heuristic Map of Myotomal Innervation in Humans Using Direct Intraoperative Nerve Root Stimulation: Clinical Article. *J. Neurosurg. Spine* **2011**, *15*, 64–70. [CrossRef]
17. Edgar, M.A.; Ghadially, J.A. Innervation of the Lumbar Spine. *Clin. Orthop. Relat. Res.* **1976**, *115*, 35. [CrossRef]
18. Thaisetthawatkul, P.; Dyck, P.J.B. Chapter 9—Cervical and Lumbosacral Radiculoplexus Neuropathies. In *Dysimmune Neuropathies*; Rajabally, Y.A., Ed.; Academic Press: Cambridge, MA, USA, 2020; pp. 199–223. [CrossRef]
19. Lumbosacral Plexus—UpToDate. Available online: https://sso.uptodate.com/contents/image/print?imageKey=NEURO%2F7 5983&source=graphics_gallery&topicKey=5278 (accessed on 15 January 2023).
20. Rubin, D.I. Brachial and Lumbosacral Plexopathies: A Review. *Clin. Neurophysiol. Pract.* **2020**, *5*, 173–193. [CrossRef] [PubMed]
21. Kawaji, Y.; Uchiyama, S.; Yagi, E. Three-Dimensional Evaluation of Lumbar Disc Hernia and Prediction of Absorption by Enhanced MRI. *J. Orthop. Sci.* **2001**, *6*, 498–502. [CrossRef] [PubMed]

22. Asch, H.L.; Lewis, P.J.; Moreland, D.B.; Egnatchik, J.G.; Yu, Y.J.; Clabeaux, D.E.; Hyland, A.H. Prospective Multiple Outcomes Study of Outpatient Lumbar Microdiscectomy: Should 75 to 80% Success Rates Be the Norm? *J. Neurosurg.* **2002**, *96* (Suppl. S1), 34–44. [CrossRef] [PubMed]
23. Fisher, C.; Noonan, V.; Bishop, P.; Boyd, M.; Fairholm, D.; Wing, P.; Dvorak, M. Outcome Evaluation of the Operative Management of Lumbar Disc Herniation Causing Sciatica. *J. Neurosurg.* **2004**, *100* (Suppl. S4), 317–324. [CrossRef]
24. Bush, K.; Cowan, N.; Katz, D.E.; Gishen, P. The Natural History of Sciatica Associated with Disc Pathology. A Prospective Study with Clinical and Independent Radiologic Follow-Up. *Spine* **1992**, *17*, 1205–1212. [CrossRef]
25. Strayer, A. Lumbar Spine: Common Pathology and Interventions. *J. Neurosci. Nurs.* **2005**, *37*, 181–193. [CrossRef]
26. Sarzi-Puttini, P.; Atzeni, F.; Fumagalli, M.; Capsoni, F.; Carrabba, M. Osteoarthritis of the Spine. *Semin. Arthritis Rheum.* **2005**, *34* (Suppl. S2), 38–43.
27. Goode, A.P.; Carey, T.S.; Jordan, J.M. Low Back Pain and Lumbar Spine Osteoarthritis: How Are They Related? *Curr. Rheumatol. Rep.* **2013**, *15*, 305. [CrossRef]
28. Laplante, B.L.; DePalma, M.J. Spine Osteoarthritis. *PM R* **2012**, *4* (Suppl. S5), S28–S36. [CrossRef]
29. Goode, A.P.; Nelson, A.E.; Kraus, V.B.; Renner, J.B.; Jordan, J.M. Biomarkers Reflect Differences in Osteoarthritis Phenotypes of the Lumbar Spine: The Johnston County Osteoarthritis Project. *Osteoarthr. Cartil.* **2017**, *25*, 1672–1679. [CrossRef] [PubMed]
30. Gellhorn, A.C.; Katz, J.N.; Suri, P. Osteoarthritis of the Spine: The Facet Joints. *Nat. Rev. Rheumatol.* **2013**, *9*, 216–224. [CrossRef]
31. Lindsey, T.; Dydyk, A.M. Spinal Osteoarthritis. In *StatPearls*; StatPearls Publishing: Treasure Island, FL, USA, 2022.
32. Gandhi, R.; Woo, K.M.; Zywiel, M.G.; Rampersaud, Y.R. Metabolic Syndrome Increases the Prevalence of Spine Osteoarthritis. *Orthop. Surg.* **2014**, *6*, 23–27. [CrossRef] [PubMed]
33. Naylor, A. Factors in the Development of the Spinal Stenosis Syndrome. *J. Bone Jt. Surg. Br.* **1979**, *61-B*, 306–309. [CrossRef]
34. Hansson, T.; Suzuki, N.; Hebelka, H.; Gaulitz, A. The Narrowing of the Lumbar Spinal Canal during Loaded MRI: The Effects of the Disc and Ligamentum Flavum. *Eur. Spine J.* **2009**, *18*, 679–686. [CrossRef] [PubMed]
35. Altinkaya, N.; Yildirim, T.; Demir, S.; Alkan, O.; Sarica, F.B. Factors Associated with the Thickness of the Ligamentum Flavum: Is Ligamentum Flavum Thickening Due to Hypertrophy or Buckling? *Spine* **2011**, *36*, E1093–E1097. [CrossRef] [PubMed]
36. Sairyo, K.; Biyani, A.; Goel, V.; Leaman, D.; Booth, R.; Thomas, J.; Gehling, D.; Vishnubhotla, L.; Long, R.; Ebraheim, N. Pathomechanism of Ligamentum Flavum Hypertrophy: A Multidisciplinary Investigation Based on Clinical, Biomechanical, Histologic, and Biologic Assessments. *Spine* **2005**, *30*, 2649–2656. [CrossRef]
37. Sakamaki, T.; Sairyo, K.; Sakai, T.; Tamura, T.; Okada, Y.; Mikami, H. Measurements of Ligamentum Flavum Thickening at Lumbar Spine Using MRI. *Arch. Orthop. Trauma Surg.* **2009**, *129*, 1415–1419. [CrossRef]
38. Sudhir, G.; Vignesh Jayabalan, S.; Gadde, S.; Venkatesh Kumar, G.; Karthik Kailash, K. Analysis of Factors Influencing Ligamentum Flavum Thickness in Lumbar Spine—A Radiological Study of 1070 Disc Levels in 214 Patients. *Clin. Neurol. Neurosurg.* **2019**, *182*, 19–24. [CrossRef]
39. Yoshiiwa, T.; Miyazaki, M.; Kawano, M.; Ikeda, S.; Tsumura, H. Analysis of the Relationship between Hypertrophy of the Ligamentum Flavum and Lumbar Segmental Motion with Aging Process. *Asian Spine J.* **2016**, *10*, 528–535. Available online: https://pubmed.ncbi.nlm.nih.gov/27340534 (accessed on 22 January 2023). [CrossRef]
40. Ito, M.; Abumi, K.; Takeda, N.; Satoh, S.; Hasegawa, K.; Kaneda, K. Pathologic Features of Spinal Disorders in Patients Treated with Long-Term Hemodialysis. *Spine* **1998**, *23*, 2127–2133. [CrossRef] [PubMed]
41. Inatomi, K.; Matsumoto, T.; Tomonaga, T.; Eto, M.; Shindo, H.; Hayashi, T.; Konishi, H. Histological Analysis of the Ligamentum Flavum of Patients with Dialysis-Related Spondyloarthropathy. *J. Orthop. Sci.* **2004**, *9*, 285–290. [CrossRef] [PubMed]
42. Dunn, B. Lumbar spondylolysis and spondylolisthesis. *J. Am. Acad. Physician Assist.* **2019**, *32*, 50–51. Available online: https://journals.lww.com/jaapa/Fulltext/2019/12000/Lumbar_spondylolysis_and_spondylolisthesis.11.aspx (accessed on 22 January 2023). [CrossRef] [PubMed]
43. Nickerson, E.K.; Sinha, R. Vertebral osteomyelitis in adults: An update. *Br. Med. Bull.* **2016**, *117*, 121–138. Available online: https://academic.oup.com/bmb/article/117/1/121/1744712 (accessed on 22 January 2023). [CrossRef] [PubMed]
44. Katz, J.N.; Harris, M.B. Lumbar Spinal Stenosis. *N. Engl. J. Med.* **2008**, *358*, 818–825. [CrossRef] [PubMed]
45. Deyo, R.A.; Mirza, S.K. Herniated Lumbar Intervertebral Disk. *N. Engl. J. Med.* **2016**, *374*, 1763–1772. [CrossRef]
46. Chiou-Tan, F.Y.; Kemp, K., Jr.; Elfenbaum, M.; Chan, K.T.; Song, J. Lumbosacral plexopathy in gunshot wounds and motor vehicle accidents: Comparison of electrophysiologic findings. *Am. J. Phys. Med. Rehabil.* **2001**, *80*, 280–285. Available online: https://pubmed.ncbi.nlm.nih.gov/11277135 (accessed on 22 January 2023). [CrossRef]
47. Kutsy, R.L.; Robinson, L.R.; Routt, M.L., Jr. Lumbosacral plexopathy in pelvic trauma. *Muscle Nerve* **2000**, *23*, 1757–1760. Available online: https://pubmed.ncbi.nlm.nih.gov/11054756 (accessed on 22 January 2023). [CrossRef]
48. Laughlin, R.S.; Dyck, P.J.B. Diabetic Radiculoplexus Neuropathies. *Handb. Clin. Neurol.* **2014**, *126*, 45–52. [CrossRef]
49. Ng, P.S.; Dyck, P.J.; Laughlin, R.S.; Thapa, P.; Pinto, M.V.; Dyck, P.J.B. Lumbosacral Radiculoplexus Neuropathy: Incidence and the Association with Diabetes Mellitus. *Neurology* **2019**, *92*, e1188–e1194. [CrossRef]
50. Ladha, S.S.; Dyck, P.J.B.; Spinner, R.J.; Perez, D.G.; Zeldenrust, S.R.; Amrami, K.K.; Solomon, A.; Klein, C.J. Isolated amyloidosis presenting with lumbosacral radiculoplexopathy: Description of two cases and pathogenic review. *J. Peripher. Nerv. Syst.* **2006**, *11*, 346–352. [CrossRef] [PubMed]
51. Jaeckle, K.A.; Young, D.F.; Foley, K.M. The Natural History of Lumbosacral Plexopathy in Cancer. *Neurology* **1985**, *35*, 8–15. [CrossRef] [PubMed]

52. Song, E.J.; Park, J.S.; Ryu, K.N.; Park, S.Y.; Jin, W. Perineural Spread Along Spinal and Obturator Nerves in Primary Vaginal Carcinoma: A Case Report. *World Neurosurg.* **2018**, *115*, 85–88. [CrossRef] [PubMed]
53. Ladha, S.S.; Spinner, R.J.; Suarez, G.A.; Amrami, K.K.; Dyck, P.J.B. Neoplastic lumbosacral radiculoplexopathy in prostate cancer by direct perineural spread: An unusual entity. *Muscle Nerve* **2006**, *34*, 659–665. Available online: https://pubmed.ncbi.nlm.nih.gov/16810682 (accessed on 22 January 2023). [CrossRef]
54. Thomas, J.E.; Cascino, T.L.; Earle, J.D. Differential Diagnosis between Radiation and Tumor Plexopathy of the Pelvis. *Neurology* **1985**, *35*, 1–7. [CrossRef]
55. Wu, L.; Cruz, R. Lumbar Spinal Stenosis. In *StatPearls*; StatPearls Publishing: Treasure Island, FL, USA, 2022.
56. Davis, D.; Maini, K.; Vasudevan, A. Sciatica. In *StatPearls*; StatPearls Publishing: Treasure Island, FL, USA, 2022.
57. Donnally, C.J., III; Butler, A.J.; Varacallo, M. Lumbosacral Disc Injuries. In *StatPearls*; StatPearls Publishing: Treasure Island, FL, USA, 2022.
58. Hall, S.; Bartleson, J.D.; Onofrio, B.M.; Baker, H.L.; Okazaki, H.; O'Duffy, J.D. Lumbar Spinal Stenosis. Clinical Features, Diagnostic Procedures, and Results of Surgical Treatment in 68 Patients. *Ann. Intern. Med.* **1985**, *103*, 271–275. [CrossRef]
59. Dyck, P.J.B.; Thaisetthawatkul, P. Lumbosacral Plexopathy. *Contin. Lifelong Learn. Neurol.* **2014**, *20*, 1343. [CrossRef]
60. Lang, E.M.; Borges, J.; Carlstedt, T. Surgical Treatment of Lumbosacral Plexus Injuries. *J. Neurosurg. Spine* **2004**, *1*, 64–71. [CrossRef]
61. Yabuki, S.; Fukumori, N.; Takegami, M.; Onishi, Y.; Otani, K.; Sekiguchi, M.; Wakita, T.; Kikuchi, S.; Fukuhara, S.; Konno, S. Prevalence of Lumbar Spinal Stenosis, Using the Diagnostic Support Tool, and Correlated Factors in Japan: A Population-Based Study. *J. Orthop. Sci.* **2013**, *18*, 893–900. [CrossRef]
62. Schmid, R.; Reinhold, M.; Blauth, M. Lumbosacral Dislocation: A Review of the Literature and Current Aspects of Management. *Injury* **2010**, *41*, 321–328. [CrossRef]
63. Falkson, S.R.; Hinson, J.W. Westphal Sign. In *StatPearls*; StatPearls Publishing: Treasure Island, FL, USA, 2022.
64. Kurt, S.; Kaplan, Y.; Karaer, H.; Erkorkmaz, U. Femoral Nerve Involvement in Diabetics. *Eur. J. Neurol.* **2009**, *16*, 375–379. [CrossRef] [PubMed]
65. Solheim, L.F.; Hagen, R. Femoral and Sciatic Neuropathies After Total Hip Arthroplasty. *Acta Orthop. Scand.* **1980**, *51*, 531–534. [CrossRef] [PubMed]
66. Poage, C.; Roth, C.; Scott, B. Peroneal Nerve Palsy: Evaluation and Management. *JAAOS J. Am. Acad. Orthop. Surg.* **2016**, *24*, 1–10. [CrossRef] [PubMed]
67. Fleischman, A.N.; Rothman, R.H.; Parvizi, J. Femoral Nerve Palsy Following Total Hip Arthroplasty: Incidence and Course of Recovery. *J. Arthroplast.* **2018**, *33*, 1194–1199. [CrossRef]
68. Distad, B.J.; Weiss, M.D. Clinical and Electrodiagnostic Features of Sciatic Neuropathies. *Phys. Med. Rehabil. Clin. N. Am.* **2013**, *24*, 107–120. [CrossRef]
69. Maravilla, K.R.; Bowen, B.C. Imaging of the Peripheral Nervous System: Evaluation of Peripheral Neuropathy and Plexopathy. *AJNR Am. J. Neuroradiol.* **1998**, *19*, 1011–1023.
70. Stynes, S.; Konstantinou, K.; Ogollah, R.; Hay, E.M.; Dunn, K.M. Clinical Diagnostic Model for Sciatica Developed in Primary Care Patients with Low Back-Related Leg Pain. *PLoS ONE* **2018**, *13*, e0191852. [CrossRef]
71. Ro, T.H.; Edmonds, L. Diagnosis and Management of Piriformis Syndrome: A Rare Anatomic Variant Analyzed by Magnetic Resonance Imaging. *J. Clin. Imaging Sci.* **2018**, *8*, 6. [CrossRef]
72. Moore, K.R.; Tsuruda, J.S.; Dailey, A.T. The Value of MR Neurography for Evaluating Extraspinal Neuropathic Leg Pain: A Pictorial Essay. *AJNR Am. J. Neuroradiol.* **2001**, *22*, 786–794.
73. Gupta, L.; Yadav, M.; Thulkar, S. 'Trident Sign' in Pelvis: Sinister Sign with Poor Prognosis. *BMJ Case Rep.* **2017**, *2017*, bcr2017220460. [CrossRef]
74. Laughlin, R.S.; Dyck, P.J.B. Electrodiagnostic Testing in Lumbosacral Plexopathies. *Phys. Med. Rehabil. Clin. N. Am.* **2013**, *24*, 93–105. [CrossRef] [PubMed]
75. Ehler, E.; Vyšata, O.; Včelák, R.; Pazdera, L. Painful Lumbosacral Plexopathy. *Medicine* **2015**, *94*, e766. [CrossRef] [PubMed]
76. Bastron, J.A.; Thomas, J.E. Diabetic Polyradiculopathy: Clinical and Electromyographic Findings in 105 Patients. *Mayo Clin. Proc.* **1981**, *56*, 725–732. [PubMed]
77. Haig, A.J.; Tong, H.C.; Yamakawa, K.S.J.; Quint, D.J.; Hoff, J.T.; Chiodo, A.; Miner, J.A.; Choksi, V.R.; Geisser, M.E. The Sensitivity and Specificity of Electrodiagnostic Testing for the Clinical Syndrome of Lumbar Spinal Stenosis. *Spine* **2005**, *30*, 2667–2676. [CrossRef]
78. Jacobs, W.C.H.; van Tulder, M.; Arts, M.; Rubinstein, S.M.; van Middelkoop, M.; Ostelo, R.; Verhagen, A.; Koes, B.; Peul, W.C. Surgery versus Conservative Management of Sciatica Due to a Lumbar Herniated Disc: A Systematic Review. *Eur. Spine J.* **2011**, *20*, 513–522. [CrossRef]
79. Garozzo, D.; Zollino, G.; Ferraresi, S. In Lumbosacral Plexus Injuries Can We Identify Indicators That Predict Spontaneous Recovery or the Need for Surgical Treatment? Results from a Clinical Study on 72 Patients. *J. Brachial Plex. Peripher. Nerve Inj.* **2014**, *9*, 1. [CrossRef]
80. Lurie, J.; Tomkins-Lane, C. Management of Lumbar Spinal Stenosis. *BMJ* **2016**, *352*, h6234. [CrossRef]

81. Ammendolia, C.; Stuber, K.; de Bruin, L.K.; Furlan, A.D.; Kennedy, C.A.; Rampersaud, Y.R.; Steenstra, I.A.; Pennick, V. Nonoperative Treatment of Lumbar Spinal Stenosis with Neurogenic Claudication: A Systematic Review. *Spine* **2012**, *37*, E609–E616. [CrossRef]

82. Shabat, S.; Folman, Y.; Leitner, Y.; Fredman, B.; Gepstein, R. Failure of Conservative Treatment for Lumbar Spinal Stenosis in Elderly Patients. *Arch. Gerontol. Geriatr.* **2007**, *44*, 235–241. [CrossRef]

83. Whitman, J.M.; Flynn, T.W.; Fritz, J.M. Nonsurgical Management of Patients with Lumbar Spinal Stenosis: A Literature Review and a Case Series of Three Patients Managed with Physical Therapy. *Phys. Med. Rehabil. Clin. N. Am.* **2003**, *14*, 77–101. [CrossRef]

84. Macedo, L.G.; Hum, A.; Kuleba, L.; Mo, J.; Truong, L.; Yeung, M.; Battié, M.C. Physical Therapy Interventions for Degenerative Lumbar Spinal Stenosis: A Systematic Review. *Phys. Ther.* **2013**, *93*, 1646–1660. [CrossRef] [PubMed]

85. Aalto, T.J.; Leinonen, V.; Herno, A.; Alen, M.; Kröger, H.; Turunen, V.; Savolainen, S.; Saari, T.; Airaksinen, O. Postoperative Rehabilitation Does Not Improve Functional Outcome in Lumbar Spinal Stenosis: A Prospective Study with 2-Year Postoperative Follow-Up. *Eur. Spine J.* **2011**, *20*, 1331–1340. [CrossRef] [PubMed]

86. Messiah, S.; Tharian, A.R.; Candido, K.D.; Knezevic, N.N. Neurogenic Claudication: A Review of Current Understanding and Treatment Options. *Curr. Pain Headache Rep.* **2019**, *23*, 32. [CrossRef] [PubMed]

87. Dowell, D.; Haegerich, T.M.; Chou, R. CDC Guideline for Prescribing Opioids for Chronic Pain—United States, 2016. *JAMA* **2016**, *315*, 1624–1645. [CrossRef] [PubMed]

88. Rutkove, S.B.; Sax, T.W. Lumbosacral Plexopathies. In *Neuromuscular Disorders in Clinical Practice*; Katirji, B., Kaminski, H.J., Ruff, R.L., Eds.; Springer: New York, NY, USA, 2014; pp. 1063–1071. [CrossRef]

89. Song, S.H.; Ryu, G.H.; Park, J.W.; Lee, H.J.; Nam, K.Y.; Kim, H.; Kim, S.Y.; Kwon, B.S. The Effect and Safety of Steroid Injection in Lumbar Spinal Stenosis: With or Without Local Anesthetics. *Ann. Rehabil. Med.* **2016**, *40*, 14–20. [CrossRef]

90. Deyo, R.A.; Mirza, S.K.; Martin, B.I.; Kreuter, W.; Goodman, D.C.; Jarvik, J.G. Trends, Major Medical Complications, and Charges Associated with Surgery for Lumbar Spinal Stenosis in Older Adults. *JAMA J. Am. Med. Assoc.* **2010**, *303*, 1259–1265. [CrossRef] [PubMed]

91. Kovacs, F.M.; Urrútia, G.; Alarcón, J.D. Surgery Versus Conservative Treatment for Symptomatic Lumbar Spinal Stenosis: A Systematic Review of Randomized Controlled Trials. *Spine* **2011**, *36*, E1335. [CrossRef]

92. Silvers, H.R.; Lewis, P.J.; Asch, H.L. Decompressive Lumbar Laminectomy for Spinal Stenosis. *J. Neurosurg.* **1993**, *78*, 695–701. [CrossRef]

93. Estefan, M.; Munakomi, S.; Camino Willhuber, G.O. Laminectomy. In *StatPearls*; StatPearls Publishing: Treasure Island, FL, USA, 2022.

94. Bydon, M.; Macki, M.; Abt, N.B.; Sciubba, D.M.; Wolinsky, J.-P.; Witham, T.F.; Gokaslan, Z.L.; Bydon, A. Clinical and Surgical Outcomes after Lumbar Laminectomy: An Analysis of 500 Patients. *Surg. Neurol. Int.* **2015**, *6* (Suppl. S4), S190–S193. [CrossRef]

95. Guha, D.; Heary, R.F.; Shamji, M.F. Iatrogenic Spondylolisthesis Following Laminectomy for Degenerative Lumbar Stenosis: Systematic Review and Current Concepts. *Neurosurg. Focus* **2015**, *39*, E9. [CrossRef]

96. Ghogawala, Z.; Dziura, J.; Butler, W.E.; Dai, F.; Terrin, N.; Magge, S.N.; Coumans, J.-V.C.E.; Harrington, J.F.; Amin-Hanjani, S.; Schwartz, J.S.; et al. Laminectomy plus Fusion versus Laminectomy Alone for Lumbar Spondylolisthesis. *N. Engl. J. Med.* **2016**, *374*, 1424–1434. [CrossRef] [PubMed]

97. Sharif, S.; Shaikh, Y.; Bajamal, A.H.; Costa, F.; Zileli, M. Fusion Surgery for Lumbar Spinal Stenosis: WFNS Spine Committee Recommendations. *World Neurosurg. X* **2020**, *7*, 100077. [CrossRef] [PubMed]

98. De Cicco, F.L.; Camino Willhuber, G.O. Nucleus Pulposus Herniation. In *StatPearls*; StatPearls Publishing: Treasure Island, FL, USA, 2022.

99. Blamoutier, A. Surgical Discectomy for Lumbar Disc Herniation: Surgical Techniques. *Orthop. Traumatol. Surg. Res.* **2013**, *99* (Suppl. S1), S187–S196. [CrossRef] [PubMed]

100. Lurie, J.D.; Faucett, S.C.; Hanscom, B.; Tosteson, T.D.; Ball, P.A.; Abdu, W.A.; Frymoyer, J.W.; Weinstein, J.N. Lumbar Discectomy Outcomes Vary by Herniation Level in the Spine Patient Outcomes Research Trial. *J. Bone Jt. Surg. Am.* **2008**, *90*, 1811–1819. [CrossRef]

101. Loupasis, G.A.; Stamos, K.; Katonis, P.G.; Sapkas, G.; Korres, D.S.; Hartofilakidis, G. Seven- to 20-Year Outcome of Lumbar Discectomy. *Spine* **1999**, *24*, 2313–2317. [CrossRef]

102. Alexandre, A.; Corò, L.; Azuelos, A. Microsurgical Treatment of Lumbosacral Plexus Injuries. *Acta Neurochir. Suppl.* **2005**, *92*, 53–59. [CrossRef]

103. Nichols, D.S.; Fenton, J.; Cox, E.; Dang, J.; Garbuzov, A.; McCall-Wright, P.; Chim, H. Surgical Interventions for Lumbosacral Plexus Injuries: A Systematic Review. *Plast. Reconstr. Surg. Glob. Open* **2022**, *10*, e4436. [CrossRef]

104. Safavi-Abbasi, S.; Feiz-Erfan, I.; Shetter, A.G. Neurolysis. In *Encyclopedia of the Neurological Sciences*, 2nd ed.; Aminoff, M.J., Daroff, R.B., Eds.; Academic Press: Oxford, UK, 2014; pp. 406–407. [CrossRef]

105. Lipinski, L.J.; Spinner, R.J. Neurolysis, Neurectomy, and Nerve Repair/Reconstruction for Chronic Pain. *Neurosurg. Clin. N. Am.* **2014**, *25*, 777–787. [CrossRef]

106. Kim, D.H.; Murovic, J.A.; Tiel, R.L.; Kline, D.G. Intrapelvic and Thigh-Level Femoral Nerve Lesions: Management and Outcomes in 119 Surgically Treated Cases. *J. Neurosurg.* **2004**, *100*, 989–996. [CrossRef]

107. Kitagawa, R.; Kim, D.; Reid, N.; Kline, D. Surgical Management of Obturator Nerve Lesions. *Neurosurgery* **2009**, *65* (Suppl. S4), A24–A28. [CrossRef]

108. Dahlin, L.B. Techniques of Peripheral Nerve Repair. *Scand. J. Surg.* **2008**, *97*, 310–316. [CrossRef] [PubMed]
109. Kornfeld, T.; Vogt, P.M.; Radtke, C. Nerve Grafting for Peripheral Nerve Injuries with Extended Defect Sizes. *Wien. Med. Wochenschr.* **2019**, *169*, 240–251. [CrossRef] [PubMed]
110. Osgaard, O.; Husby, J. Femoral Nerve Repair with Nerve Autografts. Report of Two Cases. *J. Neurosurg.* **1977**, *47*, 751–754. [CrossRef] [PubMed]
111. Cao, Y.; Li, Y.; Zhang, Y.; Li, S.; Jiang, J.; Gu, Y.; Xu, L. Different Surgical Reconstructions for Femoral Nerve Injury: A Clinical Study on 9 Cases. *Ann. Plast. Surg.* **2020**, *84* (Suppl. S3), S171–S177. [CrossRef] [PubMed]
112. Shacklock, M.; Rade, M.; Poznic, S.; Marčinko, A.; Fredericson, M.; Kröger, H.; Kankaanpää, M.; Airaksinen, O. Treatment of Sciatica and Lumbar Radiculopathy with an Intervertebral Foramen Opening Protocol: Pilot Study in a Hospital Emergency and In-Patient Setting. *Physiother. Theory Pract.* **2022**, 1–11. [CrossRef]
113. Metz, L.N.; Deviren, V. Low-Grade Spondylolisthesis. *Neurosurg. Clin. N. Am.* **2007**, *18*, 237–248. [CrossRef]
114. Abdu, W.A.; Sacks, O.A.; Tosteson, A.N.A.; Zhao, W.; Tosteson, T.D.; Morgan, T.S.; Pearson, A.; Weinstein, J.N.; Lurie, J.D. Long-Term Results of Surgery Compared With Nonoperative Treatment for Lumbar Degenerative Spondylolisthesis in the Spine Patient Outcomes Research Trial (SPORT). *Spine* **2018**, *43*, 1619–1630. [CrossRef]
115. Hennemann, S.; de Abreu, M.R. Degenerative Lumbar Spinal Stenosis. *Rev. Bras. Ortop.* **2021**, *56*, 9–17. [CrossRef]
116. Bae, J.; Lee, S.-H.; Wagner, R.; Shen, J.; Telfeian, A.E. Full Endoscopic Surgery for Thoracic Pathology: Next Step after Mastering Lumbar and Cervical Endoscopic Spine Surgery? *BioMed Res. Int.* **2022**, *2022*, 8345736. [CrossRef]
117. Kim, H.S.; Raorane, H.D.; Wu, P.H.; Heo, D.H.; Sharma, S.B.; Jang, I.-T. Incidental Durotomy During Endoscopic Stenosis Lumbar Decompression: Incidence, Classification, and Proposed Management Strategies. *World Neurosurg.* **2020**, *139*, e13–e22. [CrossRef]
118. Liang, J.; Lian, L.; Liang, S.; Zhao, H.; Shu, G.; Chao, J.; Yuan, C.; Zhai, M. Efficacy and Complications of Unilateral Biportal Endoscopic Spinal Surgery for Lumbar Spinal Stenosis: A Meta-Analysis and Systematic Review. *World Neurosurg.* **2022**, *159*, e91–e102. [CrossRef]
119. Häckel, S.; Christen, S.; Vögelin, E.; Keel, M.J.B. Exposure of the Lumbosacral Plexus by Using the Pararectus Approach: A Technical Note. *Oper. Neurosurg.* **2023**, *24*, e1–e9. [CrossRef] [PubMed]
120. Kurze, C.; Keel, M.J.B.; Kollár, A.; Siebenrock, K.A.; Klenke, F.M. The Pararectus Approach—A Versatile Option in Pelvic Musculoskeletal Tumor Surgery. *J. Orthop. Surg.* **2019**, *14*, 232. [CrossRef] [PubMed]
121. Liu, G.; Chen, J.; Liang, C.; Zhang, C.; Li, X.; Hu, Y. The Pararectus Approach in Acetabular Fractures Treatment: Functional and Radiologcial Results. *BMC Musculoskelet. Disord.* **2022**, *23*, 370. [CrossRef] [PubMed]
122. Sanuki, N.; Kodama, S.; Seta, H.; Sakai, M.; Watanabe, H. Radiation Therapy for Malignant Lumbosacral Plexopathy: A Case Series. *Cureus* **2022**, *14*, e20939. [CrossRef]
123. Cao, X.; Gao, X.-S.; Li, W.; Liu, P.; Qin, S.-B.; Dou, Y.-B.; Li, H.-Z.; Shang, S.; Gu, X.-B.; Ma, M.-W.; et al. Contouring Lumbosacral Plexus Nerves with MR Neurography and MR/CT Deformable Registration Technique. *Front. Oncol.* **2022**, *12*, 818953. [CrossRef]
124. Chen, A.M.; Hall, W.H.; Li, J.; Beckett, L.; Farwell, D.G.; Lau, D.H.; Purdy, J.A. Brachial Plexus-Associated Neuropathy After High-Dose Radiation Therapy for Head-and-Neck Cancer. *Int. J. Radiat. Oncol.* **2012**, *84*, 165–169. [CrossRef]
125. Kong, Y.; Kuss, M.; Shi, Y.; Fang, F.; Xue, W.; Shi, W.; Liu, Y.; Zhang, C.; Zhong, P.; Duan, B. Exercise Facilitates Regeneration after Severe Nerve Transection and Further Modulates Neural Plasticity. *Brain Behav. Immun. Health* **2022**, *26*, 100556. [CrossRef]
126. Chen, Y.-W.; Li, Y.-T.; Chen, Y.C.; Li, Z.-Y.; Hung, C.-H. Exercise Training Attenuates Neuropathic Pain and Cytokine Expression after Chronic Constriction Injury of Rat Sciatic Nerve. *Anesth. Analg.* **2012**, *114*, 1330–1337. [CrossRef]
127. Kami, K.; Tajima, F.; Senba, E. Exercise-Induced Hypoalgesia: Potential Mechanisms in Animal Models of Neuropathic Pain. *Anat. Sci. Int.* **2017**, *92*, 79–90. [CrossRef]
128. de Freria, C.M.; Van Niekerk, E.; Blesch, A.; Lu, P. Neural Stem Cells: Promoting Axonal Regeneration and Spinal Cord Connectivity. *Cells* **2021**, *10*, 3296. [CrossRef]
129. Kadoya, K.; Lu, P.; Nguyen, K.; Lee-Kubli, C.; Kumamaru, H.; Yao, L.; Knackert, J.; Poplawski, G.; Dulin, J.N.; Strobl, H.; et al. Spinal Cord Reconstitution with Homologous Neural Grafts Enables Robust Corticospinal Regeneration. *Nat. Med.* **2016**, *22*, 479–487. [CrossRef]
130. Dulin, J.N.; Adler, A.F.; Kumamaru, H.; Poplawski, G.H.D.; Lee-Kubli, C.; Strobl, H.; Gibbs, D.; Kadoya, K.; Fawcett, J.W.; Lu, P.; et al. Injured Adult Motor and Sensory Axons Regenerate into Appropriate Organotypic Domains of Neural Progenitor Grafts. *Nat. Commun.* **2018**, *9*, 84. [CrossRef]
131. Cofano, F.; Boido, M.; Monticelli, M.; Zenga, F.; Ducati, A.; Vercelli, A.; Garbossa, D. Mesenchymal Stem Cells for Spinal Cord Injury: Current Options, Limitations, and Future of Cell Therapy. *Int. J. Mol. Sci.* **2019**, *20*, 2698. [CrossRef]
132. Qu, J.; Zhang, H. Roles of Mesenchymal Stem Cells in Spinal Cord Injury. *Stem Cells Int.* **2017**, *2017*, 5251313. [CrossRef]
133. Zachar, L.; Bačenková, D.; Rosocha, J. Activation, Homing, and Role of the Mesenchymal Stem Cells in the Inflammatory Environment. *J. Inflamm. Res.* **2016**, *9*, 231–240. [CrossRef]
134. Huang, Y.-C.; Huang, Y.-Y. Biomaterials and Strategies for Nerve Regeneration. *Artif. Organs* **2006**, *30*, 514–522. [CrossRef]
135. George, P.M.; Lyckman, A.W.; LaVan, D.A.; Hegde, A.; Leung, Y.; Avasare, R.; Testa, C.; Alexander, P.M.; Langer, R.; Sur, M. Fabrication and Biocompatibility of Polypyrrole Implants Suitable for Neural Prosthetics. *Biomaterials* **2005**, *26*, 3511–3519. [CrossRef]

136. Lladó, J.; Haenggeli, C.; Maragakis, N.J.; Snyder, E.Y.; Rothstein, J.D. Neural Stem Cells Protect against Glutamate-Induced Excitotoxicity and Promote Survival of Injured Motor Neurons through the Secretion of Neurotrophic Factors. *Mol. Cell. Neurosci.* **2004**, *27*, 322–331. [CrossRef]

137. Huang, E.J.; Reichardt, L.F. Trk Receptors: Roles in Neuronal Signal Transduction. *Annu. Rev. Biochem.* **2003**, *72*, 609–642. [CrossRef]

138. Song, S.; McConnell, K.W.; Amores, D.; Levinson, A.; Vogel, H.; Quarta, M.; Rando, T.A.; George, P.M. Electrical Stimulation of Human Neural Stem Cells via Conductive Polymer Nerve Guides Enhances Peripheral Nerve Recovery. *Biomaterials* **2021**, *275*, 120982. [CrossRef] [PubMed]

Review

Evidence of Improvement of Lower Limb Functioning Using Hydrotherapy on Spinal Cord Injury Patients

Liliana Elena Stanciu [1], Madalina Gabriela Iliescu [1,*], Liliana Vlădăreanu [1], Alexandra Ecaterina Ciota [1], Elena-Valentina Ionescu [1] and Claudia Ileana Mihailov [2]

[1] Department of Physical Medicine and Rehabilitation, Faculty of Medicine, "Ovidius" University of Constanta, 1 University Alley, Campus–Corp B, 900470 Constanta, Romania

[2] Department of Reumatology, Faculty of Medicine, "Ovidius" University of Constanta, 1 University Alley, Campus–Corp B, 900470 Constanta, Romania

* Correspondence: iliescumadalina@gmail.com

Abstract: Background: Spinal cord injury (SCI) is a devastating problem for modern society, whether it affects young people in the most productive period of their lives or the elderly. The spinal cord injury is currently without curative treatment and the therapeutic intervention aims to minimize secondary complications and maximize residual function through rehabilitation medicine. The main objective of this scientific paper is to determine whether there is evidence in the literature regarding the importance and/or use of hydrotherapy, as part of the therapeutic management of the SCI patient, in order to decrease the degree of spasticity, of pain symptoms, increase or maintain range of motion, improve respiratory, cardiovascular, and metabolic status, as well as improve function and psychological benefits. Methods: Using Preferred Reporting Items for Systematic Reviews and Meta-Analyses (PRISMA) procedures, the following databases were analyzed between 2000 and 2021: Pub Med, Pub Med Central, Science Direct, Scopus, and SpringerLink. Initial keywords: rehabilitation treatment, spinal cord injury. Additional keywords: hydrotherapy, aqua therapy, spasticity, human. For the scientific quality of the included articles, risk of bias was assessed using the Downs and Black Appraisal Modified Scale. Results: Our research used only four publications as per PRISMA protocol, assessed with Downs and Black Scale. The study models used in the individual studies included in the research are the following: two systematic reviews, one experimental non-randomized control, and one individual semi-structured interview. Due to the low number of studies, despite two of them being reviews, there is the necessity for a more standardized methodology to prove the benefits hydrotherapy for SCI patients for the improvement of lower limb functioning. Conclusion: Hydrotherapy is an important component of the treatment of an SCI patient, despite the limited number of scientific studies that support this aspect. Clinical trials in the future are required.

Keywords: aqua therapy; hydrotherapy; human; rehabilitation treatment; spasticity; spinal cord injury

Citation: Stanciu, L.E.; Iliescu, M.G.; Vlădăreanu, L.; Ciota, A.E.; Ionescu, E.-V.; Mihailov, C.I. Evidence of Improvement of Lower Limb Functioning Using Hydrotherapy on Spinal Cord Injury Patients. *Biomedicines* **2023**, *11*, 302. https://doi.org/10.3390/biomedicines11020302

Academic Editor: Nicolas Guerout

Received: 20 December 2022
Revised: 16 January 2023
Accepted: 18 January 2023
Published: 21 January 2023

1. Introduction

Spinal cord injury is a serious pathology that can cause various aspects of a patient's life to significantly deteriorate. The primary goal of rehabilitation is to improve a patient's functional level and decrease their secondary morbidity. In this paper, we review various secondary long-term complications that can occur after a spinal cord injury (SCI). Some of these include respiratory, cardiovascular, urinary, and intestinal difficulties, as well as spasticity, pain syndromes, pressure ulcers, and fractures [1].

Most of the time, treating patients with spasticity using drugs can lead to side effects. Hydrotherapy can assist in reducing the amount of medication needed as part of rehabilitation treatment [2].

Spasticity is a major issue for SCI patients, also limiting a patient's mobility and affecting their ability to perform various activities of daily life. Spasticity represents a significant challenge, both for the patient and for the rehabilitation team [3].

Medical hydrotherapy has been prescribed since Hippocrates revealed the curative virtues of water hundreds of years ago [4].

During the 1800s, Sebastian Kneipp, the 'founder of hydrotherapy', wrote extensively about the healing effects of water. His research was immediately recognized by healthcare professionals [5].

Hydrotherapy can relieve pain for patients with various conditions [6]. It can also improve their sensory perception by blocking the nociception signals [7,8]. Additionally, warm water can help nourish the body and reduce the effects of lactic acid and other chemicals in the body. It can also decrease the pain threshold and improve muscle relaxation [9,10].

Giesecke defined various aquatic exercise goals, including decreased spasticity, improved respiratory status, increased endurance, and psychological benefits. The temperature of the water is critical because excessively hot or cold temperatures may exacerbate spasticity and can also decrease the effectiveness of certain therapeutic techniques [11]. Warm waters allow individuals with a spinal cord injury to move freely in ways that would be unpleasant and difficult to do on land, and also help avoid the fear of falling. As a result, many physical and functional aims can be accomplished easier [11].

Unfortunately, aquatic therapy is not routinely available to patients with a spinal cord injury. Many of them are vulnerable to colostomies and frequent involuntary incontinence due to neurogenic bowel dysfunction [12].

According to Recio and Cabahug, aqua-based therapy could be effective for treating spinal cord-injured patients with suprapubic and Foley indwelling catheters, tracheostomy tubes, and pressure ulcers [13]. The patients exhibited no signs of dehydration or respiratory problems during the hydrotherapy sessions. Additionally, in patients with suprapubic and indwelling catheters, there was no evidence of a catheter being pulled [13–15].

The primary goal of this study is to determine the importance of hydrotherapy as a part of the therapeutic management of SCI patients on a global scale and whether there is up-to-date literature to confirm our clinical findings.

2. Methods

To conduct the research, the following databases: PubMed, PubMed Central, Science Direct, Scopus, and SpringerLink were analyzed between 2000 and 2021, according to the PRISMA procedures recommendations for systematic reviews [16]. Initial keywords—rehabilitation treatment, spinal cord injury. Additional keywords—hydrotherapy, aqua therapy, spasticity, human. Downs and Black Appraisal Modified Scale was used to evaluate the scientific quality of the included articles.

Titles and abstracts were read and evaluated for inclusion using the following criteria:

(1) Population (patients with neuromotor or neuromuscular deficits due to spinal cord injury);
(2) Intervention (hydrotherapy/aqua therapy).

Exclusion criteria included publications from previous time periods, publications about hydrotherapy that have as an aim other pathologies than spinal cord injury and publications that do not use English as the publishing language.

The scientifical quality of the studies was evaluated through the modified Downs and Black Scale [17,18], which is a validated and reliable method for assessing the effectiveness of randomized controlled and noncontrolled trials. The modified Downs and Black Scale has 27 items that are focused on five broad categories, namely reporting (10 items), internal validity—bias (7 items), internal validity—confounding (6 items), power (1 item), and external validity. The scale's overall score goes from 0 to 28. A study of superior quality has a higher value of the maximum possible score. The overall quality of the papers was: 80% (very good), 70–79% (good), 50–69% (fair), <50% (weak) (Downs and Black 1998).

For possible addition, one author examined titles and abstracts. Another two separately identified the full text of collected articles, with differences addressed by consensus or arbitration from a third author.

The study initially started with a search in the databases mentioned above for a shorter period, namely 2015–2021. Given the small number of articles eligible for a review to demonstrate the importance of hydrotherapy in the therapeutic management of patients with spinal cord injury, the search period was extended to 2000–2021, keeping the same databases. Unfortunately, the number of items identified as eligible did not increase significantly.

The International Committee of Medical Journal Editors suggested, in 1993, that articles submitted to biomedical journals should follow the same set of guidelines.

We chose articles published in 2000 and after because we supported that during that period the International Committee of Medical Journal Editors' suggestion had been embraced by relevant researchers and had increased the quality of reports [19].

3. Results

As per Figure 1 and Table 1, the algorithm used in finding and selecting the articles shows that only four remained for the final analysis. The years of publication for the articles were 2004, 2017, 2018, and 2019. The main characteristics of the articles are described in Table 2. For further information, please refer to the bibliography of the mentioned studies.

The first study selected was a study published in 2004 by Kesictas N. et al., where researchers evaluated the effectiveness of aquatic therapy on patients with spinal cord injuries. Twenty-four participants were divided into two groups. The first group received standard aquatic exercise twice a day and the other group received oral baclofen. The participants were also assessed for their muscle spasms and the effects of the treatment on their breathing and movement. The researchers also used the Ashworth scales to measure their progress. The hydrotherapeutic intervention successfully reduced spasticity. There was a statistical improvement in Ashworth scores for both study and control groups ($p < 0.01$ and $p < 0.02$, respectively).

The participants in the hydrotherapy group exhibited a significant increase in their Functional Independence Measurement (FIM) scores. They also showed a decrease in oral baclofen consumption. Regarding spasm severity, the hydrotherapy group showed a significant decrease in spasm severity ($p < 0.02$). The control group's spasm severity was 1.4, and the hydrotherapy group's was 0.7 (respectively, $p < 0.05$, $p < 0.001$).

The researchers also noted that the use of hydrotherapy led to a drop in spasm severity. This is the only clinical trial that we found using our algorithm [1].

The second study selected was a 2017 systematic review made by Chunxiao Li et al. that evaluated the effects of aquatic exercise on the physical function and fitness of people with spinal cord injury. Four studies cited in the 2017 review showed that aquatic exercise combined with physiotherapy improved the participants' functional independence. The programs were either a 10-week program or a 16-week program. A test–retest study conducted on eight participants revealed that underwater walking sessions, three times a week for eight weeks significantly improved their walking ability. A case study also showed that aquatic and land-based exercises can improve their physical fitness. The study showed that the 8-week underwater treadmill program led to a decrease in participants' daily walking heart rate. A 15-week aquatic exercise program was found to improve participants' swimming distance and force critical capacity. However, the program did not enhance the participants' forced vital capacity (FVC) and forced expiratory flow rate. A similar study conducted on a 3-year program showed that combining swimming and physiotherapy sessions helped improve participants' cardiorespiratory efficiency [1].

Figure 1. Flow chart of the review process. N—number of records; SCI—spinal cord injury.

The third paper that met the criteria for selection in our review was a review published in 2018 by Terry J. et al. focused on the effects of aquatic exercise on patients with spinal cord injuries. The study analyzed the effects of different types of aquatic exercise on different aspects of health. It was theorized that the effects of buoyancy and hydrostatic pressure on the patients' lumbopelvic hip complex and force closure improved their underwater walking. Additionally, the effects of buoyancy on the hip swing were proven to counteract the gravity-based effects of walking. The researchers also noted that the energy expenditure of patients decreased during underwater walking sessions [20,21]. The 1998 study conducted by Zamparo and Pagliaro from this 2018 [22] review was limited by its design and did not follow the usual follow-up procedures. As a result, it was not able to provide sufficient evidence supporting the study's findings. While submerged in warm water, an exerciser's heart rate lowers and enhances the thermoregulatory response of the body, which helps prolong the participant's ability to exercise. This benefit is evidenced by the fact that water provides the ideal conductor of heat, which helps the body maintain a low core temperature. This allows the exerciser to endure longer and improve their energy-use efficiency. A study published in 2004 focused on the effects of hydrotherapy on SCI patients. The researchers discovered that a reduction in the dosage of the anti-inflammatory drug Baclofen significantly reduced muscle spasticity [22]. According to a 2009 study, the physiological effects of hydrotherapy on various conditions are still not clear. Further studies are needed to analyze the mechanism that influences this treatment's effectiveness. The effects of hydrostatic pressure on the body are similar to those of Boyle's law. Breathing underwater increases the body's pressure, which makes breathing more costly. According to Becker, a patient's vital capacity decreases by about 6% to 9% after performing underwater exercises due to hydrostatic pressure. This effect counters the inspiratory muscle action of underwater exercises. A study conducted in 1980 noted that SCI patients who participated in aquatic exercises improved their respiratory fitness. Other studies also suggest that performing aquatic exercises can improve a person's expiratory muscle strength [22].

The fourth study we selected was a 2019 study made by Andresa R et al. that discussed the use of aquatic therapy by rehabilitation professionals for patients with spinal cord injuries. The researchers stated that aquatic therapy is a useful tool for improving the adherence of patients to their treatment. It does not require a standard technique or design [23].

Table 1. Appraisal of records according to the modified Downs and Black Appraisal Scale.

Authors	Downs and Black Appraisal					
	Reporting ($n = 5$)	External Validity ($n = 3$)	Internal Validity ($n = 3$)	Power ($n = 5$)	Total ($n =16$)	Grading % = x/16 × 100
N. Kesiktas, N. Paker, N. Erdogan, G. Gülsen, D. Biçki, and H. Yilmaz (2004) [1]	5	3	2	4	14	87.5 % (very good)
C. Li, S. Khoo, A. Adnan (2017) [19]	5	3	3	4	15	93.75% (very good)
T. J. Ellapen, H. V. Hammill, M. Swanepoel, G. L. Strydom (2018) [20]	5	3	3	4	15	93.75% (very good)
A. R. Marinho-Buzelli, A. J. Zaluski, A. Mansfield, A. M. Bonnyman, K. E. Musselman (2019) [22]	5	1	3	3	11	68.75% (fair)

N, number; x, sum of Downs and Black appraisal.

Due to the low number of studies supporting the effectiveness of aquatic therapy for patients with spinal cord injuries, further studies are needed to establish a standard procedure for conducting follow-up studies.

In our research, we were able to use only four publications, assessed with Downs and Black Scale as shown in Table 1. Due to the low number of studies, there is need for further studies and standardized methodology to prove the benefits of hydrotherapy for SCI patients.

Table 2. Chronological overview of the characteristics and findings of the records (*n* = 4).

Characteristics of the Study				
Authors	**Type of Study**	**Sample**	**Method**	**Findings**
N. Kesiktas, N. Paker, N. Erdogan, G. Gülsen, D. Biçki, and H. Yilmaz (2004) [1]	Experimental non-randomised control	Hydrotherapy group: 10, mean age 32.13 ± 8.34, gender: 2 females and 8 males, injury time (months): 8.6 ± 5.5, FIM: 52 ± 14.13, Ashworth Score 3 ± 0.92, Oral Baclofen (mg) 96 ± 12, Etiology (accident) 50%. Control group: 10, mean age 33.10 ± 10.71, gender: 3 females and 7 males, injury time (months): 7.70 ± 6.06, FIM: 54.70 ± 18.8, Ashworth Score 2.50 ± 1.18, Oral Baclogen (mg) 100 ± 0, Etiology (accident) 50%.	The hydrotherapy group received 20 min of underwater exercises at 71 °F (21.6 °C) 3 times/week, and also participated in the usual rehabilitation, which included passive range of motion 2 times/day, psychotherapy and oral baclofen for 10 weeks. The control group were able to maintain their usual activities through the conventional rehabilitation program.	The hydrotherapeutic intervention successfully reduced spasticity and oral baclofen doses and raised FIM scores compared to control group.
C. Li, S. Khoo, A. Adnan (2017) [19]	Systematic review	A total of 143 participants with SCI were reported and the sample size of each study ranged from 1 to 60. More male participants were reported than female (male = 91, female = 52). Participants were adults aged between 18 and 63 years. A total of seven of eight studies reported participants' injury levels on the spinal cord (the study by Pachalski and Mekarski did not report the specific injury level). Only 4 studies provided the grade of ASIA impairment scale. There was a big range in terms of postinjury time from 7 months to 28 years. In terms of study design, 3 were controlled clinical trials, 2 single group test–retest designs, 1 randomized controlled trial, 1 single-subject design, and 1 case study.	Eight of 276 studies met the inclusion criteria, of which none showed high research quality. Four studies assessed physical function outcomes and 4 studies evaluated aerobic fitness as outcome measures. Significant improvements on these 2 outcomes were generally found. Other physical or fitness outcomes including body composition, muscular strength, and balance were rarely reported.	There is insufficient evidence to support the efficacy of aquatic exercise on increasing physical function and aerobic fitness among SCI patients. We cannot yet draw any conclusion about the effectiveness of underwater training on body composition, muscular strength, and balance among the study population.

Table 2. *Cont.*

	Characteristics of the Study			
Authors	**Type of Study**	**Sample**	**Method**	**Findings**
T. J. Ellapen, H. V. Hammill, M. Swanepoel, G. L. Strydom (2018) [20]	Systematic review	A total of 142 participants were reported (but 83 PWSCI), with sample sizes varying from 1 to 30 and participant age varying from 5 to 70 years. Five studies provided kinanthropometric characteristics, whereas 5 studies considered the number of years injured, and 10 studies described the aquatic exercise intervention. The overall quality of the studies was rated as fair (62.0%).	A literature surveillance was conducted between 1998 and 2017, through the Crossref meta-database and Google Scholar, according to the PRISMA procedures. Key search words were water-therapy, aquatic-therapy, hydrotherapy, spinal cord injury, rehabilitation, human, kinematics, underwater gait, cardiorespiratory, thermoregulation and spasticity. The quality of each paper was evaluated using a modified Downs and Black Appraisal Scale. The participants were recorded pertaining to SCI and hydrotherapy. The outcomes of interest were hydrotherapy interventions, the impact of hydrotherapy on gait kinematics, thermoregulation during water submersion, and cardiorespiratory function of PWSCI.	Hydrotherapy increases PWSCI underwater gait kinematics, cardiorespiratory and thermoregulatory responses and reduces spasticity.
A. R. Marinho-Buzelli, A. J. Zaluski, A. Mansfield, A. M. Bonnyman, K. E. Musselman (2019) [22]	Individual semi-structured interviews	None mentioned.	Six PT (2 male, 4 female), three PTA (female) and 1 KIN (female) participated. The following four themes were identified: (1) multi-system benefits from AT (e.g., from impairment to function, confidence, and enjoyment); (2) application of AT; (3) perceived barriers to implementing AT; and (4) water as an enabler to function on land. All were interviewed regarding their clinical findings while working with SCI patients in aquatic environment.	The participants reported AT was a unique and fickle approach that benefits the multi-dimensional aspects of the health of individuals with SCI/D. They inserted AT very well into their clinical practice despite the barriers professionals and clients face.

AT—Aqua therapy; PWSCI—Patient/SCI patients; SCI—Spinal cord injury; FIM—Functional independence measure; PTs—Physical therapists; PTA—PT assistants; KIN—kinesiologists; ASIA—American Spinal Cord Injury Association.

4. Discussion

The use of hydrotherapy in the treatment of the SCI patient decreases spasticity, increase underwater functioning of the lower limbs and gait kinematics, increase cardiorespiratory and thermoregulatory status, significant complications that can create a varying degree of disabilities. This therapy has a different mechanism of action on these complications through all the properties that therapeutic water can present: physical, chemical, and mechanical properties [1,20,24].

As it appears in the data from the specialized literature, spasticity can develop months or years after the acute spine cord injury and lead to severe function loss and hospitalization [25]. According to studies, 65–78% of the patients with chronic SCI ($\geq$1 year postinjury) experience spasticity symptoms [26]. Up to 5 years after the spine cord injury, one-third of all patients have difficulties with spasticity [27]. Any medical intervention that leads to an improvement in neurological symptoms is vital for patient therapeutical management. Future clinical trials are needed to make significant arguments about the role of hydrotherapy.

Risk factors for cardiovascular diseases are increased in SCI patients [28]. SCI patients have considerably reduced daily energy expenditure due to a lack of motor function, as well as fewer opportunities to participate in physical exercise [29]. Additionally, SCI is associated with abnormal blood pressure, heart rate variability, arrhythmias, and a decreased cardiovascular response to exercise, which can restrict physical activity capacity [29]. Improving cardiorespiratory status through hydrotherapy is an important element for the SCI patient's compliance in therapeutic management and the subsequent evolution of cardiac symptoms. Improving cardiovascular status through hydrotherapy is an important element for the SCI patient's compliance with treatment and subsequent cardiac and respiratory symptoms evolution.

The possible effect of increasing underwater functioning of the lower limbs and gait kinematics is an important element that supports the need for future scientific studies on the role of hydrotherapy in the SCI patient.

Design and intervention methods are required for a successful study. They play an essential role in improving the quality of the study and its results.

A significant number of clinical case reports were discovered in the context of the research articles included. Although clinical case reports are not considered objective sources of information, they still play a vital role in the development and dissemination of medical knowledge.

The importance of clinical trials is often neglected. Although they may seem simple, they are conducted through a rigorous process that is governed by ethical principles.

Randomized controlled trials are often used to evaluate the effectiveness of interventions and have been acknowledged as the gold standard because they are considered one of the most powerful types of evidence for use in the development of evidence-based clinical therapeutical programs [30]. However, randomized controlled trials have often been poorly reported in medical journals [31,32].

It is essential to find solutions for implementing randomized controlled trials in medical rehabilitation units, to support the use and effectiveness of balneal factors in the treatment of various orthopedic and neurologic diseases [32], and many literature data with different studies in various pathology can be used as model [33].

The fact that when the PRISMA algorithm was applied the number of articles that could be included in the study in the current review was minimal compared to the vast number of patients with neurological pathology studied is of importance in medical rehabilitation as well as a treatment methodology, and arguments were sought for this aspect. One of the hypotheses is that there is an actual development of other treatments, such as robotic, computerized, or immersion in virtual reality.

The exoskeleton systems represent a new therapeutic approach in the rehabilitation program of these patients and can be successfully used for gait re-education, proposing kinetic models of the exoskeleton with permanent control over the angle, and angular velocity at the level of each joint, as well as over the parameter muscles (muscle parameters), offering the patient continuous passive movement [34,35]. This new therapy, along with the safety offered to the patient during treatment, reduces the therapist's workload and therapy cost [34]. The cost level of a rehabilitation program will always represent a significant concern of any health service so that as many patients as possible can benefit from rehabilitation services. The exoskeleton has much potential to maintain good health and improve the QOL in individuals with SCI. Of course, future studies are needed in this therapeutic sector as well.

Another therapeutic class used in the treatment of spasticity in patients with spinal cord injury in current medical rehabilitation is botulinum toxin. There are standardized schemes that can also be customized for the individual patient's treatment situation and the patient's target muscles, and the treatment is more effective in the first 6 months [36–38].

The strength of the study is as follows: The number of patients with neurological pathology in the field of spinal cord injury is very high, and the disability resulting from this pathology has a significant impact on the quality of life of both the patient and his family. For this reason, it is necessary to investigate all the therapeutic possibilities used to increase independence and maintain the existing functional remainder. Hydrokinetotherapy is part of the therapeutic rehabilitation management of these pathologies, with a therapeutic impact on several systems, as discussed in the Section 1. The study was based on the PRISMA research algorithm necessary for conducting systematic reviews. The period included in the research is essential, namely 2000–2021. The databases included in the research have medical scientific value, with broad international recognition: PubMed, PubMed Central, Science Direct, Scopus, and SpringerLink. The subject is of interest both for the medical rehabilitation specialty and for neurological specialists because the therapy studied with an impact on the symptomatology that induces a high degree of disability can lead to a decrease in the doses of medication with a neurological impact, which increases the patient's compliance.

The limitations of the study are as follows: The small number of studies in specialized literature regarding the impact of hydrokinetotherapy on patients with spinal cord injury makes it impossible to carry out a meta-analysis on this subject and very difficult to carry out a review that respects all the stages imposed by the PRISMA algorithm. The fact that the review was not pre-registered in an international prospective register of systematic reviews, such as PROSPERO, and a funnel plot was not made are also a limitation of this review. In addition to the limited number of existing scientific studies, there are other drawbacks, such as heterogeneous methodologies, disparate study populations, and different training programs, which must be overcome in future studies.

5. Conclusions

Due to the low quality of the controlled studies and the lack of sufficient follow-up randomized controlled trials, there is not enough evidence supporting the importance of hydrotherapy for the functional rehabilitation of patients with spinal cord injury. Hydrotherapy's physiologic effects in orthopedic recovery and neurological disease must be investigated further. Financial support for such studies and the required clinical trials are still an issue. Despite its significant therapeutic benefits, aquatic exercise and balneal rehabilitation are underutilized.

Rehabilitation techniques must represent the main purpose of multiple and important qualitative scientific studies because medical rehabilitation represents a key role in spinal disorders such as SCI.

Author Contributions: Conceptualization, L.E.S. and M.G.I.; methodology, L.V.; software, A.E.C.; validation, E.-V.I., L.E.S. and C.I.M.; formal analysis, A.E.C.; investigation, L.V.; resources, L.E.S.; data curation, L.V.; writing—original draft preparation, L.E.S.; writing—review and editing, M.G.I.; visualization, C.I.M.; supervision, L.E.S.; project administration, L.E.S. All authors have read and agreed to the published version of the manuscript.

Funding: This research received no external funding.

Institutional Review Board Statement: The study was approved by the Hospital Ethical Committee (approval no 1735 from 2 February 2022) and complied with the revised ethical guidelines of the Declaration of Helsinki.

Informed Consent Statement: Not applicable.

Acknowledgments: The study was made within the Research Core of Balneal and Rehabilitation of Sanatorium Techirghiol, Romania.

Conflicts of Interest: The authors declare no conflict of interest.

References

1. Kesiktas, N.; Paker, N.; Erdogan, N.; Gülsen, G.; Biçki, D.; Yilmaz, H.G. The use of hydrotherapy for the management of spasticity. *Neurorehabil. Neural Repair* **2004**, *18*, 268–273. [CrossRef] [PubMed]
2. Priebe, M.M.; Sherwood, A.M.; Thornby, J.I.; Kharas, N.F.; Markowski, J. Clinical assessment of spasticity in spinal cord injury: A multidimensional problem. *Arch. Phys. Med. Rehabil.* **1996**, *77*, 713–716. [CrossRef] [PubMed]
3. Hippocrates. On Airs, Waters, and Places. 400 BC. Available online: http://classics.mit.edu/Hippocrates/airwatpl.html (accessed on 11 January 2022).
4. Kneipp, S. *My Water Cure, as Tested through more than Thirty Years, and Described for the Healing of Diseases and the Preservation of Health*, 30th ed.; William Blackwood&Sons: Edinburgh, UK, 1894.
5. Hall, J.; Swinkels, A.; Briddon, J.; McCabe, C.S. Does aquatic exercisere lieve pain in adults with neurologic or musculoskeletal disease? A systematic reviewand meta-analysis of randomized controlled trials. *ArchPhys. Med. Rehabil.* **2008**, *89*, 873–883. [CrossRef]
6. Bender, T.; Karaglle, Z.; Balint, G.P.; Gutenbrunner, C.; Balint, P.V.; Sukenik, S. Hydrotherapy, balneotherapy, and spa treatment in pain management. *Rheumatol. Int.* **2005**, *25*, 220–224. [CrossRef]
7. Yamazaki, F.; Endo, Y.; Torii, S.; Sagawa, S.; Shiraki, K. Continuous monitoring of change in hemodilution during water immersion in humans: Effect of water temperature. *Aviat. Space Environ. Med.* **2000**, *71*, 632–639. [PubMed]
8. Gabrielsen, A.; Ek, V.A.; Johansen, L.B.; Warberg, J.; Christensen, N.J.; Pump, B.; Norsk, P. Forearm vascular and neuroendocrine responses to graded water immersion in humans. *Acta PhysioScand* **2000**, *169*, 87–94. [CrossRef] [PubMed]
9. Fam, A.G. Spa treatment in arthritis: A rheumarologist's view. *J. Rheumatol.* **1991**, *18*, 1775–1777. [PubMed]
10. Giesecke, C. Aquatic rehabilitation of clients with spinal cord injury. In *Aquatic Rehabilitation*; Ruoti, R.G., Morris, D.M., Cole, A.J., Eds.; Lippincott, Williams andWilkins: Hagerstown, MD, USA, 1997; pp. 125–150.
11. Stiens, S.A.; Biener-Bergman, S.; Goetz, L.L. Neurogenic boweldys function after spinal cord injury: Clinical evaluation and rehabilitation management. *ArchPhys. Med. Rehab.* **1997**, *78*, S86–S102. [CrossRef]
12. Recio, A.C.; Cabahug, P. Safety of aquatic therapy for adults with complex medical conditions among chronic spinal cord injury. In Proceedings of the ASCIP Annual Meeting, Nashville, TN, USA, 4–7 September 2016; pp. 568–569.
13. Stiens, S.A.; Shamberg, S.; Shamberg, A.; Guistini, A. Environ mental barriers: Solutions to participation, collaboration and to getherness. In *Physical Medicine and Rehabilitation Secrets*, 3rd ed.; O'Young, B.J., Young, M.A., Stiens, S.A., Eds.; Mosby: St. Louis, MI, USA, 2007; pp. 76–85.
14. Stiens, S.A.; O'Young, B.J.; Young, M.A. Person-centered rehabilitation: Interdisciplinary intervention to enhance patient enablement, physical medicine and rehabilitation secrets. In *Physical Medicine and Rehabilitation Secrets*, 3rd ed.; O'Young, B.J., Young, M.A., Stiens, S.A., Eds.; Mosby: St. Louis, MI, USA, 2007; pp. 118–125.
15. Liberati, A.; Altman, D.G.; Tetzlaff, J.; Mulrow, C.; Gøtzsche, P.C.; Ioannidis, J.P.A.; Clarke, M.; Devereaux, P.J.; Kleijnen, J.; Moher, D. The PRISMA Statement for Reporting Systematic Reviews and Meta-Analyses of Studies That Evaluate Health Care Interventions: Explanation and Elaboration. *PLoS Med.* **2009**, *6*, e1000100. [CrossRef]
16. Downs, S.H.; Black, N. The feasibility of creating a checklist for the assessment of the methodological quality both of randomized and nonrandomized studies of health care interventions. *J. Epidemiol. Community Health* **1998**, *52*, 377–384. [CrossRef]
17. Machado, M.; Bajcar, J.; Guzzo, G.C.; Einarson, T.R. Sensitivity of patient out comes to pharmacist interventions. Part II: Systematic reviewand meta-analysis in hypertension management. *Ann. Pharmacother.* **2007**, *41*, 1770–1781. [CrossRef] [PubMed]
18. Kamioka, H.; Tsutani, K.; Okuizumi, H.; Mutoh, Y.; Ohta, M.; Handa, S.; Okada, S.; Kitayuguchi, J.; Kamada, M.; Shiozawa, N.; et al. A systematic review of nonrandomized controlled trials on the curative effects of aquatic exercise. *Int. J. Gen. Med.* **2011**, *4*, 239–260. [CrossRef]
19. Li, C.; Khoo, S.; Adnan, A. Effects of aquatic exercise on physical function and fitness among people with spinal cord injury. *Medicine* **2017**, *96*, e6328. [CrossRef] [PubMed]
20. Ellapen, T.J.; Hammill, H.V.; Swanepoel, M.; Strydom, G.L. Strydom, The benefits of hydrotherapy to patients with spinal cord injuries. *Afr. J. Disabil.* **2018**, *7*, 450. [CrossRef] [PubMed]
21. Zamparo, P.; Pagliaro, P. The energy cost of level walking before and after hydro-kinesi therapy in patients with spastic paresis. *Scand. Med. Sci. Sport* **2007**, *8*, 222–228. [CrossRef] [PubMed]
22. Marinho-Buzelli, A.R.; Zaluski, A.J.; Mansfield, A.; Bonnyman, A.M.; Musselman, K.E. The use of aquatic therapy among rehabilitation professionals for individuals with spinal cord injury or disorde. *J. Spinal. Cord. Med.* **2019**, *42* (Suppl. 1), 158–165. [CrossRef] [PubMed]
23. Wall, T.; Falvo, L.; Kesten, A. Activity-specific aquatic therapy targeting gait for a patient with incomplete spinal cord injury. *Physiother. Theory Pract.* **2017**, *33*, 331–344. [CrossRef]
24. Munteanu, C. *Therapeutic Mineral Waters*; Balneary: Bucharest, Romania, 2013; ISBN 978-606-93550-6-0.

25. Rekand, T.; Hagen, E.; Grønning, M. Marit Grønning, Spasticity following spinal cord injury. *Tidsskr. Nor. Legeforen.* **2012**, *132*, 970–973. [CrossRef]
26. Adams, M.M.; Hicks, A.L. Spasticity after Spinal Cord Injury. *Spinal Cord* **2005**, *43*, 577–586. [CrossRef]
27. Holtz, K.A.; Lipson, R.; Noonan, V.K.; Kwon, B.K.; Mills, P.B. The prevalence and Effect of Problematic Spasticity After Traumatic Spinal Cord Injury. *Arch. Phys. Med. Rehabil.* **2017**, *98*, 1132–1138. [CrossRef]
28. Cragg, J.J.; Noonan, V.K.; Krassioukov, A.; Borisoff, J. Cardiovascular disease and spinal cord injury. *Neurology* **2013**, *81*, 723–728. [CrossRef] [PubMed]
29. Myers, J.; Lee, M.; Kiratli, J. Cardiovascular Disease in Spinal Cord Injury An Overview of Prevalence, Risk, Evaluation, and Management. *Am. J. Phys. Med. Rehabil.* **2007**, *86*, 142–152. [CrossRef] [PubMed]
30. Moher, D.; Dulberg, C.; Wells, G. Statistical power, sample size and their reporting in randomized controlled trials. *J. Am. Med. Assoc.* **1994**, *272*, 122Y124. [CrossRef]
31. Schulz, K. Randomized trials, human nature, and reporting guidelines. *Lancet* **1996**, *348*, 562. [CrossRef] [PubMed]
32. Day, S.; Altman, D. Blinding to clinical trials and other studies. *Br. Med. J.* **2000**, *321*, 504. [CrossRef] [PubMed]
33. Gasmi, A.; Bjorklund, G.; Mujawdiya, P.K.; Semenova, Y.; Peana, M.; Dosa, A.; Piscopo, S.; Benahmed, A.G.; Costea, D.O. Micronutrients deficiencies in patients after bariatric surgery. *Eur. J. Nutr.* **2021**, *61*, 55–67. [CrossRef]
34. Głowiński, S.; Ptak, M. A kinematic model of a humanoid lower limb exoskeleton with pneumatic actuators. *Acta Bioeng. Biomech.* **2022**, *24*, 145–157. [CrossRef]
35. Pons, J.L. *Wearable Robots*; John Wiley & Sons, Ltd.: Chichester, UK, 2008; ISBN 9780470987667.
36. Stevenson, V.L. Rehabilitation in practice: Spasticity management. *Clin. Rehabil.* **2010**, *24*, 293–304. [CrossRef]
37. Little, J.W.; Micklesen, P.; Umlauf, R.; Britell, C. Lower extremitiy manifestations in spasticity in chronic spinal cord injury. *Am. J. Phys. Med. Rehabil.* **1989**, *68*, 32–36. [CrossRef]
38. Horga Parte, J.F.; Pareés, M.I. Toxina botulínica: Origen, estructura, actividad farmacológica y cinética. In *Toxina Botulínica*; López, d.V.L.J., Castor, G.A., Eds.; Elsevier España: Barcelona, Spain, 2010; pp. 3–17.

Review

Designing a Clinical Trial with Olfactory Ensheathing Cell Transplantation-Based Therapy for Spinal Cord Injury: A Position Paper

Ronak Reshamwala [1,2,3], Mariyam Murtaza [1,2,3,4], Mo Chen [1,2,3], Megha Shah [1,2,3], Jenny Ekberg [1,2,3], Dinesh Palipana [5], Marie-Laure Vial [1,2,3], Brent McMonagle [1,5] and James St John [1,2,3,4,*]

[1] Menzies Health Institute Queensland, Griffith University, Southport, QLD 4222, Australia
[2] School of Pharmacy and Medical Sciences, Griffith University, Southport, QLD 4222, Australia
[3] Clem Jones Centre for Neurobiology and Stem Cell Research, Griffith University, Brisbane, QLD 4111, Australia
[4] Griffith Institute for Drug Discovery, Griffith University, Brisbane, QLD 4111, Australia
[5] Gold Coast University Hospital, Southport, QLD 4215, Australia
* Correspondence: j.stjohn@griffith.edu.au

Abstract: Spinal cord injury (SCI) represents an urgent unmet need for clinical reparative therapy due to its largely irreversible and devastating effects on patients, and the tremendous socioeconomic burden to the community. While different approaches are being explored, therapy to restore the lost function remains unavailable. Olfactory ensheathing cell (OEC) transplantation is a promising approach in terms of feasibility, safety, and limited efficacy; however, high variability in reported clinical outcomes prevent its translation despite several clinical trials. The aims of this position paper are to present an in-depth analysis of previous OEC transplantation-based clinical trials, identify existing challenges and gaps, and finally propose strategies to improve standardization of OEC therapies. We have reviewed the study design and protocols of clinical trials using OEC transplantation for SCI repair to investigate how and why the outcomes show variability. With this knowledge and our experience as a team of biologists and clinicians with active experience in the field of OEC research, we provide recommendations regarding cell source, cell purity and characterisation, transplantation dosage and format, and rehabilitation. Ultimately, this position paper is intended to serve as a roadmap to design an effective clinical trial with OEC transplantation-based therapy for SCI repair.

Keywords: neurosurgery; olfactory glia; translational health research; regenerative medicine; biomedical engineering

Citation: Reshamwala, R.; Murtaza, M.; Chen, M.; Shah, M.; Ekberg, J.; Palipana, D.; Vial, M.-L.; McMonagle, B.; St John, J. Designing a Clinical Trial with Olfactory Ensheathing Cell Transplantation-Based Therapy for Spinal Cord Injury: A Position Paper. *Biomedicines* **2022**, *10*, 3153. https://doi.org/10.3390/biomedicines10123153

Academic Editor: Nicolas Guerout

Received: 8 November 2022
Accepted: 3 December 2022
Published: 6 December 2022

Publisher's Note: MDPI stays neutral with regard to jurisdictional claims in published maps and institutional affiliations.

1. Introduction

Spinal cord injury (SCI) is a devastating life-altering condition and there are currently no effective treatments. The loss of sensorimotor and autonomic function that follows an injury has an overwhelming effect on the individual, carers, society, and the healthcare system in general. Apart from paralysis, SCI leads to widespread systemic impact including inflammatory reaction, respiratory issues, cardiovascular complications, compromised immunity and bone densities, muscle wasting and several other complications to the individual's mental, physical as well as financial health. According to a report by Spinal Cure Australia, there are over 20,800 people living with spinal cord injury in Australia currently, and the lifetime cost of healthcare alone is estimated to be AUD 3.3 billion, with the total lifetime socioeconomic burden being over AUD 75 billion [1]. A conservative estimate suggests that a small significant functional recovery in a fraction of the people living with spinal cord injury can result in savings of AUD 3.5 billion, potentially up to AUD 10 billion [1]. This further highlights the urgent need for a clinically available reparative therapy for SCI, as the current standard clinical practice can only provide damage control and strategies to mitigate complications.

To meet this need, several different approaches are under investigation (Figure 1). Briefly, the core mechanism of action for the approaches used at different stages of injuries include (1) a combination of drugs and anti-inflammatory strategies for damage control and mitigating secondary degeneration in immediate and acute phases—which is currently clinically available [2–4]; (2) cell transplantation in the acute, or subacute phase to replace lost glia [5,6]; once the paralysis sets in: (3) robotics to simulate motor repairs [7–9]; and (4) biological approaches to restore the natural motor, sensory and autonomic function, which can include cell transplantation, immunomodulation, and growth factor supplementation [10,11]. The biological approaches offer the opportunity to replace the lost neural tissue mass and mitigate the damage by shrinking the defect size. Out of all the stem and non-stem cell types explored for cell transplantation, olfactory ensheathing cells (OECs) stand out as promising candidates for neural repair due to their unique properties [12,13]. The OECs are the primary glial cells of the olfactory nerve, where they play a crucial role in replacing up to 1–3% of olfactory neurons daily and guide them to their intended targets in the olfactory bulb throughout life [14]. OECs from the olfactory mucosal tissue are relatively easy to access via intranasal endoscopy without causing long-term issues, making OECs an excellent candidate for clinical translation, as evidenced by numerous pre-clinical animal trials [15–20] as well as several safety/efficacy trials conducted in recent decades [21–31].

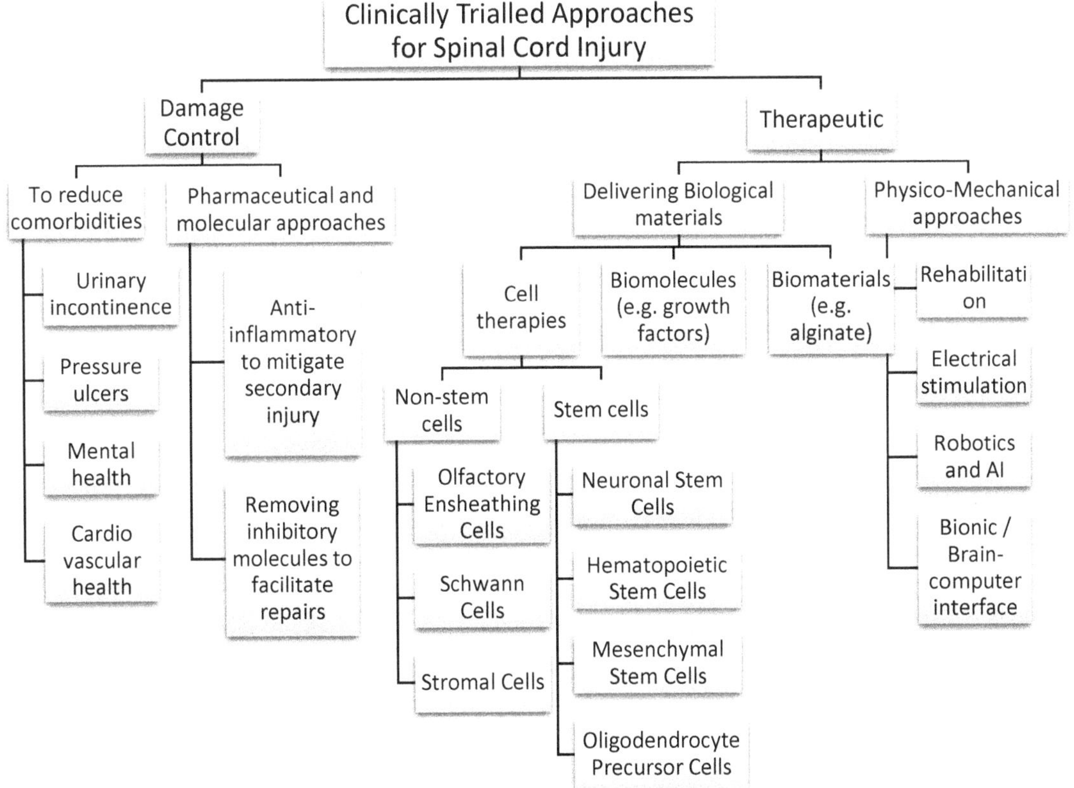

Figure 1. An overview of approaches under exploration for spinal cord injury treatment.

Despite this promising evidence, the fact that cell transplantation therapy with OECs (or any other cells) has not yet made it to wider clinical use raises some concerns for caution for future clinical trials. To improve translational outcomes from OEC-based cell therapies, there is a need to identify potential barriers to successful translation and to discuss the preclinical-to-clinical translational approaches with an appreciation of fundamental biological properties of OECs. In this position paper, we discuss the distinct aspects of clinical trial design from donor recruitment, clinical-grade cell production to assessment regimes for us to obtain reliable data from OEC transplantation therapies, and to improve the likelihood of neural repair.

2. What to Translate to a Clinical Trial?

Clinical trials are typically translated from pre-clinical animal studies, where there is robust evidence of the safety and efficacy of a treatment modality. In recent years, nearly 20% of the cell transplantation-based clinical trials for SCI were conducted around the use of OECs. OECs have been proven safe for transplantation in animal spinal cord injury models and they have shown varying degrees of success at restoring sensory, motor, and autonomic functions following treatments in rodents [18–20,32–37], canine [15], and primate models [17]. Based on the evidence from pre-clinical studies, OECs are a promising, low-risk therapeutic candidate for clinical translation with a high chance of success.

2.1. What Makes OECs Suitable Candidates for the SCI Repair?

OECs are the supporting cells of the olfactory nerve and are present throughout the course of the nerve from olfactory mucosa to the olfactory bulb. In their natural environment, OECs envelope or ensheathe the olfactory axons and provide them with ongoing support. In this way, they are similar to other glial cells such as Schwann cells and astrocytes. However, unlike the other glial cells, OECs have physiologically developed proficiency in migration and phagocytosis to clean up cellular debris following injuries [38,39], promotion, and augmentation of axonal regrowth by secreting neurotrophic factors [40,41], axonal guidance [42], and modulation of inflammatory profile of their immediate vicinity by expressing different specific cytokine profiles and other molecular markers such as macrophage migration inhibitory factor [43,44]. This can be primarily attributed to their intrinsic environment being a region of high turn-over for axonal damage and repair [5].

Importantly, OECs are shown to retain these direct and indirect mechanisms of inducing and supporting axonal repair when they are transplanted to a neural injury site such as the spinal cord injury [5,10,45]. Additionally, unlike other glia such as Schwann cells, OECs are able to interact with the injury site and the reactive astrocytes that form the scar tissue due to their unique heparin sulphate expression profile [45–47]. Thus, the OECs show a natural tendency towards axonal repair. Several types of stem cells are also reported to possess similar properties for nerve repair and secretion of neurotrophic factors [48–50]. However, OECs have some distinct advantages over the stem cells for their use in cell-transplantation-based therapy. OECs are differentiated and functionally mature cells, and therefore, they do not need to be differentiated into a functional cell type. In some instances, the OECs have been used as a delivery mechanism for therapeutic transgenes in addition to providing the usual nerve repairs in a spinal cord injury site [51]. Another critical advantage from a safety point of view is that the OECs possess negligible-to-no tumorigenic potential, which is a crucial consideration for a cell transplantation therapy [52]. Being non-stem cells, OECs, especially mucosal OECs, do not face the same logistical, ethical, or moral concerns and controversies for their source as stem cells (except for allografted foetal OECs obtained from abortifacients).

Thus, the natural cellular properties supporting nerve repair, the robust safety profile, and remarkable evidence of efficacy in the animal trials make OECs highly suitable for spinal cord injury repair trials.

2.2. What about the Trials so Far?

There have been several clinical trials using OECs in different formats and using different modes of administration. While there is sizeable evidence of safety, the evidence of efficacy lacks consistency, and these trials have not progressed beyond phase IIa. The trials have also identified challenges to a successful and widespread translation of the therapy, with various aspects described below.

Since 2005, there have been around fifty clinical trials attempting cell transplantation-based therapies for SCI across the world. Out of these, eleven trials used OECs as their therapeutic cell population and were focussed on feasibility (pilot), safety (phase I), or evidence of efficacy (phase IIa), and were conducted in many different countries such as Australia, Portugal, India, China, USA, and Poland. However, 1 of the 11 trials was a longer-term follow up of a previous trial (having the same patients) [23], and another trial compared the role of intense rehabilitation in SCI patients who received a cell transplant with those who did not [27], and included patients from a separate trial [28]. Thus, for this review, we have considered the trial design and surgical specifics for only nine trials, although we have included the original individual findings of all eleven trials. Additionally, some case studies involving stand-alone reporting of interesting individual cases related to the use of OECs for spinal cord injury repair were reviewed and are discussed. However, the difference must be noted between the patient-focussed case studies, and the clinical trials where the focus is to answer specific clinical research questions (feasibility, safety, and efficacy) by recruiting patients.

2.3. Anatomical Origin of Cells Used in Trials

OECs can be obtained either from the olfactory mucosa via a minimally invasive biopsy from within the nasal cavity, or from the outer layer of the olfactory bulb via an invasive procedure from within the cranial cavity. Some researchers prefer harvesting cells from the olfactory bulb as it can lead to higher purity cultures, but there is a higher risk with the procedure and likely to be permanent damage to the integrity of the olfactory bulb, which may impact on the functioning of the sense of smell. Out of the nine human trials, seven used OECs from the olfactory mucosa, of which three trials used OECs isolated and cultured from mucosal tissues [21,28,29], three used autografts of minced whole olfactory mucosa [22,24,25], and one trial used autologous lamina propria (the part of the mucosa after removing mucosal epithelial lining) for transplantation [31]. The other two trials used OECs obtained from the olfactory bulb [26,30]. Table 1 contains a summary of the details regarding cell origin used in each clinical trial.

Interestingly, a few case studies which were performed outside of a formal clinical trial have also tested OEC transplantation. In a popularly reported case, an autologous olfactory bulb (unilateral) was used as the source of OECs because the patient's mucosa was rendered inaccessible due to extensive nasal polyps [53]. Another recent case report used foetal olfactory bulbs as the source of OECs [54].

Thus, OECs from both anatomical sources have been tried and tested in clinical settings. Pre-clinical trials have indicated that OECs from both sources have comparable reparative properties [55] despite being two distinctly different sub-populations [56]. The transplantation of cells from both the sources has been shown to be feasible and safe in clinical trials, however, determining the purity of mucosal OECs and their purification have been identified as challenges, thus, raising some serious safety concerns and must be addressed, as discussed later.

Table 1. A summary of details of interest regarding clinical trials design and protocols.

Author, Year	Transplantation Format	Cell Characterisation	Purity	Adverse/Severe Adverse Events	Safety Established	Efficacy
Féron et al., 2005 [21]	Isolated mucosal OECs (autograft)	GFAP + S100; p75NTR	>95%; 76–88%	None	Yes	Not assessed
Lima et al., 2006 [22]	Olfactory mucosa pieces (autograft)	N/A	N/A	None	Yes	Modest improvement in ASIA scores
Mackay-Sim et al., 2005 [23]	Isolated mucosal OECs (autograft)	N/A	N/A	Meningitis, CSF leakage, IBS, new visceral pain	Yes	No significant functional improvements
Chhabra et al., 2009 [24]	Olfactory mucosa pieces (autograft)	N/A	N/A	1 syrinx; 1- more sensory loss recovering gradually	Yes	No significant improvements, but statistically significant improvements in SCIM, BDI and ISCIS
Lima et al., 2010 [25]	Olfactory mucosa pieces (autograft)	N/A	N/A	None	Yes	Possible with post-operative rehabilitation
Wu et al., 2012 [26]	Foetal OB OECs (allograft)	GFAP and S100 Immuno-staining		One patient had reduced ASIA sensory with pain and tingling, transient pain resolving with analgesics. No SAEs	Yes	Moderate sensory and spasticity improvements, minimal locomotor improvements
Larson et al., 2013 [27]	Olfactory mucosa pieces (autograft)	N/A	N/A	N/A	Yes	Motor recovery was observed, no sensory improvement. Recovery was not significantly greater compared to the control group
Tabakow et al., 2013 [28]	Isolated mucosal OECs (autograft)	S100, p75NTR	>5%	Some immediate adverse events over post-operative phase, resolving within 3–4 days. No AE or SAE over 1-year follow up.	Yes	2 of the 3 patients improved ASIA scores, third patient had some neurological recovery without ASIA score improvement
Rao et al., 2013 [29]	Isolated mucosal OECs (autograft)	morphology	Not reported	No SAEs	Yes	3/8 patients had substantial sensorimotor recovery; 2/8 had bladder function restored
Chen et al., 2014 [30]	Foetal OB OECs (allograft)	p75NTR, S100 for OECs; S100 for SCs	94%	One patient had fever, No SAEs	Yes	4/5 treated patients showed significant electrophysiological improvements, 5/5 showed some functional improvement
Wang et al. [31]	Olfactory lamina propria pieces (autograft)	N/A	N/A	No SAEs	Yes	Limited functional recovery; 2/8 patients had ASIA score improvement

ASIA = American Spinal Injury Association, BDI = Beck Depression Inventory, ISCIS = International Spinal Cord Injury Scale, SCIM = Spinal Cord Independence Measure, OB = Olfactory bulb, OECs = Olfactory ensheathing cells, AE = Adverse event, SAE = Serious adverse event. N/A = Not applicable.

2.4. Autologous vs. Allogenic Cell Source

Due to the accessibility of OECs from the olfactory nerve, most trials have used autologous transplantation. However, treatments that require early intervention for acute spinal cord injury or when large numbers of cells are required may consider allogeneic donor cells. Two trials used foetal OECs as allografts for treatments, one of which progressed to a phase II study. The use of human foetal OECs does raise numerous ethical issues, and thus, if donor cells are required, the use of adult donor cells would avoid many ethical complications.

In addition to the accessibility of cell source, cell survival and integration after transplantation are important considerations when deciding between these two cell source types [5]. Some pre-clinical trials have indicated that the transient survival of OECs may be sufficient to induce neural repair in SCI, and therefore, immuno-incompatibility may not be critical factor, at least not in the rodent models [16]. However, neuroinflammation and further complications that ensue after a graft rejection, make the autograft a more logical and prudent option.

2.5. Cell Purity and Characterisation

OECs can be difficult to purify and cultures of OECs often contain large proportions of other cells, such as fibroblasts. In addition, cultured cells can lose the expression of markers and may potentially alter the function.

Perhaps due to these reasons, the trials using olfactory mucosa (three trials) and olfactory lamina propria autografted the small pieces of intact tissues rather than isolated cells. In this way, the cells were retained within their natural niche, but considerable unwanted cells were also transplanted. However, by using intact tissue, there was no way to characterise the treatment cell population or their purities and no means to define the cell numbers used for treatments in each patient.

Interestingly, most of the remaining trials did not comment on the quality or purity of the transplanted cells in each individual patient—rather, an overview of the cell population was given. For example, the foetal OECs were reported to be ~94% pure in one trial [30]. In another phase I trial, where adult autologous mucosal OECs were used, it was reported that >95% cells expressed the GFAP and S100 markers for OECs and 76–88% cells expressed the marker p75NTR [21]. Conversely, in one trial, the acceptable purity threshold was kept as low as 5% and their treatment cell population was defined as the combination of OECs and the olfactory nerve fibroblasts, which are the most common accompanying cells for OECs of mucosal origin [28]. One trial confirmed their cell populations as OECs by visualising their morphology, but not by immunocytochemistry [29]. Specific details of interest regarding cell purity and characterisation are summarised in Table 1.

Thus, there has been no uniform method or threshold for the acceptability of the transplantation cell population in the clinical trials. This may be an important factor responsible for inconsistent efficacy outcomes. Defining a robust characterisation method and setting a high quality but realistic acceptance threshold is a crucial quality control measure moving forward.

2.6. Cell Dosage

The amount of cells transplanted, total treatment volume and the mode of transplantation all have a bearing on the outcomes of the surgical intervention [5]. There is a limit to the amount of the cells and total volume of the treatments that can be transplanted safely at the injury site, however, it is also crucial to maintain the critical mass of the treatment for the therapeutic effect to take place.

As mentioned earlier, cell quantity or treatment volumes were not possible to determine for the four trials that used mucosal pieces or lamina propria grafts. The three trials using autologous OECs prepared in the form of single cell suspensions injected the cells in the cord parenchyma adjacent to the injury in varying doses. The injected cell doses varied widely from 12, 24, and 28 million cells in one trial [21], to 1.8, 1.92, and 21.2 million in another trial (with 30,000–200,000 cells/μL concentration) [28]. In the third trial, a fixed dose of 1 million cells in 2 mL suspension was injected [29]. Two trials using foetal OECs used 500,000 cells in a total volume of 5 μL [26], and a fixed dose of 1 million cells in a total volume of 50 μL [30], respectively. In the case reports mentioning cell suspension injections, 500,000 cells in 48 μL [53] and 1 million cells in 60 μL volume [54] were injected. Thus, the total number and volume of cell suspension injections varied considerably. Surprisingly, the total injection volumes varied from 5 μL to 2 mL in different trials, however, none of the trials reported any links between the treatment volume and final outcomes or adverse events. One complicating factor in linking the dose and outcomes is that cell survival may be affected by the volumes or concentrations that are used.

Understandably, there are considerable differences between the transplantation methods depending on the format in which cells are transplanted. The trials using intact tissue transplanted the treatments directly in the injury site. For this, injury site manipulation, cord untethering, and partial scar debridement were necessary. Conversely, the cell suspension injections were made into the cord parenchyma around the injury site, where scar debridement was optional and not necessary. The injections were generally made as low flow-rate micro-injections at multiple locations. In some cases, all the treatments were carried out identically, where all the patients received 1.1 μL injections in a 3×5 grid at 4 different depths [21], or a total volume of 2 mL in 6 injections at the caudal end of the injury site [29]; however, in one trial, patients received 60, 64, and 120 μL by the use of 120, 128, and 212 micro-injections with a fixed 0.5 μL volume with a flow rate of 2 μL/min [28]. The same protocol was used to inject 500,000 cells in a 48 μL volume over 96 total micro-injections in a case report by the same team [53]. The procedural details for these cell transplantations are summarised in Table 2. As mentioned, the injections were made in the relatively healthy and intact cord parenchyma around the injury site where each injection is potentially a separate microinjury. In this manner, both types of approaches are markedly different in their execution and carry their own set of risks and benefits. Even though all the trials concluded that their approaches were safe, a novel approach that combines the benefits of both methods while mitigating the risks may enhance the efficacy of the treatments.

Table 2. Procedural detail summary for cell transplantation in the clinical trials using cell suspension injections.

Author, Year	Cells Transplanted	Cell Concentration	Injection Volume	Flow Rate	Number of Total Injections
Féron et al., 2005 [21]	12 million, 24 million, and 28 million, respectively, injected in 3 patients	not mentioned	1.1 μL/injection, ~132 μL total	not mentioned	4 depths in a 3×5 grid, both rostrally and caudally = 120 injections
Wu et al., 2012 [26]	500,000 cells	100,000 cells/μL	5 μL	not mentioned	2 injections, 1 each rostrally and caudally from the injury
Tabakow et al., 2013 [28]	1.8 million, 1.9 million, and 21 million cells, respectively, in 3 patients	30,000–200,000 cells/μL	60, 64, and 106 μL, respectively	2 μL/min	120, 128 and 210 injections, respectively, 0.5 μL per injection
Rao et al., 2013 [29]	1 million cells	50,000 cells/μL	2 mL	not mentioned	6 injections total
Chen et al., 2014 [30]	1 million cells	20,000 cells/μL	50 μL	not mentioned	2 injections, 1 each rostrally and caudally from the injury

2.7. Patient Recruitment

The SCI is a highly variable condition where no two injuries are ever the same, which is why the inclusion of different injury types in the clinical trial may impact on the trial outcomes. Injury level, degree of severity or completeness, and time since injury are some critical factors that may influence the clinical outcomes.

All the clinical trials enrolled patients with at least 6 months after the initial SCI. However, one trial specifically included chronic patients with a neurological profile of the injury that had stabilised for at least 6 months. The included patient ages ranged from 16 to 65 for all trials, with most trials including patients aged 18 or above. Thus, the surgical interventions are attempted in "settled" injury sites in adult patients, where the chances of spontaneous regeneration are null clinically. The acute phase injuries are thus far not preferred for the experimental surgical intervention in the trials which are assessing the safety of the procedure in the first instance.

Most trials included patients with American Spinal Injury Association (ASIA) grade A or B, however, one trial also included ASIA-C patients; one trial used the Frankel scale and only included "complete" injuries on the scale (similar to ASIA-A). The neurological level of the injuries varies widely from cervical to low thoracic levels. Thus, the selection of patients apparently favours complete or more severe injuries in most clinical trials, conceivably to avoid doing any harm in the pilot or phase I trials where the safety of the intervention is still being tested. However, a wider variety of injury types must be considered for the efficacy trials in phase II and further.

2.8. Trial Design

Three of the clinical trials were designated "Pilot" studies with 5, 7, and 8 patients enrolled, respectively [22,24,29]. The remaining studies were either designated phase I/IIa trials [23,25], or they started with only phase I and progressed to a phase II study with a 3-year follow-up window [26,31]. Thus, depending on the background pre-clinical work that the trials aim to build on, they may be designed as pilot, phase I, or phase I/IIa trials.

2.9. Rehabilitation as an Adjuvant Intervention

Rehabilitation following neurotrauma or other sudden accidental central nervous system events (such as stroke) has been recognised as an important intervention for functional recovery. Rehabilitation is even recognised as a stand-alone treatment modality for spinal cord injury. Recent studies also show that sufficient and sustained rehabilitation may be able to induce limited nerve repairs on its own [27]. This makes rehabilitation a great synergistic adjuvant to any cell transplantation treatment.

Importantly, 5 of the 11 trials, as well as both case reports reviewed here, had a component of rehabilitation associated with the cell transplantation protocol, while 1 of the trials assessed neurorehabilitation as the primary intervention [27]. Notably, three of the trials included a pre-operative component of intense rehabilitation ranging from 3 months [28] and 6 months [30] to 8 months (35 weeks) [25]. These trials also had the post-operative intense rehabilitation of 24 months [25] and 21 months [28]; however, information was not specified in the third trial protocol. Another trial included only the post-operative component of rehabilitation lasting from 2.5 to 4.5 months [27], whereas the remaining trial instructed the participants to perform home-based rehabilitation [31]. The dose of the rehabilitation also varied considerably, ranging from an average of 8.5 h a week [27], to nearly 20 h a week (4–5 h per day, 3–5 days a week) [28], with the highest amount reported between 25 and 39 h a week [25].

It is important to note that the trial by Lima et al. included the highest weekly hours and longest total duration of rehabilitation, however, no patient dropouts were mentioned [25]. Conversely, the trial by Chen et al. had a mandatory 6 months of pre-operative rehabilitation included but only 28 of the original 64 participants were able to complete the intensive rehabilitation program; however, only 7 of these 28 patients were able to receive the transplants due to limited resources [30]. Thus, neurorehabilitation, akin to the several other specific parts of the protocol, showed marked differences across different trials. Most commonly, the availability of resources, such as funding and patient adherence, were the determining factors in this regard.

2.10. Outcome Measures

Most of the studies focussing on the safety profile of the intervention primarily defined their outcome measure as patient safety, which is monitored by recording any adverse events (AE) or severe adverse events (SAE) as well as any regression of the neurological function/worsening of existing symptoms. The patients were followed up for at least 1-year post-intervention in most trials, with only one exception, where the follow-up period was 6 months. The maximum follow-up period was 48 months in one trial and a few more trials had up to 3-year follow ups.

Almost all the trials reported no SAEs and minimal AEs. However, one trial reported a patient with reduced ASIA sensory grade with pain and tingling over 18 months of follow up, and another trial reported a case of meningitis, cerebrospinal fluid leak, and new occurrence of visceral pain over the follow up of 48 months.

The efficacy of the interventions was assessed using several different clinical assessment tools across different clinical trials as the secondary outcome measures, such as ASIA grading, electrophysiological assays such as electromyography, somatosensory-evoked potentials (SSEP), motor-evoked potentials (MEP), autonomic function such as bowel and bladder control, imaging changes as seen on MRI and diffusion tensor imaging (DTI), monitoring of neuropathic pain, functional independence measure (FIM), activities of daily living rating (ADL), walking index for spinal cord injury (WISCI), modified Ashworth's scale for spasticity (MAS), and international association of neurorestoratology—spinal cord injury functional rating scale (IANR–SCIFRS). Thus, there are numerous assessments that can be carried out to determine functional and psychosocial outcomes.

2.11. Safety Concerns

In addition to the AEs and SAEs reported in the clinical trial outcomes (see Table 1), there are a few further safety concerns. Outside of the reviewed clinical trials and case reports here, there have been additional case reports where improper cell harvesting techniques and/or surgical protocols have resulted in disastrous outcomes for the patients several years after the intervention.

The earliest reported case was that of a young female patient presenting with a cystic mass at the transplantation site 8 years after the intervention [57]. More recently, another case of a 38-year-old male was also reported with a similar presentation, 12 years after the intervention [58]. Both these patients presented with severe back pain and needed a subsequent surgery to remove the mass. Surprisingly, the masses were found to have mucus-producing cells from the respiratory mucosa. Both these unfortunate incidents occurred from the transplantation of mucosal pieces that demonstrate that transplanting un-quantified, uncharacterised, and uncultured tissue grafts is not a desirable transplantation approach. This further highlights the importance of having a thorough understanding of the relevant cell types and their physiological properties as well as having robust isolation, purification, and characterisation protocols in place for the transplantation cell population.

2.12. Conclusions of the Past Clinical Trials

All the studies unequivocally concluded that the therapy was safe and feasible, and most studies found that the therapy was effective with varying levels of functional recoveries recorded, as summarised in Table 1. However, none of the therapies have progressed beyond this level of clinical testing, which may be due to a combination of limited resources and variable outcomes in the clinical trials. There are numerous potential factors contributing to variable outcomes, which if addressed, could improve the efficacy of the treatment. For example, the past trials transplanted cells either as single cell suspension via injections, or as small pieces of tissue where the quality and quantity of the cells cannot be controlled or confirmed, likely leading to a high variation in outcomes. Trials opting to inject cell suspensions did not address the scar tissue, whereas the trials transplanting mucosal pieces performed varying degrees of partial scar removal. The scar removal was shown to be safe, however, much debate still persists regarding the wisdom of scar removal. One important aspect of the past trials was the use of fixed volume/dose treatments for different injuries. Different injuries with different volumes, shapes, and sizes are likely to respond differently, and thus, the cell dose needs to be tailored to suit the injury.

3. How to Design a Clinical Trial?

A clinical trial must be strategically planned around the primary aims of the trial, specifically as to avoid any confounding factors from the skewing trial outcomes. It is also critical to incorporate measures to address and overcome the limitations and issues identified during past clinical trials in order to successfully progress further if the trial's aims are met sufficiently.

In our proposed approach, we have identified the following strategies to address and overcome these challenges, thereby strengthening the approach for OEC transplantation to repair SCI.

3.1. Olfactory Mucosa as a Source of the Cells

It is clear from the clinical trials that the preferred source of autologous OECs is the olfactory mucosa from the nasal cavity. In the only instance where autologous cells were obtained from the olfactory bulb, it was carried out because the mucosa was rendered inaccessible [53]. The OECs of mucosal origin are easily accessible with minimally invasive means using local anaesthesia, or with general anaesthesia if the surgeon prefers to gain deeper access to the nasal cavity. As the nasal mucosa is the only tissue that is involved, there is little risk to the patient during the harvest procedure. From the point of view of the integrity of the sense of smell, harvesting a small region of the nasal mucosa does not affect the ability to smell. In contrast, harvesting OECs from the olfactory bulb requires invasive access into the cranial cavity and is likely to result in considerable damage to the nerve fibre layer of the olfactory bulb, with likely ongoing perturbation to the functional capacity of the sense of smell. Importantly, the mucosal OECs are known to express a unique combination of proteins that are developmentally relevant for nerve repair and are not expressed by the OECs from the bulb [59,60].

3.2. Comprehensive Assessment of Cell Purity and Function

To improve the clinical outcomes from OEC-based cell transplantation therapies, there is a need to standardize the cell isolation, expansion, and good manufacturing practice (GMP) of cell production aspects. There is no consensus on the methods for isolating OECs from the olfactory tissues or markers to identify OECs and the other cells obtained in the primary culture [59,61]. The cells obtained from the olfactory mucosa for clinical purposes should be tested at various points during the GMP production process and should be reported for every patient. At a minimum, the starting material, the intermediate expansion product, and the final cellular product must be analysed. Cell assessments should involve an analysis of cell size, shape, morphology, growth characteristics, and the evaluation of cell surface markers. The cells should be analysed for the cell surface marker expression of a combination of positive and negative markers by quantitative immunofluorescence staining or flow cytometry. These assessments will help determine the quantity and purity of the cells, as well as identify subpopulations of cells. For the clinical trial, the research team must define the specified ranges of cell surface marker expression and viability, and this pre-determined acceptance criteria must be used prior to the release of OEC cells for clinical transplantation.

In the previous clinical trials conducted for OECs, the biological activity of the transplanted cells has not been shown prior to cell transplantation. However, the measured biological activity must be related to the intended biological effect as there is the potential for the culturing conditions to alter cell function. Therefore, there is a need for assays to pre-determine the biological effect of OECs as a potential predictor of clinical outcomes and to control the quality of the cell therapy product. However, this can prove to be challenging—it can be difficult to produce large quantities of cells from the olfactory biopsy and diverting cells for functional testing may adversely impact the dosage available for clinical use. The development of assays must, therefore, keep these limitations in mind and use small numbers of cells. Further, the assays should be simple and cost-effective for widespread adoption in clinical practice.

3.3. A 3D Construct Is Warranted

Cell integration and survival is one the major factors affecting the outcomes of any cell transplantation-based therapy including OECs. Several factors such as inflammation and hostile milieu of the SCI site adversely affect the ability of OECs to survive and integrate with the injury site. Additionally, the cells have historically been transplanted in a suspension form via an injection in the healthy cord parenchyma around the injury site. These needle tracks essentially inflict further injury to an already injured spinal cord, and there is a limited volume that can be applied to the spinal cord.

To overcome this, some trials in the past have opted to transplant pieces of olfactory mucosa or lamina propria directly in the SCI site, without manipulating the surrounding healthy cord tissue. This provides the transplanted cells their own native tissue scaffolding, and thus, improves their chances of survival and integration; however, this approach has its own significant draw backs. The lamina propria has several other cell types present and the OECs can neither be purified nor quantified prior to transplantation. Thus, unknown quantities and quality of the transplanted cells are transplanted which can be a significant confounding factor leading to widely varying outcomes. The approach also has some associated risks as it involves non-purified tissue autograft. The most significant risk is the possibility of accidentally transplanting tissues collected from sub-optimal sites which may contain respiratory mucosa or cells other than OECs, thus, potentially leading to disastrous outcomes [58].

An alternative cell preparation approach is to create three-dimensional constructs in which cells are embedded within supporting structures or gels [62,63], thus, enabling them to be deposited into the injury site.

3.4. Cell Dose and Treatment Volume

As mentioned before, there is a huge disparity amongst the trial with regard to the cell dose and treatment volumes. Assuming the highest reported numbers to be the maximum safe treatment thresholds, the highest used treatment volume was 2 mL and in a separate trial, the highest used cell dose was ~21 million cells. Thus, the maximum dosage that can be performed safely via injections is twenty million cells in 2 mL volume. However, using the alternative 3D preparation approach presents another critical edge in this regard, as the volume/amount of cells can be tailored to suit the size of the cavity, with surgeons able to customise the dosage to fill the cavity with the 3D cell preparation.

3.5. Surgical Approach for Transplantation

It has been well established that the scar at the site of SCI is unique in its cellular and molecular make up. The scar also plays a vital role in the protection and stabilisation of the injury site up to a certain point; however, after the injury is stabilised, the scar tissue acts as a physical barrier to any potential repairs. The molecules such as chondroitin sulphate proteoglycans (CSPG) also emit inhibitory signals for the reparative processes.

Thus, strategic manipulation of the scar tissue is crucial for a successful transplantation intervention. Leaving the scar tissue untouched can impede the cellular repairs and axonal growth through the injury zone; however, the removal or over-dissection of the scar can further destabilise the injury site and trigger adverse neuroinflammatory signals. One of the advantages of using OECs is that they can interact and permeate through the dense glial scar of the SCI [64,65], and therefore, a complete or partial removal of scar may not be necessary. We propose that a tactical minimal debridement of the scar tissue aimed at merely mobilising the adhesions and securing an approach to deposit the 3D preparation of cells within the defect would be best suited.

3.6. Patient Recruitment Should Be Carried out Based on the Clinical Trial Phase

3.6.1. Complete or Incomplete Injuries

Patient safety is a priority. For this reason, chronic injuries in which the neurological function has stabilised are the safer option for cell transplantation clinical trials at this point. In addition, as most phase I and IIa clinical trials have done in the past, patients with a more complete injury profile should be selected as there is a reduced risk of causing further neurological harm due to the intervention. Similarly, the lower neurological level injuries are best suited to test for the safety, however, a more lenient range can be set for including patients with the different neurological levels of injury depending on access to the patient pool and patient enrolments. It is worth considering that the patients with incomplete injuries have a larger portion of cord tissue spared with less extensive scarring, and therefore, have a higher chance of benefitting from the therapy. Once the use of OEC transplantation has been shown to be safe, patients with less complete injuries could be considered for potential treatment.

3.6.2. Time since Injury

Another aspect for the SCI profile is the time since injury. While 'the sooner the better' stands true for regaining neurological function after SCI, there are several factors to be considered in this regard. Neuroinflammation is a complex process which significantly affects the survival and integration of transplanted cells. For some patients, there may be a spontaneous regain of function following an injury, which an early intervention can impede. Depending on the extent of the initial injury, the injury might be deemed too unstable for an invasive experimental intervention such as cell transplantation. Considering all such factors, patients with injuries at least 12 months old should be included for the safety trial. The injury site is fairly stable at this point, and the probability of spontaneous repairs becomes considerably reduced after the first few months of injury. Thus, there is less chance of harm by intervention at this stage. The additional advantage of including chronic injuries is that the enrolment process can be expedited since there is no need for prospective enrolment, which can be time consuming and limited. Once the safety of the cell transplantation treatment is confirmed, acute/subacute injuries can be included in future trials.

3.7. Adaptive Trial Design May Be Beneficial

With the clinical translation of such critical therapies, the clinical benefits and urgency must be balanced against patient safety and pragmatism. Thus, we propose that new clinical trials should primarily be aimed at assessing the safety and feasibility despite the abundant evidence supporting the safety of OECs. Although this approach uses the same cells, new purification and identification techniques and functional assessments, as well as the ability to prepare cells in a 3D format, will warrant another phase I trial. However, efficacy can be pursued as a secondary objective and a phase I/IIa trial can be designed accordingly. In the likely event of a successful phase I trial outcome, an adaptive trial design can be employed for the subsequent studies to expedite the further translation. For example, with an adaptive design, the cell dosage and timing can be tested and revised as feedback from the outcomes of the first patients is obtained. In this way, the optimal findings can promptly be re-incorporated in the trial.

3.8. Outcome Measures Must Be Selected Strategically

The primary outcome measure must remain to be patient safety as monitored by the adverse events and severe adverse events, as well as the worsening of any existing symptoms. However, for the secondary outcome measures, a wide variety of clinical assessment tools can be used as reported by the past clinical trials. Nevertheless, it is critically important that none of the therapeutic impacts go undetected. Therefore, the clinical assessment tools must be employed strategically to guarantee that the assessments can detect any change in the patients' condition after intervention, while ensuring that the outcomes are not over-interpreted. This is also necessary for a cost–benefit evaluation. For example, AIS grades, the most commonly used clinical tool, may be too crude to pick up small-scale improvements that can be detected by the spinal cord independence measure (SCIM), minimal clinically important difference (MCID), or a reduction in neurological level of injury (NLI) [11]. A clinically small significant improvement is defined as a four-point improvement in SCIM, which may or may not be reflected in AIS grading at all; however, such a change may drastically improve the quality of life for the patient. Additionally, a conservative economical estimate by the Australian government suggests that a therapy that can consistently yield such a small significant improvement may result in savings of over AUD 3.5 billion. On the other hand, including too many assessment tools can be counterproductive by increasing the costs of the trial and reducing patient adherence or causing distress to the patients.

3.9. Rehabilitation Is Crucial for Functional Recovery

The reviewed clinical trials and case reports all definitively conclude that rehabilitation is crucial for functional recovery following the transplantation of OECs. There is some debate regarding if the rehabilitation alone is sufficient for any significant recovery, and that combining the cell transplantation with rehabilitation makes it difficult to link any potential recovery to cell transplantation. However, the aim of this position paper is to derive clinically relevant information for developing therapeutic approaches, and as such, the recommendation is clearly in favour of combining cell transplantation with rehabilitation. The overarching aim here must be to develop an approach that helps clinical patients regain their lost functions, and intense rehabilitation combined with the transplantation of OECs offers the best chance to achieve that.

Ideally, the amount of rehabilitation should be kept uniform (albeit, not necessarily identical) for all participants, where they would spend similar hours on distinct aspects of training such as posture, balance, pre-gait, and gait trainings as well sensory training in each session. The specifics of the sessions, however, such as the weights and intensities of each different training session, should be customised from patient to patient to maintain feasibility to continue over a long time, while avoiding stagnation in their recovery. Having a personalised approach to rehabilitation is critical as each patient's journey will be different. It is, therefore, advisable to have a minimum target for the amount of rehabilitation, but allow for some flexibility for participants in the trial to adjust their rehabilitation regimen to suit their individual needs and capacities.

Obtaining sufficient funding for rehabilitation has been a major limiting factor in past trials. Several distinct aspects of the prolonged rehabilitation such as accessibility, availability of the rehabilitation at multiple geographic locations, and a balance between site-based as well as home-based rehabilitation programs would rely on access to sufficient funding. Considering the importance of rehabilitation in complementing the cell transplantation, it is critical that the trial is funded sufficiently to enable participants to complete the rehabilitation program.

3.10. Prehabilitation Is Indicated

Several of the past trials included a pre-operative neuro-rehabilitation component. Some trials explained the rationale of this intervention, as it is important to see if there was any opportunity for spontaneous recovery. Additionally, pre-operative rehabilitation (or prehabilitation) is likely to prime the participants for post-operative rehabilitation—which is of key importance—prepare them with what they can expect after the treatment surgery, and overall enhance the patient adherence, thus, improving the likelihood of success overall. Therefore, prehabilitation with the same intense rehabilitation regime is indicated.

4. Conclusions

Olfactory ensheathing cell transplantation offers a promising therapy for repairing spinal cord injury. Clinical trials of OEC transplantation have shown that it is safe and feasible, however, past clinical trials also highlight challenges that must be overcome to complete a successful and widespread clinical translation of the therapy. Future clinical trials should be designed to incorporate a range of aspects that will increase the likelihood of success. The cells sourced from olfactory mucosa represent the safest clinical approach, however, complete and robust cell characterisation and quality control is necessary to ensure that the appropriate treatment cell population is transplanted. Importantly, transplanting cells in a three-dimensional format is the most suitable way to overcome the adversities of surgically transplanting cells into the spinal cord. While patient recruitment will likely involve people living with chronic injuries, planning for treating acute injuries can be incorporated into an adaptive trial design which can also test changes in dose. Finally, combining cell transplantation with neurorehabilitation provides the best chance of functional recovery for the trial participants.

Author Contributions: R.R. Conceptualization, literature review and writing; M.M. writing; M.C. writing, M.S. literature review, J.E. Supervision; D.P. clinical insights, M.-L.V. rehabilitation insights; B.M. clinical insights; J.S.J. supervision, project administration, funding acquisition. All authors have read and agreed to the published version of the manuscript.

Funding: This work was funded by the Motor Accident Insurance Commission, the Clem Jones Foundation, and the Perry Cross Spinal Research Foundation.

Institutional Review Board Statement: Not applicable.

Informed Consent Statement: Not applicable.

Data Availability Statement: Not applicable.

Conflicts of Interest: The authors declare no conflict of interest.

References

1. Spinal Cure Australia. *Spinal Cord Injury in Australia: The Case for Investing in New Treatments*; Spinal Cure Australia: Sydney, Australia, 2020.
2. Bracken, M.B.; Holford, T.R. Neurological and functional status 1 year after acute spinal cord injury: Estimates of functional recovery in National Acute Spinal Cord Injury Study II from results modeled in National Acute Spinal Cord Injury Study III. *J. Neurosurg.* **2002**, *96*, 259–266. [CrossRef] [PubMed]
3. Silva, N.A.; Sousa, N.; Reis, R.L.; Salgado, A.J. From basics to clinical: A comprehensive review on spinal cord injury. *Prog. Neurobiol.* **2014**, *114*, 25–57. [CrossRef] [PubMed]
4. Chio, J.C.T.; Xu, K.J.; Popovich, P.; David, S.; Fehlings, M.G. Neuroimmunological therapies for treating spinal cord injury: Evidence and future perspectives. *Exp. Neurol.* **2021**, *341*, 113704. [CrossRef]
5. Reshamwala, R.; Shah, M.; St John, J.; Ekberg, J. Survival and Integration of Transplanted Olfactory Ensheathing Cells are Crucial for Spinal Cord Injury Repair: Insights from the Last 10 Years of Animal Model Studies. *Cell Transplant.* **2019**, *28*, 132S–159S. [CrossRef] [PubMed]
6. Ahuja, C.S.; Mothe, A.; Khazaei, M.; Badhiwala, J.H.; Gilbert, E.A.; Kooy, D.; Morshead, C.M.; Tator, C.; Fehlings, M.G. The leading edge: Emerging neuroprotective and neuroregenerative cell-based therapies for spinal cord injury. *Stem Cells Transl. Med.* **2020**, *9*, 1509–1530. [CrossRef] [PubMed]
7. Kressler, J.; Thomas, C.K.; Field-Fote, E.C.; Sanchez, J.; Widerstrom-Noga, E.; Cilien, D.C.; Gant, K.; Ginnety, K.; Gonzalez, H.; Martinez, A.; et al. Understanding therapeutic benefits of overground bionic ambulation: Exploratory case series in persons with chronic, complete spinal cord injury. *Arch. Phys. Med. Rehabil.* **2014**, *95*, 1878–1887.e1874. [CrossRef] [PubMed]
8. Hochberg, L.R.; Serruya, M.D.; Friehs, G.M.; Mukand, J.A.; Saleh, M.; Caplan, A.H.; Branner, A.; Chen, D.; Penn, R.D.; Donoghue, J.P. Neuronal ensemble control of prosthetic devices by a human with tetraplegia. *Nature* **2006**, *442*, 164–171. [CrossRef]
9. Pizzolato, C.; Saxby, D.J.; Palipana, D.; Diamond, L.E.; Barrett, R.S.; Teng, Y.D.; Lloyd, D.G. Neuromusculoskeletal Modeling-Based Prostheses for Recovery After Spinal Cord Injury. *Front. Neurorobot.* **2019**, *13*, 97. [CrossRef]
10. Gilmour, A.D.; Reshamwala, R.; Wright, A.A.; Ekberg, J.A.K.; St John, J.A. Optimizing Olfactory Ensheathing Cell Transplantation for Spinal Cord Injury Repair. *J. Neurotrauma* **2020**, *37*, 817–829. [CrossRef]
11. Fehlings, M.G.; Pedro, K.; Hejrati, N. The Management of Acute Spinal Cord Injury: Where have we been? Where are we now? Where are we going? *J. Neurotrauma* **2022**, *ahead of print*. [CrossRef]
12. Willison, A.G.; Smith, S.; Davies, B.M.; Kotter, M.R.N.; Barnett, S.C. A scoping review of trials for cell-based therapies in human spinal cord injury. *Spinal Cord* **2020**, *58*, 844–856. [CrossRef] [PubMed]
13. Oieni, F.; Reshamwala, R.; St John, J. Olfactory Ensheathing Cells for Spinal Cord Injury: The Cellular Superpowers for Nerve Repair. *Neuroglia* **2022**, *3*, 139–143. [CrossRef]
14. Ekberg, J.A.; Amaya, D.; Mackay-Sim, A.; St John, J.A. The migration of olfactory ensheathing cells during development and regeneration. *Neuro-Signals* **2012**, *20*, 147–158. [CrossRef] [PubMed]
15. Carwardine, D.; Prager, J.; Neeves, J.; Muir, E.M.; Uney, J.; Granger, N.; Wong, L.F. Transplantation of canine olfactory ensheathing cells producing chondroitinase ABC promotes chondroitin sulphate proteoglycan digestion and axonal sprouting following spinal cord injury. *PLoS ONE* **2017**, *12*, e0188967. [CrossRef] [PubMed]

16. Li, Y.; Li, D.; Raisman, G. Functional Repair of Rat Corticospinal Tract Lesions Does Not Require Permanent Survival of an Immunoincompatible Transplant. *Cell Transplant.* **2016**, *25*, 293–299. [CrossRef]
17. Radtke, C.; Akiyama, Y.; Brokaw, J.; Lankford, K.L.; Wewetzer, K.; Fodor, W.L.; Kocsis, J.D. Remyelination of the nonhuman primate spinal cord by transplantation of H-transferase transgenic adult pig olfactory ensheathing cells. *FASEB J. Off. Publ. Fed. Am. Soc. Exp. Biol.* **2004**, *18*, 335–337. [CrossRef]
18. Ramon-Cueto, A.; Cordero, M.I.; Santos-Benito, F.F.; Avila, J. Functional recovery of paraplegic rats and motor axon regeneration in their spinal cords by olfactory ensheathing glia. *Neuron* **2000**, *25*, 425–435. [CrossRef]
19. Ramon-Cueto, A.; Nieto-Sampedro, M. Regeneration into the spinal cord of transected dorsal root axons is promoted by ensheathing glia transplants. *Exp. Neurol.* **1994**, *127*, 232–244. [CrossRef]
20. Ziegler, M.D.; Hsu, D.; Takeoka, A.; Zhong, H.; Ramon-Cueto, A.; Phelps, P.E.; Roy, R.R.; Edgerton, V.R. Further evidence of olfactory ensheathing glia facilitating axonal regeneration after a complete spinal cord transection. *Exp. Neurol.* **2011**, *229*, 109–119. [CrossRef]
21. Féron, F.; Perry, C.; Cochrane, J.; Licina, P.; Nowitzke, A.; Urquhart, S.; Geraghty, T.; Mackay-Sim, A. Autologous olfactory ensheathing cell transplantation in human spinal cord injury. *Brain* **2005**, *128*, 2951–2960. [CrossRef]
22. Lima, C.; Pratas-Vital, J.; Escada, P.; Hasse-Ferreira, A.; Capucho, C.; Peduzzi, J.D. Olfactory Mucosa Autografts in Human Spinal Cord Injury: A Pilot Clinical Study. *J. Spinal Cord Med.* **2006**, *29*, 191–203. [CrossRef] [PubMed]
23. Mackay-Sim, A.; Féron, F.; Cochrane, J.; Bassingthwaighte, L.; Bayliss, C.; Davies, W.; Fronek, P.; Gray, C.; Kerr, G.; Licina, P.; et al. Autologous olfactory ensheathing cell transplantation in human paraplegia: A 3-year clinical trial. *Brain* **2008**, *131*, 2376–2386. [CrossRef] [PubMed]
24. Chhabra, H.S.; Lima, C.; Sachdeva, S.; Mittal, A.; Nigam, V.; Chaturvedi, D.; Arora, M.; Aggarwal, A.; Kapur, R.; Khan, T.A. Autologous olfactory [corrected] mucosal transplant in chronic spinal cord injury: An Indian Pilot Study. *Spinal Cord* **2009**, *47*, 887–895. [CrossRef] [PubMed]
25. Lima, C.; Escada, P.; Pratas-Vital, J.; Branco, C.; Arcangeli, C.A.; Lazzeri, G.; Maia, C.A.; Capucho, C.; Hasse-Ferreira, A.; Peduzzi, J.D. Olfactory mucosal autografts and rehabilitation for chronic traumatic spinal cord injury. *Neurorehabil. Neural Repair* **2010**, *24*, 10–22. [CrossRef]
26. Wu, J.; Sun, T.; Ye, C.; Yao, J.; Zhu, B.; He, H. Clinical observation of fetal olfactory ensheathing glia transplantation (OEGT) in patients with complete chronic spinal cord injury. *Cell Transplant.* **2012**, *21* (Suppl. 1), S33–S37. [CrossRef]
27. Larson, C.A.; Dension, P.M. Effectiveness of intense, activity-based physical therapy for individuals with spinal cord injury in promoting motor and sensory recovery: Is olfactory mucosa autograft a factor? *J. Spinal Cord Med.* **2013**, *36*, 44–57. [CrossRef]
28. Tabakow, P.; Jarmundowicz, W.; Czapiga, B.; Fortuna, W.; Miedzybrodzki, R.; Czyz, M.; Huber, J.; Szarek, D.; Okurowski, S.; Szewczyk, P.; et al. Transplantation of Autologous Olfactory Ensheathing Cells in Complete Human Spinal Cord Injury. *Cell Transplant.* **2013**, *22*, 1591–1612. [CrossRef]
29. Rao, Y.; Zhu, W.; Liu, H.; Jia, C.; Zhao, Q.; Wang, Y. Clinical application of olfactory ensheathing cells in the treatment of spinal cord injury. *J. Int. Med. Res.* **2013**, *41*, 473–481. [CrossRef]
30. Chen, L.; Huang, H.; Xi, H.; Zhang, F.; Liu, Y.; Chen, D.; Xiao, J. A prospective randomized double-blind clinical trial using a combination of olfactory ensheathing cells and Schwann cells for the treatment of chronic complete spinal cord injuries. *Cell Transplant.* **2014**, *23* (Suppl. 1), S35–S44. [CrossRef]
31. Wang, S.; Lu, J.; Li, Y.A.; Zhou, H.; Ni, W.F.; Zhang, X.L.; Zhu, S.P.; Chen, B.B.; Xu, H.; Wang, X.Y.; et al. Autologous Olfactory Lamina Propria Transplantation for Chronic Spinal Cord Injury: Three-Year Follow-Up Outcomes from a Prospective Double-Blinded Clinical Trial. *Cell Transplant.* **2016**, *25*, 141–157. [CrossRef]
32. Chehrehasa, F.; Windus, L.C.E.; Ekberg, J.A.K.; Scott, S.E.; Amaya, D.; Mackay-Sim, A.; St John, J.A. Olfactory glia enhance neonatal axon regeneration. *Mol. Cell. Neurosci.* **2010**, *45*, 277–288. [CrossRef] [PubMed]
33. Takeoka, A.; Jindrich, D.L.; Munoz-Quiles, C.; Zhong, H.; van den Brand, R.; Pham, D.L.; Ziegler, M.D.; Ramon-Cueto, A.; Roy, R.R.; Edgerton, V.R.; et al. Axon regeneration can facilitate or suppress hindlimb function after olfactory ensheathing glia transplantation. *J. Neurosci.* **2011**, *31*, 4298–4310. [CrossRef] [PubMed]
34. Collins, A.; Li, D.; Liadi, M.; Tabakow, P.; Fortuna, W.; Raisman, G.; Li, Y. Partial Recovery of Proprioception in Rats with Dorsal Root Injury after Human Olfactory Bulb Cell Transplantation. *J. Neurotrauma* **2018**, *35*, 1367–1378. [CrossRef] [PubMed]
35. Delarue, Q.; Robac, A.; Massardier, R.; Marie, J.P.; Guerout, N. Comparison of the effects of two therapeutic strategies based on olfactory ensheathing cell transplantation and repetitive magnetic stimulation after spinal cord injury in female mice. *J. Neurosci. Res.* **2021**, *99*, 1835–1849. [CrossRef] [PubMed]
36. Khankan, R.R.; Griffis, K.G.; Haggerty-Skeans, J.R.; Zhong, H.; Roy, R.R.; Edgerton, V.R.; Phelps, P.E. Olfactory Ensheathing Cell Transplantation after a Complete Spinal Cord Transection Mediates Neuroprotective and Immunomodulatory Mechanisms to Facilitate Regeneration. *J. Neurosci.* **2016**, *36*, 6269–6286. [CrossRef] [PubMed]
37. Liu, T.; Ji, Z.Q.; Ahsan, S.M.; Zhang, Y.; Zhang, P.; Fan, Z.H.; Shen, Y.X. Intrathecal transplantation of olfactory ensheathing cells by lumbar puncture for thoracic spinal cord injury in mice. *J. Neurorestoratol.* **2017**, *5*, 103–109. [CrossRef]
38. Nazareth, L.; Lineburg, K.E.; Chuah, M.I.; Tello Velasquez, J.; Chehrehasa, F.; St John, J.A.; Ekberg, J.A. Olfactory ensheathing cells are the main phagocytic cells that remove axon debris during early development of the olfactory system. *J. Comp. Neurol.* **2015**, *523*, 479–494. [CrossRef]

39. Nazareth, L.; Tello Velasquez, J.; Lineburg, K.E.; Chehrehasa, F.; St John, J.A.; Ekberg, J.A. Differing phagocytic capacities of accessory and main olfactory ensheathing cells and the implication for olfactory glia transplantation therapies. *Mol. Cell. Neurosci.* **2015**, *65*, 92–101. [CrossRef]

40. Gu, M.; Gao, Z.; Li, X.; Guo, L.; Lu, T.; Li, Y.; He, X. Conditioned medium of olfactory ensheathing cells promotes the functional recovery and axonal regeneration after contusive spinal cord injury. *Brain Res.* **2017**, *1654*, 43–54. [CrossRef]

41. Wright, A.A.; Todorovic, M.; Tello-Velasquez, J.; Rayfield, A.J.; St John, J.A.; Ekberg, J.A. Enhancing the Therapeutic Potential of Olfactory Ensheathing Cells in Spinal Cord Repair Using Neurotrophins. *Cell Transplant.* **2018**, *27*, 867–878. [CrossRef]

42. Yao, R.; Murtaza, M.; Velasquez, J.T.; Todorovic, M.; Rayfield, A.; Ekberg, J.; Barton, M.; St John, J. Olfactory Ensheathing Cells for Spinal Cord Injury: Sniffing out the Issues. *Cell Transplant.* **2018**, *27*, 879–889. [CrossRef] [PubMed]

43. Xie, J.; Li, Y.; Dai, J.; He, Y.; Sun, D.; Dai, C.; Xu, H.; Yin, Z.Q. Olfactory Ensheathing Cells Grafted into the Retina of RCS Rats Suppress Inflammation by Down-Regulating the JAK/STAT Pathway. *Front. Cell. Neurosci.* **2019**, *13*, 341. [CrossRef] [PubMed]

44. Wright, A.A.; Todorovic, M.; Murtaza, M.; St John, J.A.; Ekberg, J.A. Macrophage migration inhibitory factor and its binding partner HTRA1 are expressed by olfactory ensheathing cells. *Mol. Cell. Neurosci.* **2020**, *102*, 103450. [CrossRef] [PubMed]

45. Gómez, R.M.; Sánchez, M.Y.; Portela-Lomba, M.; Ghotme, K.; Barreto, G.E.; Sierra, J.; Moreno-Flores, M.T. Cell therapy for spinal cord injury with olfactory ensheathing glia cells (OECs). *GLIA* **2018**, *66*, 1267–1301. [CrossRef]

46. Lakatos, A.; Franklin, R.J.; Barnett, S.C. Olfactory ensheathing cells and Schwann cells differ in their in vitro interactions with astrocytes. *GLIA* **2000**, *32*, 214–225. [CrossRef]

47. Santos-Silva, A.; Fairless, R.; Frame, M.C.; Montague, P.; Smith, G.M.; Toft, A.; Riddell, J.S.; Barnett, S.C. FGF/Heparin Differentially Regulates Schwann Cell and Olfactory Ensheathing Cell Interactions with Astrocytes: A Role in Astrocytosis. *J. Neurosci.* **2007**, *27*, 7154–7167. [CrossRef]

48. Lu, P.; Jones, L.L.; Snyder, E.Y.; Tuszynski, M.H. Neural stem cells constitutively secrete neurotrophic factors and promote extensive host axonal growth after spinal cord injury. *Exp. Neurol.* **2003**, *181*, 115–129. [CrossRef]

49. Teli, P.; Kale, V.; Vaidya, A. Extracellular vesicles isolated from mesenchymal stromal cells primed with neurotrophic factors and signaling modifiers as potential therapeutics for neurodegenerative diseases. *Curr. Res. Transl. Med.* **2021**, *69*, 103286. [CrossRef]

50. Trzyna, A.; Banaś-Ząbczyk, A. Adipose-Derived Stem Cells Secretome and Its Potential Application in "Stem Cell-Free Therapy". *Biomolecules* **2021**, *11*, 878. [CrossRef]

51. Carvalho, L.A.; Teng, J.; Fleming, R.L.; Tabet, E.I.; Zinter, M.; de Melo Reis, R.A.; Tannous, B.A. Olfactory Ensheathing Cells: A Trojan Horse for Glioma Gene Therapy. *J. Natl. Cancer Inst.* **2019**, *111*, 283–291. [CrossRef]

52. Murtaza, M.; Chacko, A.; Delbaz, A.; Reshamwala, R.; Rayfield, A.; McMonagle, B.; St John, J.A.; Ekberg, J.A.K. Why are olfactory ensheathing cell tumors so rare? *Cancer Cell Int.* **2019**, *19*, 260. [CrossRef] [PubMed]

53. Tabakow, P.; Raisman, G.; Fortuna, W.; Czyz, M.; Huber, J.; Li, D.; Szewczyk, P.; Okurowski, S.; Miedzybrodzki, R.; Czapiga, B.; et al. Functional regeneration of supraspinal connections in a patient with transected spinal cord following transplantation of bulbar olfactory ensheathing cells with peripheral nerve bridging. *Cell Transplant.* **2014**, *23*, 1631–1655. [CrossRef] [PubMed]

54. Chen, D.; Xi, H.; Tan, K.; Huang, H. Recovering Voiding and Sex Function in a Patient with Chronic Complete Spinal Cord Injury by Olfactory Ensheathing Cell Transplantation. *Case Rep. Neurol. Med.* **2022**, *2022*, 9496652. [CrossRef] [PubMed]

55. Mayeur, A.; Duclos, C.; Honore, A.; Gauberti, M.; Drouot, L.; do Rego, J.C.; Bon-Mardion, N.; Jean, L.; Verin, E.; Emery, E.; et al. Potential of olfactory ensheathing cells from different sources for spinal cord repair. *PLoS ONE* **2013**, *8*, e62860. [CrossRef]

56. Ekberg, J.A.; St John, J.A. Olfactory ensheathing cells for spinal cord repair: Crucial differences between subpopulations of the glia. *Neural Regen. Res.* **2015**, *10*, 1395–1396. [CrossRef]

57. Dlouhy, B.J.; Awe, O.; Rao, R.C.; Kirby, P.A.; Hitchon, P.W. Autograft-derived spinal cord mass following olfactory mucosal cell transplantation in a spinal cord injury patient: Case report. *J. Neurosurg. Spine* **2014**, *21*, 618–622. [CrossRef]

58. Woodworth, C.F.; Jenkins, G.; Barron, J.; Hache, N. Intramedullary cervical spinal mass after stem cell transplantation using an olfactory mucosal cell autograft. *CMAJ Can. Med. Assoc. J. J. De L'association Med. Can.* **2019**, *191*, E761–E764. [CrossRef]

59. Reshamwala, R.; Shah, M.; Belt, L.; Ekberg, J.A.K.; St John, J.A. Reliable cell purification and determination of cell purity: Crucial aspects of olfactory ensheathing cell transplantation for spinal cord repair. *Neural Regen. Res.* **2020**, *15*, 2016–2026. [CrossRef]

60. Au, E.; Roskams, A.J. Olfactory ensheathing cells of the lamina propria in vivo and in vitro. *GLIA* **2003**, *41*, 224–236. [CrossRef]

61. Kawaja, M.D.; Boyd, J.G.; Smithson, L.J.; Jahed, A.; Doucette, R. Technical Strategies to Isolate Olfactory Ensheathing Cells for Intraspinal Implantation. *J. Neurotrauma* **2009**, *26*, 155–177. [CrossRef]

62. Geissler, S.A.; Sabin, A.L.; Besser, R.R.; Gooden, O.M.; Shirk, B.D.; Nguyen, Q.M.; Khaing, Z.Z.; Schmidt, C.E. Biomimetic hydrogels direct spinal progenitor cell differentiation and promote functional recovery after spinal cord injury. *J. Neural Eng.* **2018**, *15*, 025004. [CrossRef] [PubMed]

63. Tukmachev, D.; Forostyak, S.; Koci, Z.; Zaviskova, K.; Vackova, I.; Vyborny, K.; Sandvig, I.; Sandvig, A.; Medberry, C.J.; Badylak, S.F.; et al. Injectable Extracellular Matrix Hydrogels as Scaffolds for Spinal Cord Injury Repair. *Tissue Eng. Part A* **2016**, *22*, 306–317. [CrossRef] [PubMed]

64. Gueye, Y.; Ferhat, L.; Sbai, O.; Bianco, J.; Ould-Yahoui, A.; Bernard, A.; Charrat, E.; Chauvin, J.P.; Risso, J.J.; Féron, F.; et al. Trafficking and secretion of matrix metalloproteinase-2 in olfactory ensheathing glial cells: A role in cell migration? *GLIA* **2011**, *59*, 750–770. [CrossRef] [PubMed]

65. Roet, K.C.; Verhaagen, J. Understanding the neural repair-promoting properties of olfactory ensheathing cells. *Exp. Neurol.* **2014**, *261*, 594–609. [CrossRef] [PubMed]

MDPI AG
Grosspeteranlage 5
4052 Basel
Switzerland
Tel.: +41 61 683 77 34

Biomedicines Editorial Office
E-mail: biomedicines@mdpi.com
www.mdpi.com/journal/biomedicines

www.ingramcontent.com/pod-product-compliance
Lightning Source LLC
LaVergne TN
LVHW070951170726
843515LV00005B/962